Cell Line Development

AF793323

Cell Engineering

Volume 6

Series Editor

Professor Mohamed Al-Rubeai
UCD School of Chemical and Bioprocess Engineering
University College Dublin
Dublin, Ireland

Editorial Board

Dr Hansjorg Hauser
Helmholtz Centre for Infection Research
Braunschweig
Germany

Professor Michael Betenbaugh
Johns Hopkins University
Baltimore, USA

Professor Martin Fussenegger
Swiss Federal Institute of Technology
Zurich, Switzerland

Professor. Nigel Jenkins
National Institute for Bioprocessing Research and Training
Dublin, Ireland

Dr Otto-Wilhelm Merten
A.F.M.-Genethon 11
Gene Therapy Program
Evry, France

For other titles published in this series, go to
www.springer.com/series/5728

CELL ENGINEERING

Vol. 6: Cell Line Development

Edited by

Mohamed Al-Rubeai

University College Dublin, Ireland

Springer

Editor
Prof. Dr. Mohamed Al-Rubeai
School of Chemical and Bioprocess Engineering
University College Dublin
Belfield, Dublin 4
Ireland

ISBN 978-94-007-3040-3 e-ISBN 978-90-481-2245-5
DOI 10.1007/978-90-481-2245-5
Springer Dordrecht Heidelberg London New York

© Springer Science+Business Media B.V. 2009
Softcover reprint of the hardcover 1st edition 2009
No part of this work may be reproduced, stored in a retrieval system, or transmitted in any form or by any means, electronic, mechanical, photocopying, microfilming, recording or otherwise, without written permission from the Publisher, with the exception of any material supplied specifically for the purpose of being entered and executed on a computer system, for exclusive use by the purchaser of the work.

Printed on acid-free paper

Springer is part of Springer Science+Business Media (www.springer.com)

Cell Engineering – Overview

The development of cell lines has undergone several advances over the years, essentially to meet the requirement to cut the time and costs associated with using such complex hosts as production platforms. This book reviews the aspects involved in the development of cell lines and the cell engineering approach that can be employed to enhance productivity, improve cellular metabolism, control proliferation and apoptosis, and reduce instability.

Cell engineering is a new research approach which began in the early 1990s, coinciding with an increasing interest in apoptosis. This approach of manipulation of apoptotic regulatory functions in cells was highlighted in an article appearing in 1995 in "Trends in Biotechnology", in which we suggested that through the manipulation of cell lines by transfecting them with anti-apoptotic genes, one may be able to enhance the robustness and survival of those cells in culture. Since then several papers have been published which demonstrate the effectiveness of this approach; this has since evolved to embrace methodologies offered by molecular biology for the development of cell lines, which could provide platforms to improve recombinant protein production and the efficiency of industrial culture processes. The underlying principles of cell engineering are very simple, essentially involving the identification of a gene of interest, expression of it into a cell, or knocking down its function and development of a new cell line. Today, cell engineering covers several topics, including expression engineering, and involves the use of technologies such as functional genomics, proteomics and metabolomics which have become an integral part of cell engineering. Its scope has also become broader but the underlying principle remains the same. It involves various strategies of adding and deleting genetic elements such as single and multiple gene insertion, gene disruption (genetic deletions), gene silencing, mutagenesis and directed evolution. The figure below is a diagrammatic representation, which is useful in explaining the interaction between cell engineering and "omics" technologies.

It is undeniable that integrating genomics and proteomics data with vast amounts of bioprocess data will improve significantly the analysis of the biological pathways involved, with a consequent discovery of lead genes that may serve as candidates for cell engineering. Ultimately, new cell engineering strategies should provide greater insight into regulatory networks within cells in a bioprocess environment, thus greatly advancing our understanding of cellular mechanics in conjunction with

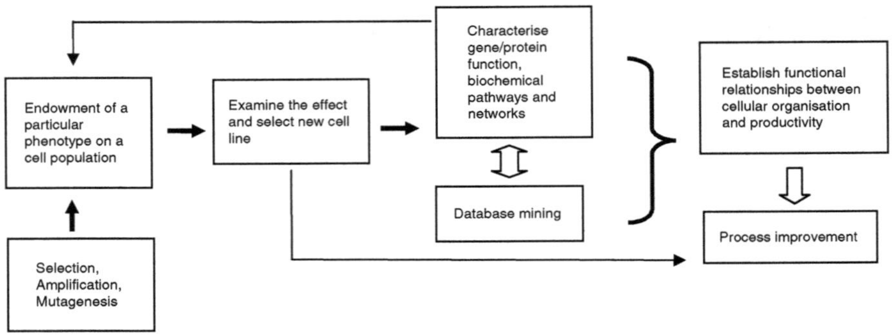

improvements in overall productivity in the manufacturing environment; this will lead ultimately to efficient and safe processing of protein products.

Cellular engineering approaches can be used to integrate bioprocess improvements like high cell density cultures, operational strategies like fed-batch and perfusion cultures and high productivity (Qp) at a cellular/molecular level. However, this approach has only met with limited success, as very little is known about the cellular dynamics related to productivity. At a cellular level, Qp has been suggested to be dependent on events downstream of transcription, thus modifying the translational and secretory pathways that would overcome limitations of protein folding and assembly reactions.

In the pursuit of increasing productivity, molecular analysis of recombinant protein production at different organisational levels within the cells (transcript, polypeptide, assembly and secretion), together with metabolite data, in effective combination with gene expression data, should piece together a more comprehensive picture about the adaptive state of productivity. The increase in productivity will not only require important transcription and mRNA stability, but also a simultaneous increase in the post-translational capacity of the cell, including N-glycan biosynthesis and secretory functions. Energy metabolism appears to become a limiting factor, with high AMP levels and high oxidative stress taking place in highly productive cells. These changes are likely to be controlled at the gene transcription level, suggesting that a cell engineering approach can alter cellular organisation to support substantial increases in yield.

In conclusion, "direct" cell engineering approaches of manipulating apoptosis, proliferation, metabolism, glycosylation and secretion have resulted in several current and potential improvements in cell lines of biopharmaceutical importance. With the advent of genomic and proteomic tools, "indirect" cell engineering is becoming a useful strategic approach for the improvement of biopharmaceutical cell lines. In combination with advances in expression engineering, clone selection and media development, it is hoped to decrease development times and dramatically increase cell line productivity, thus reducing overall costs for the next generation of approved products.

Mohamed Al-Rubeai
19 March 2009

Contents

Use of MAR Elements to Increase the Production of Recombinant Proteins

Cori Gorman, Salina Arope, Mélanie Grandjean, Pierre-Alain Girod, and Nicolas Mermod

Abstract The biopharmaceutical industry continues to face the challenge of producing large amount of recombinant proteins for use as therapeutics, and eighty percent of protein therapeutics in clinical development are produced in mammalian cell systems. Approaches to increase production addressing growth conditions, such as the improvement of media composition and process control, or transcription of the recombinant gene via the use of strong promoters/enhancers and amplification of gene copy number, have increased the yields obtained from mammalian cells considerably over the past decades. However these processes remain laborious, and extensive screening of clones is often required, as stable cell line and/or protein production is not always obtained. Unstable or variable expression is linked to the location of transgene integration site, the regulation of gene expression, the silencing of genes, and the loss of gene copies. Genetic elements that may remodel chromatin to maintain the transgene in an active configuration are now being employed increasingly to improve protein production using mammalian cells. Here we will review how one type of such elements, the MARs, may increase transgene integration into the cell genome and decrease silencing effects to reduce expression variability. We also illustrate how inclusion of these elements in expression vectors leads to increased specific productivities ranging from 20 to 100 picograms per cell and per day (p/c/d), resulting in protein titers above 5 g/l.

C. Gorman and P.-A. Girod
Selexis SA, 18 chemin des Aulx, 1228 Plan-les-Ouates, Switzerland

C. Gorman
DNA Gateway International, Inc. 55 New Montgomery St. Ste 605
San Francisco CA, 94105, USA

S. Arope, M. Grandjean, and N. Mermod (✉)
Laboratory of Molecular Biotechnology, University of Lausanne, 1015, Lausanne, Switzerland
e.mail: nicolas. mermod@unil.ch

M. Al-Rubeai (ed.), *Cell Line Development*, Cell Engineering 6,
DOI 10.1007/978-90-481-2245-5_1, © Springer Science+Business Media B.V. 2009

1 Introduction to Epigenetic Silencing Issues in the Generation of Production Cell Lines

During the past two decades, cultured mammalian cells have become the most widely used platform for producing recombinant therapeutic proteins. Improving yield and stability of protein expression are, therefore, of considerable value to the industry. Previously, such improvements have mainly originated from optimizing downstream production processes and media development. Despite recent advances in the field of cell line generation, expression levels in mammalian cells are relatively low and often unstable over time; these events result in high development and production costs for therapeutic proteins. The average reported yields in mammalian cells (usually 0.5–2 g/l) are several fold lower than the yields from bacteria and yeast systems, while even the highest reproducible yields, 5 g/l, remain at least three fold lower than obtained in more simple production systems (Kwaks and Otte, 2006).

Despite significant effort, the current schemes for cell line development remain, to a large extent, empirical. There is a considerable degree of variability, and our understanding of the sources of variability in the mammalian cell line development process remains limited. This process is laborious, and extensive screening of clones, often spanning over several months, is still widely practiced in industry. Obtaining cell lines that maintain stable protein production is of utmost importance, particularly for industrial use where the goal is to commercialize the protein being produced. Problems with stability can impact the time and effort required to generate working and master cell banks. The loss of productivity between the initial cell isolates and the end-of-production cells can compromise regulatory approval and, in the worst-case scenario, may result in rejection of a particular cell line after months of development efforts. Unfortunately, the method of selecting cell lines relies heavily on a degree of chance. In part, this high degree of variability is due to the effect that random transgene integration into host cell chromosome exerts on transgene transcription often resulting in silencing. Another cause of variability results from the integration of a varying number of transgene copies from one clone to the next. Finally, a last source of variability is variegation, a phenomenon that results in the cycling of cells between productive and non-productive phases, which again may affect differently distinct cell clones.

The insertion of genes into certain areas of chromatin can lead to the so-called "position effect" (Wilson et al., 1990). Silencing of transfected genes in mammalian cells is a fundamental problem that probably involves the relatively inaccessible status of the DNA when it is imbedded in chromatin. Transgene integration, when it occurs through a random process, can either occur in highly condensed, silenced region of the chromatin, heterochromatin, or in more open and active chromatin, euchromatin (Eissenberg, 1989; Eissenberg et al., 1992; Zahn-Zabal et al., 2001). Integration into heterochromatin may result in minimal or no transgene expression. Because a large proportion of the genome is in the form of heterochromatin, the chance that a transgene integrates in, or close to, heterochromatin, and consequently is silenced or repressed, is high.

Other regions of the chromosome may be subjected to slow silencing effects (Pankiewicz et al., 2005). Because this slow silencing may not be readily apparent, it can lead to a gradual loss of productivity that suddenly appears after an initial phase of seemingly stable expression during clone isolation and characterization (see chapter by Alan Dickson in this book). A related process, known as position effect variegation (PEV), is thought to result from the stochastic spread or retreat of heterochromatin towards or away from the gene location (Eissenberg et al., 1992; Volfson et al., 2006). PEV typically leads to clones that possess heterogeneous levels of expression when comparing distinct cells in the monoclonal population. Often, this heterogeneous expression is not apparent when determining the titer of secreted proteins such immunoglobulins in the cell supernatant, but nevertheless may limit the yield. The combined effects of random transgene incorporation with these chromatin-mediated epigenetic effects collectively result in only a small percentage (less than 1%) of the initially isolated cells being capable of producing high amounts of the desired protein (Girod and Mermod, 2003). Therefore, lengthy selection and screening procedures are often required to select and identify those cells with the proper growth, transgene expression, long-term stability, and protein quality properties required for large-scale production.

For several years, Barnes et al. (2001, 2003, 2004, 2006, 2007; Barnes and Dickson, 2006) have examined the process of loss of protein production in mammalian cell lines. Despite repeated rounds of cloning, the cell lines they derived showed a wide variation in terms of maximum obtainable cell densities, rates of growth, and accumulation of secreted recombinant protein (Barnes et al., 2001). Several lines of data suggest that rapid phenotypic drift may be occurring during culture and, therefore, the cells derived from a single cell, as a result of cloning, soon diverge to become a mixed population. In this context, the term 'stable cell lines' refers to cell populations that retain stability of expression during prolonged culture (Barnes et al., 2001).

Though a variety of mammalian cells have been used for recombinant protein production, including mouse myeloma derived (NS0), human embryonic kidney (HEK-293), baby hamster kidney (BHK) and more recently, the human retina derived (PerC6) cells, the most commonly used host cell lines remain the CHO cells. The popularity of CHO based expression is largely due to the ability to use DNA amplification techniques in these cells to increase transgene copy number. However, both the DHFR and GS methods of amplifying transgenes can result in genetic instability. CHO cells typically undergo genomic rearrangements and amplifications of the locus of DNA integration resulting in increased copy numbers for both DHFR and the protein of interest. Often, clones containing several hundred copies of the vector construct can be found following amplification. This high copy number does not however lead to uniform high expression or to stable production. It is commonly reported that recombinant protein production can drop significantly within the 2 months following high-producing clone selection, particularly when the selection pressure is removed (Fann et al., 2000; Jun et al., 2006; Kim et al., 1998a; Strutzenberger et al., 1999). For instance, Strutzenberger et al. (1999) have shown that when dhfr amplification with methotrexate was used, over 75% of the integrated transgenes were lost once the drug was removed. In the absence of selective pressure, expression is lost just weeks following selection. In one study, the relative decrease in specific productivity varied among subclones, ranging from 30% to 80% (Kim et al., 1998b; Kim et al.,

2001). Southern and Northern blot analyses showed that this decreased productivity resulted mainly from the loss of amplified antibody gene copies and their respective cytoplasmic mRNAs (Barnes et al., 2006).

Overall, it is clear from the examples presented above that there is a commercial need within the biotechnology industry to understand the problem of instability of protein production associated with recombinant mammalian cell lines. Loss of recombinant gene copy number and the overgrowth of non-producing populations of cells may result in low production; however, there are several other factors that may affect expression levels and stability of production. The possibility of improving cell-line stability and decreasing the variability associated with the generation of a cell line is now possible through the use of techniques that assure that the transgenes are actively transcribed following integration and that the chromatin surrounding the transgene remains capable of active transcription. The potential of these techniques for saving time as well as human and financial resources is extremely promising. This will also open up avenues for more rapid and effective use of additional types of mammalian cells, beyond CHO cells, as expression hosts (Barnes et al., 2004, 2007).

2 Causes of Instability During Mammalian Cell Production

The consistency of growth, productivity, or product characteristics with each successive generation of the cell line defines cell line stability and these factors contribute to the overall process consistency. Some of the issues that lead to instability of protein production in mammalian cells include gene amplification, loss of genetic material, methylation and the location of integration. Gene amplification occurs through the mechanism of chromosomal rearrangement, which involves chromosomal breaks (Andrulis et al., 1983; Melton et al., 1982). Amplification can result in decreased stability of transgene expression due to such breaks and rearrangements (Flintoff et al., 1984; Yoshikawa et al., 2000). CHO cells are known to have an unstable karyotype, with chromosome rearrangements arising from translocations and recombination events, as in the amplification procedures (Yoshikawa et al., 2000). As discussed above, the predominant use of CHO cells has been paralleled with gene amplification selection methods; however, loss of protein production following amplification has been reported for several proteins including interferon, tissue plasminogen activator, and antibodies. In some cases, production levels can reach a stable value after an initial decrease during the first 30 to 50 days of culture (Kim et al., 1998a). Even in the presence of selective pressure, Jun et al. (2006) found that the stability of antibody producing subclones was very poor. Furthermore, the specific secretion rate decreased by 50% after 100 passages even with selective pressure. This might be explained by the selective silencing of the transgenes at chromosomal sites that are prone to epigenetic regulation (position effect), as the amplified gene arrays are often scattered at multiple loci in the host genome. Thus instability may be a concern in the development of CHO cell lines with DHFR and GS-mediated gene amplification (Kim et al., 1998a; Kim and Lee, 1999; Fann et al., 2000; Jun et al., 2006).

In addition to the potential loss of transgenes following amplification, the large number of repetitive gene sequences that results from this process may induce methylation of DNA sequences, thus preventing transcription (Fouremana et al., 1998). More recently, RNAi-mediated chromatin remodeling linked to the occurrence of repetitive sequences, such as the hundreds of copies that result from amplification, has been shown to contribute to gene silencing (Almeida and Robin, 2005; Morris, 2008). These observations provide a likely explanation for the fact that expression following amplification does not increase proportionally with the transgene copy number. Mechanisms by which repeated sequences such as inverse repeat transgene arrays and RNAi may trigger silent chromatin assembly include physical pairing of homologous sequences and/or DNA–RNA or RNA–RNA interactions (Selker, 1999; Matzke et al., 2001). The connection between RNAi and heterochromatin assembly has suggested a model for the RNA-mediated epigenetic structuring of the eukaryotic genomes. Double-stranded RNA is processed into small RNAs, which in turn provide specificity for targeting histone-modifying activities and epigenetic modification of the genome through homology recognition (Fig. 1; Grewal and Moazed, 2003).

Instability of expression can be due to the regulation of gene expression, the silencing of genes, and the loss of gene copies. However, it must be stressed that these mechanisms are not mutually exclusive, and often the regulation of gene expression and the occurrence of instability of expression involve interplay between different mechanisms. DNA is highly condensed into the chromatin structures, and this condensation often hinders the accessibility of DNA to transcription complexes (Felsenfeld, 1992, 1996; Woodcock and Dimitrov, 2001). Activation of transcription

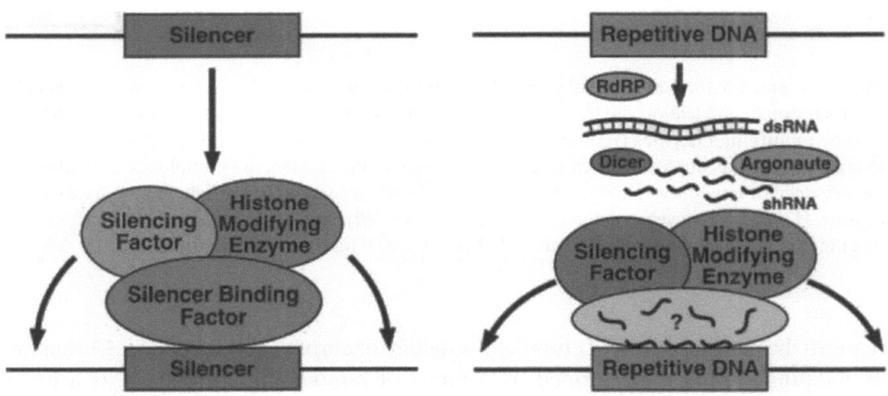

Fig. 1 Mechanisms for the initiation of heterochromatin. Heterochromatic structures can be nucleated by specific cis-acting sequences, called silencers, which are recognized by DNA binding proteins (*left*). Transcripts generated by repetitive DNA are processed into siRNAs by a mechanism requiring components of the RNAi machinery (from Grewal and Moazed, 2003, reprinted with permission from AAAS)

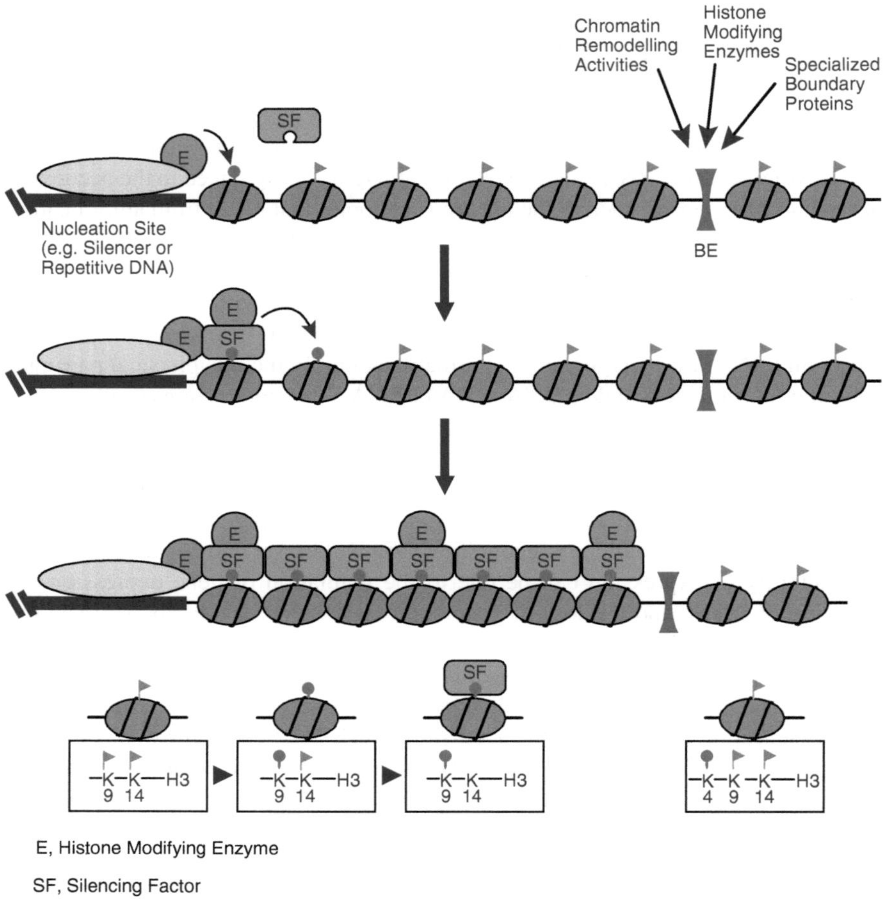

Fig. 2 Model for formation of silenced chromatin domains. After the recruitment to a specific heterochromatin nucleation site by proteins that directly bind DNA or are targeted by way of RNAs, histone modifying enzymes (E) such as deacetylases and methyltransferases modify histone tails to create a binding site for silencing factors (SF). Spreading of silencing complexes is blocked by the presence of boundary elements (BE). The modifications associated with the amino terminus of histone H3 in fission yeast heterochromatin (*bottom left*) and euchromatin (*bottom right*) are illustrated as an example (from Grewal and Moazed, 2003, reprinted with permission from AAAS)

requires the rearrangement of chromatin structure or chromatin remodeling. Chromatin remodeling, which is performed by a range of remodeling complexes, is a loose term used to define any event that alters the nuclease sensitivity of a DNA region (West et al., 2002; Grewal and Moazed, 2003) (Fig. 2).

The highly condensed heterochromatin domains are interspersed along with relatively decondensed euchromatic regions (Fig. 2; Grewal and Moazed, 2003). Given that heterochromatin structures, once nucleated, can spread in *cis*, resulting in epigenetic silencing of adjacent genes, cells have evolved antagonistic mechanisms that protect active regions from the repressive effects of nearby heterochromatin.

Chromatin and proteins important for the control of transcription can undergo a variety of modifications, such as methylation, acetylation, phosphorylation, and ubiquitinylation. It has been reported that DNA methylation of transfected DNA can play a major role in the regulation of expression. Reports have suggested that methylation causes the repression of gene expression and hypomethylated DNA around the promoter region of genes is often associated with elevated transcriptional activity. Acetylation is also an important step in transcriptional control. As a general rule, transcriptionally active genes usually exhibit acetylation, whereas transcriptionally inactive genes do not. Finally, ubiquitination has been suggested to lead to transcriptionally active DNA by disrupting higher order chromatin structures, hindering internucleosomal interactions, and/or by disrupting the association of linker histones with nucleosomes (Esteller, 2008; Feinberg, 2008).

3 Use of MAR Elements to Boost and Stabilize Expression

Typically, the stability of recombinant cell lines is determined by monitoring cell growth and protein production for several months. For some cell lines, however, protein productivity diminishes over time, usually as a result of changes in the regulation of transgene expression (Strutzenberger et al., 1999). Regulation of higher order chromatin structure is directly coupled with regulation of the expression and integrity of the genetic information of eukaryotes and is likely to be a major force in the origin and evolution of genes, chromosomes, genomes, and organisms. Some of these problems are caused by gene silencing at the level of chromatin – so-called epigenetic gene silencing. Specialized DNA elements known as boundary elements have been shown to mark the borders between adjacent chromatin domains and to serve as barriers against the effects of silencers and enhancers from the neighboring regions (West et al., 2002; Labrador and Corces, 2002; Fig. 3). Perhaps boundaries

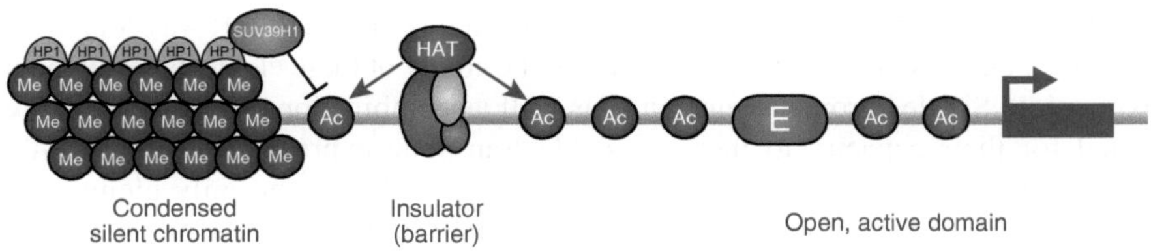

Fig. 3 A model for the barrier activity of insulators. A schematic diagram based on the example of the upstream boundary of the chicken beta-*globin* locus. Insulator proteins constitutively recruit histone acetyltransferases that acetylate flanking nucleosomes (*red spheres*). Acetylation serves to inhibit histone modifications required for the propagation of transcriptionally silent condensed chromatin (*packed bluespheres*). Barriers act to terminate the chain of repressive chromatin by competing in the histone-modification process (West et al., 2002)

delimit structural domains by interacting with each other or with some other nuclear structure (Labrador and Corces, 2002). When these human genetic elements are included in the expression vectors, the chromatin structure flanking the transgene is maintained in an active configuration. Here, we describe the inclusion of a specific type of DNA element that has been used during the past years to interfere with epigenetic gene silencing, with the aim of enhancing and stabilizing transgene expression.

One method to overcome positional dependent inactivation is the use of vectors that include a matrix or scaffold attachment region (MAR or SAR) that repress silencing. The anti-silencing effect observed in the presence of MAR may be mediated by chromatin modifications such as histone hyperacetylation at the site of chromosomal transgene integration locus (Recillas-Targa et al., 2002; Yasui et al., 2002) or changes in a specific subnuclear localization (Bode et al., 2003; Hart and Laemmli, 1998). In addition, the general increase in transgene expression can be explained in several ways. For example, transcription of transgenes can be improved, either directly or indirectly, by an activation of the transgene promoter or enhancer by MARs. MARs may also favor integration in a permissive locus within the chromosome, or they may increase the number of integrated transgene copies.

The MAR element associated with the lysozyme gene in chicken is one of the most studied elements (Zahn-Zabal et al., 2001; Girod and Mermod, 2003; Girod et al., 2005). The chicken lysozyme locus contains a 3 kb regulatory region known as the A element. This element was originally used as a MAR in a series of experiments on the effect of MARs on gene expression (Stief et al., 1989). These experiments were exciting because transgenes flanked by the A element exhibited expression that was proportional to gene copy number ('copy number-dependent'), suggesting that the element had been able to insulate transgene expression from gene silencing or position effects. The intact element has been shown to contain both enhancer and matrix-binding activities. When the intact element was divided into 1.32 and 1.45 kb pieces, both were able to confer copy number-dependent transgene expression. However, when smaller fragments were tested, the portion of the A element that bound to the nuclear matrix no longer conferred copy number dependence (Phi-Van and Strätling, 1996), and the possibility must be considered that at least some of the original effects were attributable to the enhancer portion of the element rather than the matrix-binding portion (Allen et al., 1996).

In the Zahn-Zabal et al. (2001) study, the chicken lysozyme MAR was compared with other chromatin elements with respect to the ability of these elements to augment expression. Single chromatin elements, as well as combinations of elements, were tested for their capacity to increase stable transgene expression in industrially relevant CHO cells. The chicken lysozyme 5′ MAR was the only element to significantly enhance reporter expression in pools of stable clones. While increased expression in pools of stable clones is indicative of an overall positive effect of the chicken lysozyme MAR on transgene expression, it does not provide information as to the probability of isolating a high producer clone. In order to address this issue, individual colonies were isolated and the level of expression of the transgene was measured. CHO cells were transfected with luciferase expression vectors containing no, one, or two MARs, and 15 individual colonies were randomly isolated and analyzed for each construct. Consistent with the results obtained with pools of stable clones, the average expression level of the clones analyzed increases with the number of MARs present on the construct. The use of the MAR elements also increases the proportion of high-producing clones, thus reducing the number of clones that need to be screened. Thus, MARs have been used to improve the expression of transgenes, in cells cultured *in vitro* and *in vivo* (Girod et al., 2005; Gutierrez-Adan and Pintado, 2000; Zahn-Zabal et al., 2001). Furthermore, the expression level of the most productive clones was found to be higher for constructs bearing MARs; therefore, fewer clones needed to be picked and analyzed to identify a high-level production clone when MARs were present on the expression plasmid.

Other types of epigenetic regulatory elements have also been studied, such as chromatin sequences associated with the β-globin gene (Forrester et al., 1989). The β-globin gene locus control region is comprised of five DNase I hypersensitive sites (Ostermeier et al., 2003). Expression vectors containing four of the DNaseI hypersensitive motifs have been shown to increase β-globin mRNA levels 8- to 13-fold following transfection into mouse erythroleukemia cells, while vectors containing just two motif sites increased globin expression to a lesser extent. These first β-globin sequences were seen to display cell-type specificity in that no effect was seen when the constructs were assayed in 3 T3 fibroblasts. More recent studies have also characterized the human β-globin MAR element. Kim et al. (2004) showed that the human β-globin MAR improves transgene expression in CHO cells. They constructed various deletion constructs with different orientations and examined their effects on the frequency of β-Gal positive colonies and on transgene expression levels. The enhancing effects of the human β-globin MAR depended on the integrity of the full-length fragment (regardless of the orientation) as all of the deletion constructs were much less active. Furthermore, there was no effect of the MAR on transient expression (Kim et al., 2004).

Two groups have studied the MAR/SAR elements associated with the human β-interferon gene. Klehr et al. (1991) transfected DNA corresponding to the complete chromatin domain of human β-interferon gene into mouse L cells. When the transgene is flanked by SARs, the gene's transcription was enhanced 20–30-fold with respect to DNAs containing only the immediate regulatory elements. To elucidate the role of SAR elements in the transcriptional enhancement, the position of the genomic element was varied relative to several artificial promoter-gene combinations. The data showed that SARs enhance general promoter functions in an orientation- and partially distance-independent manner; the effect of these elements is restricted to the integrated state of transfected templates. Similar to the results seen by Kim et al. (2004), when studying the β-globin MAR, the SAR elements studied by Klehr et al. (1991) were generally found to have an antagonizing effect during transient expression. Kim et al. (2005a) analyzed the frequency of positive colonies by in situ β-galactosidase staining when the human β-interferon SAR element is included in the vector. Two copies of the human β-interferon SAR element enhanced the frequency of positive colonies only by nearly 40% versus that obtained using one copy of human β-interferon SAR element, although the gene expression was enhanced twofold. The frequencies of positive colonies obtained from two copies of human interferon-beta SAR element and from one copy of human β-globin MAR element are about the same, although the expression of β-galactosidase gene with two copies of human β-interferon SAR element was about 50% greater than with one copy of the human β-globin MAR element.

These data suggest that the additional copy of the human β-interferon SAR element at the flanking region of the β-galactosidase expression unit affects transgene expression more than the frequency of positive colonies. In the case of the expression of recombinant genes in CHO cells, applications of MAR/SAR elements have been reported for the chicken lysozyme MAR element (Zahn-Zabal et al., 2001) and for the human β-globin MAR element (Kim et al., 2004). In a previous study Kim et al. (2004) demonstrated that the human β-interferon SAR element is less effective than the human β-globin MAR element; however, in the 2005 study, the human β-interferon SAR element was more effective than the human β-globin MAR element when two flanking human β-interferon SAR element were used (Kim et al., 2005). Interestingly, the chicken lysozyme MAR element was most effective when the two flanking MAR elements were used, but this was not case for the human β-globin MAR element (Zahn-Zabal et al., 2001; Kim et al., 2004). Therefore, it appears that the enhancing effects of MAR/SAR elements on the expression of recombinant genes require their proper configurations.

4 Identification of MAR DNA Sequences that Mediate Increased Expression

Association of the MARs in the chromosomal DNA with the nuclear matrix organizes the higher order structure of the genome, forming looped structures that are likely to be equivalent to active chromatin domains in terms of transcription as well as replication. The nuclear matrix was originally described as a framework of the nucleus that remains insoluble after selective extraction of histones and DNA in the chromatin loops (Sjakste and Sjakste, 2001; Girod and Mermod, 2003). The MAR sequences are generally AT-rich at 70% and possess potential of DNA bending (Yamasaki et al., 2007). These AT rich regions of MARs, which are composed of either a tract of homopolymeric adenine (dA) or a stretch of adenine.thymine dinucleotides (dA.dT), are thought to play a significant role in MARs functions. However, it has not been clearly elucidated how these unique DNA sequences regulate MAR activities. The binding of regulatory proteins to these A+T sequence motifs as well as the structural features of the A+T rich regions, which include curved DNA configuration (Homberger, 1989), a strong potential for strand separation (Bode et al., 1992), narrow minor groove width (dictated by oligo d.A tracts) may altogether mediate the functional activity of MARs in chromatin remodeling and gene expression.

The A+T rich elements have been shown to have transcriptional activation capacity in stable transformants of both plant and animal cells (Nowak et al., 2001; Bode et al., 1992). They have regions where base pairs tend to break under an unwinding stress (base unpairing region: BUR), centered at a sequence ATATAT that are referred to as BUR nucleation sequence. The tendency of base unpairing in the MAR DNA was shown to be essential in binding to the nuclear matrix and enhancing the promoter activity (Yamasaki et al., 2007).

The correlation between DNA curvature and transcriptional activation has been demonstrated by Ohyama and colleagues (Nishikawa et al., 2003), whereby a 36 bp left hand curved DNA segment activated transcription from the herpes simplex virus thymidine promoter (HSV tk) in transiently transfected COS-7 cells. The curved DNA segment was referred as T4, containing four tracts of 4 oligo dA (5′ GTGAAAAACATGGAAAAACATGAAAAACATGAAAAAC-3′), designed to have a specific left hand rotation. The T4 left hand curved DNA is predicted to have high affinity for histone core and indeed it was shown to associate with nucleosomes. This led the authors to conclude that T4 activated transcription by forming part of the nucleosome, thus, arranging the TATA box of the promoter outwards and therefore, facilitating the initiation of transcription (Nishikawa et al., 2003). Several years later, Ohyama and workers further tested the effects of longer left hand curved DNA segment comprised of 2 to 40 tandem repeats of the T4 segment on transgene activation in COS-7 and HeLa cells (Sumida et al., 2006). All the left hand curved T4 tandem repeats activated HSV tk promoter in transient assays in COS cells. The effect of right hand curved DNA was also tested but had very little effect on promoter activity, at least in the two cell lines tested. The degree of transcriptional activation correlated with the length of the curved DNA. In particular, the T32 segment was the most effective curved DNA segment, activating HSV tk promoter 150-fold relative to control construct with straight DNA fragment.

The effect of curved DNA on transcription was also tested in the context of genomic chromatin in HeLa cells. The T20 segment was shown to activate transcription of reporter gene regardless whether the construct was integrated in intergenic or coding region of a gene. By contrast, the transgene expression was extinguished in the control HeLa cell lines in which the curved DNA was removed from the reporter gene. The results of this study were important in demonstrating that this left hand curved DNA segment minimizes silencing and increases transcription of reporter gene regardless of the locus of integration. The transcriptional activation by T20 functions perhaps in a similar mechanism to that accounted for T4 curved segment but having a more dramatic effect since the T20 segment is longer with higher density of histones. The ability of T20 to "capture" and reposition histones may facilitate the accessibility of the promoter to transcriptional machinery (Kamiya et al., 2007). Nevertheless, it is possible that other regulatory proteins, such as high mobility group (HMG)-1 non-histone chromosomal protein (Landsman and Bustin, 1993) and SATB1 (Bode et al., 1992), both bind to curved DNA structures and further change DNA conformation upon binding.

Naturally occurring A+T sequence motifs derived from the MAR at the 3′ end of immunoglobulin heavy chain enhancer and 5′ upstream of the human β-interferon have also been investigated for their ability to activate transcription. Multimerization of synthetic oligonucleotides containing an AATATATTT sequence motif derived from the two MARs mentioned above, were demonstrated to be potent in increasing SV40 promoter activity in stably transformed mouse L cells, almost comparable to the transcriptional activation levels by the full 2.2 kb hu β-interferon gene MAR (Bode et al., 1992). However, whether the AATATAAT sequence motif may form a complex with nucleosomes is unknown. The transcriptional activity of the multimerized A+T sequence motif may be also related to the structural features of this DNA element and to its ability to bind to the nuclear scaffold. Indeed, the AATATAAT sequence motif was shown to have high potential for base-unpairing and nuclear matrix binding. There seems to be a correlation between the unwinding potential of this sequence and the potency of nuclear-matrix binding as well as transcriptional activation. Mutations of this DNA motif resulted in loss of unwinding property of the MAR, reduced affinity to the nuclear scaffold and loss of capability to enhance the transcriptional activity (Bode et al., 1992). Work by others have also shown that the decrease in the thermodynamic stability of MARs is correlated with enhanced strength in binding to the nuclear scaffold *in vitro* as well as in the ability to activate transcription *in vivo* (Allen et al., 1996; Schubeler et al., 1996). Specific A+T sequence motif from these two MARs may favor transcriptionally active complex by keeping the transcriptional domain in an "open" and "relaxed" conformation since MARs are thought to separate chromatin into strained loop domains. This nucleation site for DNA unwinding may also accept released histones (Clark and Felsenfeld, 1991) from the region and recruit topoisomerases (Gilmour et al., 1986) to prevent condensation of the transcriptional domain (Bode et al., 1996). However, it must be noted that the nuclear scaffold binding strength does not necessarily correlate to the potency of MAR to activate gene expression (Girod et al., 2007). A summary of the ways that the A+T sequence motif of a MAR element may modulate expression is given in Table 1.

Although there are experimental results outlining a role of A+T rich elements in several processes, we still do not understand the underlying mechanisms related to

Table 1 Summary of the mechanisms by which the A+T rich element of MARs may exert their transcriptional activation effect

Feature of A+T Motif	Mechanism	Reference
Maintaining a local "open" chromatin domain	Due to inherent curvature of the DNA	Bode et al. (1992, 1996, 2006)
Organizing chromatin structure into loop domains	By attaching to the nuclear scaffold	Bode et al. (1992, 2006)
Facilitating the unwinding and base-unpairing of DNA sequences	Due to inherent curvature of the DNA	Bode et al. (1992, 2006)
Increasing the accessibility of TATA box and *cis*-DNA elements of the promoter	Functioning as a "trap" for histones	Girod et al. (2007); Kamiya et al. (2007)
Increasing concentration of functional proteins	Serving as target sequences for transcription factors and chromatin modifiers	Gilmour et al. (1986)
Facilitating chromatin remodeling and transactivation	Increasing the concentration of chromatin proteins in a particular region	Gilmour et al. (1986)

the transcriptional initiation and the anti-silencing effects by the A + T rich elements. Further investigations of the role of MAR-associated A + T rich elements are required and may lead to useful applications of this DNA element in "chromatin engineering" (Kamiya et al., 2007). Sequences derived from MARs or synthetically designed oligomers with A + T rich sequence motifs can be a useful tool to increase and maintain high transgene expression for research applications, recombinant protein productions and gene-based therapy. Furthermore, due to their small size, they can be practical when constructing viral or non-viral vectors where the size could be a constraint.

5 Identification of the MAR–Binding Proteins as Mediators of Increased Expression

MAR/SAR activity is unlikely to arise uniquely from the intrinsic properties of its DNA motifs. Rather, the ability to protect against position effect and to regulate transcription may depend on the contribution of the protein factors that bind these motifs (Liebich et al., 2002). Transcription factors binding-sites found in the nuclear matrix are extremely diverse. This is not surprising, as MARs constitute important regulatory elements of genes, involved in DNA replication, transcription, repair, and recombination. Below, we will discuss three known MAR transcription factors: SATB1, CTCF and HMGA family of proteins.

5.1 Special AT-Rich Binding Protein (SATB1)

SATB1 binds to AT rich base-unpairing sequences (Kohwi-Shigematsu et al., 1998) (referred as 'ATC sequence context'), where one strand consists of adenine, thymidine, cytosine but not guanine (Dickinson et al., 1992). SATB1 controls gene expression by anchoring DNA sequences to the nuclear scaffold resulting in the formation of "cage-like structures" that separates heterochromatin from euchromatin (Cai et al., 2003), in a cell-type specific manner, as it is predominately expressed in thymocytes. In addition, SATB1, serves as a docking site for recruiting chromatin remodeling proteins such as ACF, ISWI, HAT and HDAC, and these chromatin modifiers were suggested to activate or suppress gene expression through nucleosome remodeling histone acetylation or deacetylation at SATB1 bound MARs (Yasui et al., 2002; Kumar et al., 2005).

The ability of SATB1 to recruit either HAT (coactivator) or HDAC (corepressor) appears to be mediated by the phosphorylation state of SATB1 (Kumar et al., 2006). A study conducted in T cells demonstrated that the phosphorylation of SATB1 by protein kinase C (PKC) was followed by the recruitment of HDAC1 to the IL-2 promoter, resulting in repression of IL-2 transcription. Dephosphorylation of

SATB1 exerted the opposing effects, whereby; the interaction of SATB1 with PCAF causes the derepression of the IL-2 gene. Kumar et al. suggest that a similar mechanism involving the posttranslational modification of SATB1 may be involved in the global regulation of gene expression.

An elegant study by Cai et al. (2006) demonstrated how SATB1 regulates long-range intrachromosomal interactions by changing the chromatin loop landscape to coordinate the expression of several genes (Il4, Il5, Il13) in T-helper 2 (TH-2) cells. Two essential methods were used in their study: (1) chromatin conformation capture (3C) assay to determine whether two remote genomic sequences interact, and (2) CHIP-loop assay to determine chromatin loops that are attached at their bases with a specific protein. Their results showed that upon TH-2 cells activation, SATB1 expression was induced to assemble a transcriptionally active chromatin structure at the cytokine locus. The "cage-like structure" was made up of numerous loops, all attached to SATB1 at their bases. In addition, histone H3 acetylated at Lys9 and Lys 14, c-maf (a transcription factor in TH-2 cells important for Il13 expression), chromatin remodeling enzyme Brg1 and RNA polymerase II are all bound within this 200-kb region. When SATB1 expression was reduced using RNAi, the TH-2 cells did not form a dense loop structure (same structure as found in inactivated TH-2 cells) and as a result, Il4, Il5, Il13 were not expressed. Therefore, this study has provided an insight to how SATB1 may coordinate the expression of multiple genes in a cluster by bringing them to closer proximity. This would allow for a more efficient interaction between the promoter of these genes and transcriptional regulatory factors.

In another study, Kumar et al. (2007) described how SATB1 organizes the gene rich region of MHC-1 class locus into several chromatin loops by anchoring the MARs to the nucleus at specific distances to ensure proper expression of genes within the locus. Promyelocytic leukemia (PML) oncoprotein, a protein associated with the nuclear body was identified as a SATB1 interacting protein. Together, SATB1 and PML formed a functional complex with putative MARs at the base of the chromatin loops. They mapped five chromatin loops within the MHC-I locus in Jurkat cells (Kumar et al., 2007) spanning 300 kb of the MHC-I locus containing HCG-9 and HLA-F genes in the presence and absence of SATB1. SATB1 appears to be involved in the activation of HCG-9 and repression of the expression of most other genes on the MHC-I locus (see Fig. 4). Using SATB1 RNAi, the chromatin architecture of MHC-1 locus underwent a reorganization taking chromatin structure similar to that of γIFN treated Jurkat cells, leading to upregulation of HCG4, HCG4P6 as well as HCG-9. The expression of HCG-9 was enhanced when the gene became part of the giant loop.

Although SATB1 is predominantly expressed in thymocytes, its expression in other cell-types also affects chromatin organization and the regulation of many genes. Recently, Kohwi-Shigematsu and workers (Han et al., 2008) showed that the SATB1 expression in breast cancer cells is associated with these cells to become metastatic by reprogramming the chromatin organization, thus resulting in the upregulation of metastasis-associated gene. When SATB1 expression was abolished by RNA-interference in highly aggressive cancer cells (MDA-MB-231),

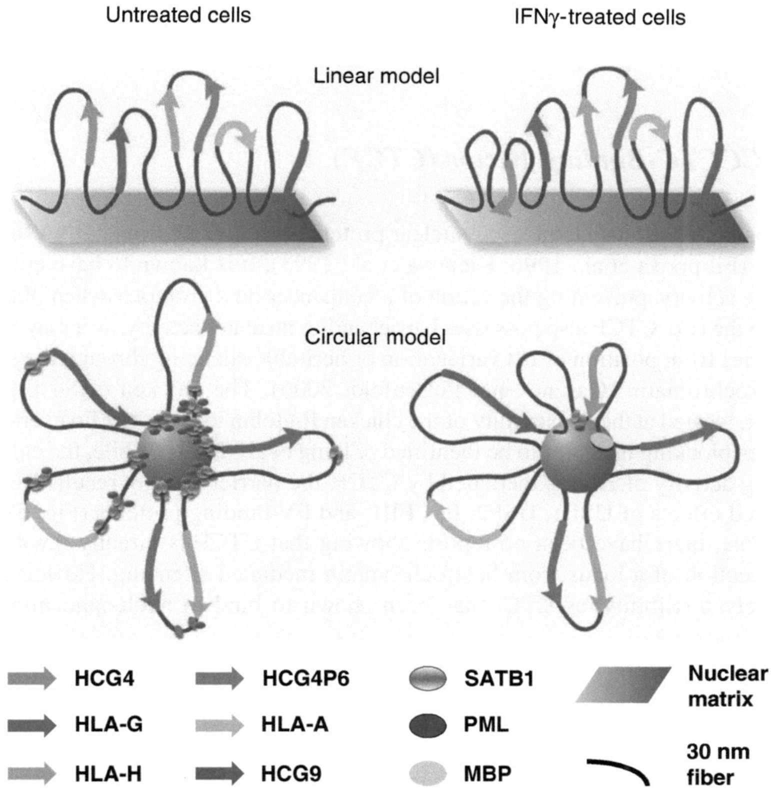

Fig. 4 Schematic representation of the chromatin loop structure of MHC-1 locus in control cells and γIFN treated cells as determined by ChIP-loop assay (details of the experiment and results are described in Kumar et al., 2006). Diagrams on *top* depict the loop landscape in linear fashion while those at the *bottom* indicate the same in a circular manner. On the *lower left*, the non-random distribution of SATB1 and PML across the MHC-I locus is exhibited by depicting the occupancy of the two proteins deduced by chromatin immunoprecipitation assay. On *lower right side*, the diagram depicts only the major changes upon γIFN treatment, notably the chromatin loop containing HCG-9 that becomes larger and extends out from the core of the chromosome, resulting in enhanced HCG-9 gene expression as well as the replacement of SATB1 by another MAR-binding protein (MBP, depicted by a *yellow ellipse*) (modified from Galande et al., 2007)

expression of more than 1,000 genes were altered and reversed the tumor growth and metastasis *in vivo*. SATB1 expression in non-aggressive (SKBR3) cells induced the expression of many genes that are associated with aggressive-tumor phenotypes, causing these cells to acquire the ability to metastasis *in vivo*. HMGA, another MAR associated transcription factor that binds to the minor groove of AT rich regions, is also implicated in breast cancer cell progression (Reeves et al., 2001) and may cooperate with SATB1 to promote cell growth and differentiation (Han et al., 2008). The authors propose that SATB1 may be used as a "molecular indicator" to predict the progression of breast tumors and future studies should investigate SATB1 as a therapeutic target for metastatic breast diseases.

5.2 CCCTC-Binding Factor (CTCF)

CTCF is a ubiquitously expressed nuclear protein with 11-zinc finger DNA binding domain (Filippova et al., 1996; Klenova et al., 1993). It is known to have enhancer blocking activity, preventing the action of an enhancer on a promoter when placed in between the two. CTCF also possesses barrier and/or insulator activity, as it may protect transgenes from position effect variegation or heritable silencing through the spread of heterochromatin (Gaszner and Felsenfeld, 2006). The chicken cHS4 β-globin insulator, located at the 5′ extremity of the chicken β-globin locus, is the first vertebrate enhancer-blocking insulator to be identified (Chung et al., 1993). While, the enhancer blocking activity of HS4 is mediated by CTCF, the barrier activity results from the combined effects of USF1, USF2, FI-, FIII- and FV-binding proteins (Fig. 5).

To date, there have been no reports showing that CTCF is directly involved in the protection of a locus from heterochromatin mediated silencing. However, this is a likely possibility, as CTCF has been shown to bind to nucleophosmin (also

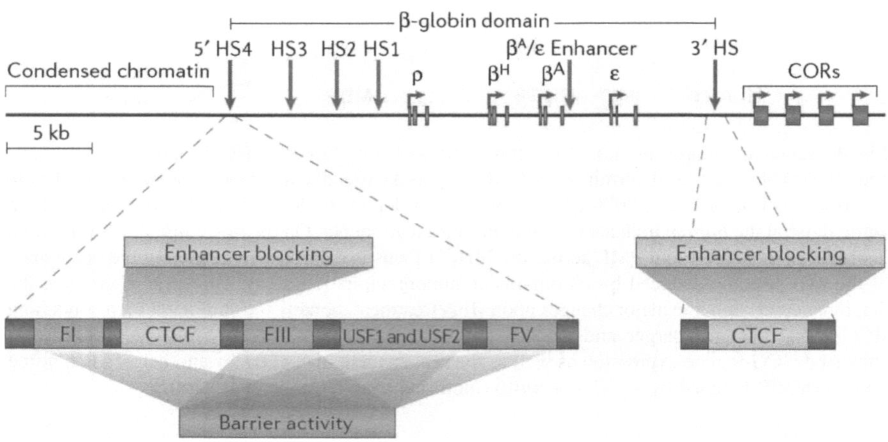

Fig. 5 In the chicken β-globin gene locus, the 5′HS4 and 3′HS insulator elements define the limits of a chromatin domain that encompasses the developmentally regulated beta-globin gene cluster and its locus-control region (LCR), which is comprised of the HS1–3 and β A/epsilon enhancers. The HS4 element possesses both enhancer blocking and barrier activity, presumably to prevent the LCR from inappropriately activating genes outside the domain and at the same time protecting the globin cluster against silencing that emanates from the flanking condensed-chromatin region. Enhancer blocking is mediated by CTCF, whereas barrier activity results from the combined effect of USF1 and USF2 and the as yet uncharacterized FI-, FIII- and FV-binding proteins. 3′ HS binds CTCF and functions only as an enhancer-blocking insulator (reprinted by permission from Macmillan Publishers Ltd: Nat. Rev. Genet., Gaszner and Felsenfeld, 2006, copyright 2006)

known as B32), suggesting that CTCF may anchor to the nuclear matrix, creating an independent "loop" domain that would shield the transgene from silencing. Work by Rincon-Arano et al. (2007) demonstrated that the chicken cHS4 β-globin insulator protected transgene from silencing by the telomeric heterochromatin. In this study, EGFP transgene integration was targeted to the telomere of the chicken cell line HD3 with or without cHS4 β-globin insulator. The cHS4 β-globin insulator sustained transgene expression of a single-copy integrant for over 100 days. By contrast, the un-insulated single copy clones showed a rapid extinction of the transgene expression. RNAi-mediated knockdown of USF1 did not alter cHS4 protection of the transgene from telomeric silencing, demonstrating that cHS4 insulation of the transgene is not dependent on USFI. There was no direct evidence for the role of CTCF in the protective effect. Recruitment of CTCF as a fusion to the GAL4 DNA binding domain did not protect from telomeric silencing (Esnault et al., 2009), suggesting that other nuclear factors must be recruited to the cHS4 to play a role in the protection of the transgene against telomeric position effect (TPE).

The role of CTCF as an enhancer-blocking insulator in the regulation of gene imprinting and monoallelic gene expression (Fedoriw et al., 2004; Ling et al., 2006) is well characterized in the imprinted IGF-2 (insulin-like growth factor 2)-H19 locus. IGF-2 is only expressed from the paternal allele, whereas H19 is expressed from the maternal allele, both genes sharing the same enhancers. The imprinting control region (ICR) is located at the 5′ flank of H19 gene and its deletion results in biallelic expression of both IGF-2 and H19, suggesting the role of H19-ICR to repress the maternal IGF-2 allele (Thorvaldsen et al., 1998). CTCF binding to the H19-ICR is thought to be required for the IGF-2 repression (Kaffer et al., 2000). Work by Kurukuti et al. (2006) demonstrated that the IGF-2 promoter on the paternal chromosome interacts with the enhancers. By contrast, on the maternal allele, this interaction is prevented by CTCF binding to the maternal ICR. CTCF binding to the ICR regulates its interaction with matrix attachment region 3 (MAR3) and differentially methylated region (DMR) 1 at IGF-2 gene, forming a condensed loop around the maternal IGF-2 locus. As a result, the interactions between of IGF-2 promoter and the H19 enhancers are prevented, leading to the silencing of IGF-2 expression (Fig. 6).

Since the initial discovery of CTCF, there has been great interest in identifying potential binding sites for CTCF in the eukaryotic genome as this knowledge is essential to understand how *cis*-regulatory elements coordinate expression of target genes. Kim et al. (2007) identified over 13,000 novel putative CTCF-binding sequences as well as confirmed CTCF binding sites in the human genome, using chromatin immunoprecipitation followed by genome-tiling microarrays methodology (Kim et al., 2005b). They found that most of the putative CTCF binding sites are located far from the transcriptional start sites and that their distribution is strongly correlated with genes. Interestingly, CTCF localization appears to be similar in different cell types, as determined by their analysis on the primary human fibroblast cells and hematopoietic progenitor cell line U937. In some cases, CTCF binding sites were located at the boundaries of distinct chromatin structures, but this was not a general phenomenon, as many other binding sites did not coincide with boundaries. Again this evidence points to the fact that if CTCF contributes to the

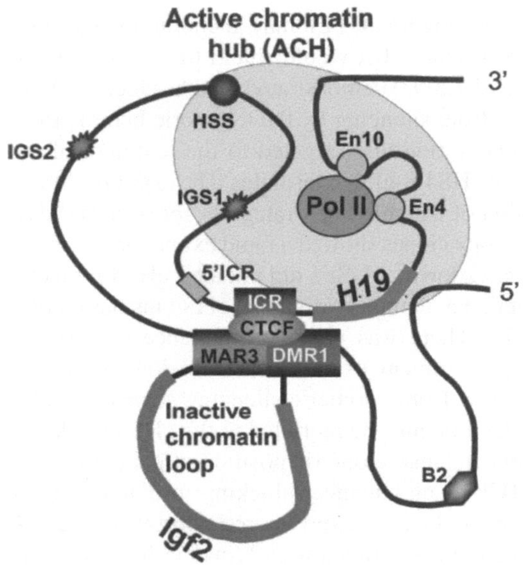

Fig. 6 Model showing contacts established within the maternal allele at the IGF-2/H19 region in neonatal liver. The model suggests a mechanism of how CTCF controls the repression of maternal IGF2 gene (located within inactive chromatin loop). CTCF binding to the H19 ICR, regulates its interaction with the matrix attachment region (MAR)3 and differentially methylated region (DMR)1, forming a condensed loop around the IGF-2 gene, restricting the H19 enhancers (en4 and en10) access to the IGF-2 promoter. This model is based on results from neonatal liver only and may not apply to other tissues (Kurukuti et al., 2006; copyright 2006 National Academy of Sciences, USA)

establishment of chromatin boundaries by elements such as MARs, other as yet unidentified activities must also contribute to the boundary effect.

5.3 High Mobility Group (HMGA)

The high mobility group A family of proteins comprises of 3 proteins HMGA1a, HMGA1b, and HMGA2 (previously known as HMGI, HMGY, and HMGI-C respectively) (Sgarra et al., 2004). These proteins contain three positively highly charged regions called the AT-hook since they bind to the minor groove of AT rich sequences of the promoter regions and MARs. HMGA1 protein has been found to co-localize with the enzyme topoisomerase II and histone H1 (Saitoh and Laemmli, 1993; Saitoh and Laemmli, 1994) suggesting that it acts as a regulator of gene transcription by controlling the structure of chromatin. It has been demonstrated that HMGA proteins can serve as transcriptional activators in the context of chromatin by displacing histones H1 from MAR sequences (Zhao et al., 1993).

Earlier footprinting studies showed that HMGA proteins preferentially bind to a stretch of five or six AT base pairs (Solomon et al., 1986). However, more recent studies have shown that HMGA proteins have more sequence specificity, requiring two or three appropriately spaced AT rich sequences as a single multivalent binding site (Maher and Nathans, 1996). For example, HMGA proteins simultaneously bind to two or three runs of AT base pairs in the regulatory regions on human β-interferon enhancer (Thanos and Maniatis, 1992), and the promoter regions of interleukin-2 (Baldassarre et al., 2001) and interleukin-2 receptor α–chain gene (John et al., 1995). Using a PCR-base systematic evolution of ligands by exponential enrichment (SELEX), Cui and Leng (2007) identified two consensus sequences for HMGA2: 5′-ATATTCGCGAWWATT-3′ and 5′-ATATTGCGCAWWATT-3′, where W represents A or T. These sequences can be divided into three segments: the first segment has five base pairs that is AT rich, the middle segment has four base pairs that GC-rich and the last segment has six base pairs that is AT-rich. All three segments are required for HMGA2 binding.

Indirect evidence for the role of HMGA in MAR functional activity was provided by studies performed using aggressive breast carcinoma cell lines, showing elevated expressions of HMGA1a and HMGA1b as compared to non-metastatic cells (Liu et al., 1999). South-western blot analysis using whole protein extracts from these tumor cells exhibited strong binding of these HMGA proteins to a synthetic MAR probe composed of multimer containing the 25-bp sequence derived form a MAR 3′ of the IgH enhancer. This 25-bp sequence of MAR is a base-unpairing region (BUR) and it binds to the nuclear matrix with high affinity. Western blot and protein sequencing analysis confirmed that these BUR-binding proteins were indeed HMGA proteins. By contrast, the HMGA proteins were shown to bind poorly to a mutated MAR probe that is still AT-rich but has lost the unwinding propensity. Therefore, HMGA proteins appeared to strictly bind to base-unpairing sequences, one of the key structural element of MARs, and they may participate in gene regulation to trigger metastatic phenotype in breast cancer cells. Similarly to SATB1, HMGA proteins may thus be used as a biomarker for tumor progression. Whether the implication of these proteins in cancer progression may result from their proposed contribution to MAR activity is an interesting but as yet unestablished possibility.

Other MAR transcription factors include B-cell specific protein called BRIGHT (Herrscher et al., 1995), NMP4 proteins known to bind to minor groove of homopolymeric (dA:dT) sites in the core unwinding regions of MARs (Torrungruang et al., 2002) and scaffold attachment factor-A (SAF-A), a multifunctional matrix specific factor that recognizes AT-rich DNA sequences (Romig et al., 1992). These proteins may also contribute to mediating some of the conformational and/or chromatin structure effects of the MAR, but their specific contribution(s) to these effects remain to be identified.

6 Effects of MARs on the Copy Number of Integrated Transgenes

In the examples given above, it is clear that MAR elements can enhance and maintain long-term expression by acting on the structure of chromosomes and of chromatin, and that these effects can lead to increased transcription of the transgenes.

In some studies, the MAR elements also appear to reduce the variability within a polyclonal cell population. Thus, MARs may provide more consistent and elevated transcription to each integrated transgene copy. However, other mechanisms such as those that relate to the transgene copy number may also concur to increased expression. In this section, we discuss the role that MAR elements may take to augment transgene integration in the host genome, thereby yielding increased transgene copy number and overall expression.

Several studies demonstrated that MAR elements increase the number of integrated transgene copies in transfected plants and mammalian cells. For instance, it was found that the inclusion of the human MAR 1-68 in transfected plasmids significantly enhances the number of copies integrated in the host genome, as compared to cells transfected without MAR. Indeed, quantitative PCR assays performed either on stable cell populations or clones confirmed a 3–4-fold higher transgene copy number in cells transfected with the MAR 1-68 (Girod et al., 2007; Galbete et al., 2009), in agreement with previous observations (Kim et al., 2004; Girod et al., 2005). In Kim et al. (2004), the authors observed higher transgene copy numbers when these genes are co-transfected with the human β-globin MAR in CHO cells. Girod et al. (2005) achieved similar results in CHO cell clones transfected with the chicken lysozyme MAR. Furthermore, fluorescent *in situ* hybridization analysis of metaphase chromosomes of stable polyclonal populations showed generally much greater intensity of a fluorescent probe in cells transfected with MARs, therefore confirming the increase of transgene integration (Girod et al., 2007).

Similarly, many examples showed that MAR elements renhanced expression in a copy-number dependent manner. For example, transgenic mice carrying multiple copies of a reporter gene flanked by the chicken lysozyme MAR expressed the gene at levels proportional to copy number, indicating that a complete gene locus, as defined by its chromatin structure, functions as an independent regulatory unit when introduced into a heterologous genome (Bonifer et al., 1990, 1994). In addition, the presence of a MAR from the chicken lysozyme locus reduced variability and conferred a copy number-dependent increase in transgene expression in transgenic rice plants (Oh et al., 2005). However, Park and Kay (2001) observed that the chicken lysozyme MAR did not improve the number of proviral DNA copies integrated in mouse hepatocytes whereas the immunoglobulin-kappa MAR exhibited a 2.5-fold augmentation.

In contrast, a study by Wang et al. (2007) revealed that the expression of the CAT enzyme in stably transformed lines of the microalgae *Dunaliella salina* was not significantly proportional to the gene copy numbers, suggesting that the effects of MARs on transgene expression may not be through increasing transgene copies. In addition, it was shown that in preimplantation mouse embryos, flanking SARs stimulated transgene expression in a copy-dependent manner. But in the differentiated tissues of newborn and adult mice, correlation with copy number was lost (Thompson et al., 1994). Furthermore, Baur et al. (2004) demonstrated that even a single gene copy might also result in a variegated expression, as show by the spontaneous changes of expression of a luciferase reporter gene integrated near HeLa cell telomeric heterochromatin. Thus, there is a clear benefit in including MAR elements in the transfection vector to increase transgene expression.

However, the ability of a MAR to confer a copy number-dependent expression to the transgene by insulating them to prevent gene silencing or position effect is less clear. Differences in the effect noted by various experimenters, in some occasions working with the same MAR element, may result from other factors influencing expression, such as the promoters and vector backbones as well as the cell lines and transfection methods used.

MAR elements appear to be able to counteract silencing effects, as exemplified when comparing stable cell populations transfected with or without MAR. Transgene copy number and cell fluorescence levels were shown to correlate well in the presence of MAR, indicating that the increase of transgene expression results from a similar increase in transgene integration (M. Grandjean and N. Mermod, unpublished data). In contrast, the normalization of EGFP mRNA levels relative to the gene copy number from stable cell clones indicated that the MAR increased gene expression by twofold on average (Galbete et al., 2009). Thus, increased transgene expression observed with MAR is likely to result both from the integration of more transgene copies in the genome of cells and from MAR-mediated inhibition of epigenetic silencing events that are associated with the integration of tandem gene copies.

It is known that mammalian cells in culture contain the enzymatic machinery required to mediate recombination between newly introduced plasmid DNA molecules and that the frequency of homologous recombination or non-homologous end-joining between co-injected plasmid molecules to form concatemers is extremely high, approaching 100% (Folger et al., 1985). However, integration of one of these concatemers into one of the chromosome is a relatively rare event in mammalian cells (Folger et al., 1985). As a result, multiple copies of the transfected gene are not scattered throughout the host genome, but they co-integrate as concatemer at a single locus in the host chromosome, usually in tandem head-to-tail orientation (Folger et al., 1982), the integration site being different in independent transformants (Robins et al., 1981). However, recombination between the newly introduced DNA from transfection and its homologous chromosomal sequence occurs exceedingly rarely in mammalian cells, at a frequency of 1:1,000 cells receiving DNA (Thomas et al., 1986).

The high copy number of transgenes integrated in the genome of the cells with a MAR does not result from a more efficient plasmid import into the nucleus during transfection, or from the occurrence of multiple chromosomal integration events (Girod et al., 2007; Grandjean et al., personal communication). This effect might rather be linked to an effect of MAR on increased DNA concatemerization and/or facilitated transgene integration. Indeed, MARs may play a role as DNA recombination signals. It was previously shown that MAR elements could regulate recombination processes such as immunoglobulin gene rearrangement (Xu et al., 1996). Breakpoints of recurrent deletions and translocations in leukemia were found to occur at MARs, thus facilitating their illegitimate recombination at the nuclear matrix (Iarovaia et al., 2004). Finally, retroviruses showed a strong preference for integration in the vicinity of MARs (Johnson and Levy, 2005).

How MARs may increase transgene integration is currently unknown. Because they mediate a permissive chromatin structure, MARs could improve homologous recombination between transfected plasmids, thus allowing the formation of larger concatemers, yielding the observed increased number of gene copies that integrate within the genome of cultured cells without leading to multiple integration sites within a transfected cell. Alternatively, but not exclusively, MAR may interact with proteins of the repair machinery that are known to contribute also to homologous recombination and non-homologous end-joining events.

7 Effects of MARs on Transgene Expression Variegation

In addition to the effects of MARs on transgene silencing and integration, an additional role of these elements in preventing variegation is currently being uncovered. The high variability among independent transformants in stable expression is thought to depend on the site of transgene integration in the chromosome (Kalos and Fournier, 1995; Recillas-Targa et al., 2002). Indeed, transgene integration may be influenced by the fortuitous presence of regulatory elements at the random integration locus in the host genome. In addition, transgene expression is thought to reflect particular chromatin structure coming from adjacent chromosomal domains (Robertson et al., 1995; Henikoff, 1996; Wakimoto, 1998). However, variability of expression can also be noted from distinct cells within a monoclonal population that have the transgene integrated at the same chromosomal locus. This effect, described as variegation, is most clearly seen when individual cells express transgenes with easily detectable products such as short half-life fluorescent proteins. However, the extent of this effect will itself vary when assessing individual clones, and some integration sites may thus be more prone to variegation than others.

The human MAR 1-68 was found to decrease variegation in addition to its effect to improve transgene expression, as cells within individual colonies showed similar levels of expression (Girod et al., 2007). The localization of transgene integration sites, as assessed by fluorescent *in situ* hybridization, did not show any multiple integration events and transgenes did not appear to be targeted to any specific chromosomal sites or particular chromosomal structures in cells transfected with the MAR 1-68 (Derouazi et al., 2006; Girod et al., 2007). Time-lapse microscopy of GFP expression in single cells indicated that MAR 1-68 mediated constant transgene expression, while cells generated without this element would cycle between states of high expression and silent states within a time frame of hours and days (Galbete et al., 2009). Thus, in addition to their long-term effects on the inhibition of heterochromatin formation, MAR can also act positively to mediate constant gene transcription, as opposed to expression cycling usually obtained from transgenes devoid of these epigenetic regulators. The MAR effect on expression variegation was discovered recently, and the molecular mechanisms that oppose a variegated expression pattern remain uncharacterized, but it may conceivably be linked to the action of MARs on chromatin structure and/or on the assembly or firing of transcription initiation complexes at promoters.

8 Isolation of Potent MAR Elements via Bioinformatics to Generate Producer Cell Lines

In the above section, we have discussed MAR elements that have been identified associated with specific genes: chicken lysozyme, β-globin and β-interferon, for example. Here we discuss the isolation of MAR elements on a genome-wide approach using bioinformatics. Since no unique consensus sequence for MARs has been found (Boulikas, 1993; Kramer and Krawetz, 1995), identifying nuclear matrix attachment DNA regions *in silico* based on high A+T percentage has proven feasible (Evans et al., 2007; Girod et al., 2007). Girod et al. (2007) designed a computational method to predict MARs from human genomic sequences based on the specific characteristics of the A+T rich region such as the low melting temperature, high curvature, deep major groove depth and wide minor groove width of the DNA, as well as from the occurrence of binding sites for particular transcription factors (Girod et al., 2007). They identified 1,566 high-scoring putative MAR sequences when the algorithm was set with very stringent parameters. Out of these, they selected seven putative MAR elements for further analysis based on the presence of A+T rich core region and putative binding sites of transcription factors. All the selected seven potential MARs contain a long stretch of DNA (200 bp to 1.5 kb in length) made up of approximately 70–85% AT dinucleotides, almost devoid of any guanine and cytosine nucleotides. They assessed the ability of each MAR to activate transgene expression and found that all but one of the seven newly identified MAR elements augmented substantially EGFP transgene expression in stably transfected CHO cells. One of these MARs significantly increased IgG production in CHO cells and maintained high expressions of erythropoietin transgene from an inducible doxycycline promoter in mice. Whether or not the AT rich regions of these MARs play a significant role in minimizing silencing and activating transgene transcription in CHO or animal model is to be further investigated.

Computational analysis on the 3-kilo base pair chicken lysozyme 5′ MAR elements showed three regions within this MAR that contain potentially curved DNA structures, a deep major groove and low DNA melting temperatures (Fig. 7). Within these regions there are short A+T rich sequence motifs composed of stretches of 6 to 10 oligo dA predicted to mediate nucleosome positioning and curved DNA configuration. There appear to be a correlation between the distribution of nucleosome positioning motifs and the sequences that increase EGFP expression levels. However, it is not clear if the A+T rich elements alone confer most of the MAR effects in sustaining EGFP transgene expression in CHO.

These novel human elements overcome the need for amplification, and assure that all copies of the transgenes are actively expressed. Girod et al. (2007) assessed the effect of MAR 1-68 on antibody expression in CHO cells. In this study a comparison of the MAR 1-68 and the chicken lysozyme MAR was made with respect with the ability of each element to augment protein production. The highest antibody expression occurred using MAR 1-68, with one clone secreting over 70 picograms of antibody per cell and per day (p/c/d). This compares favorably with

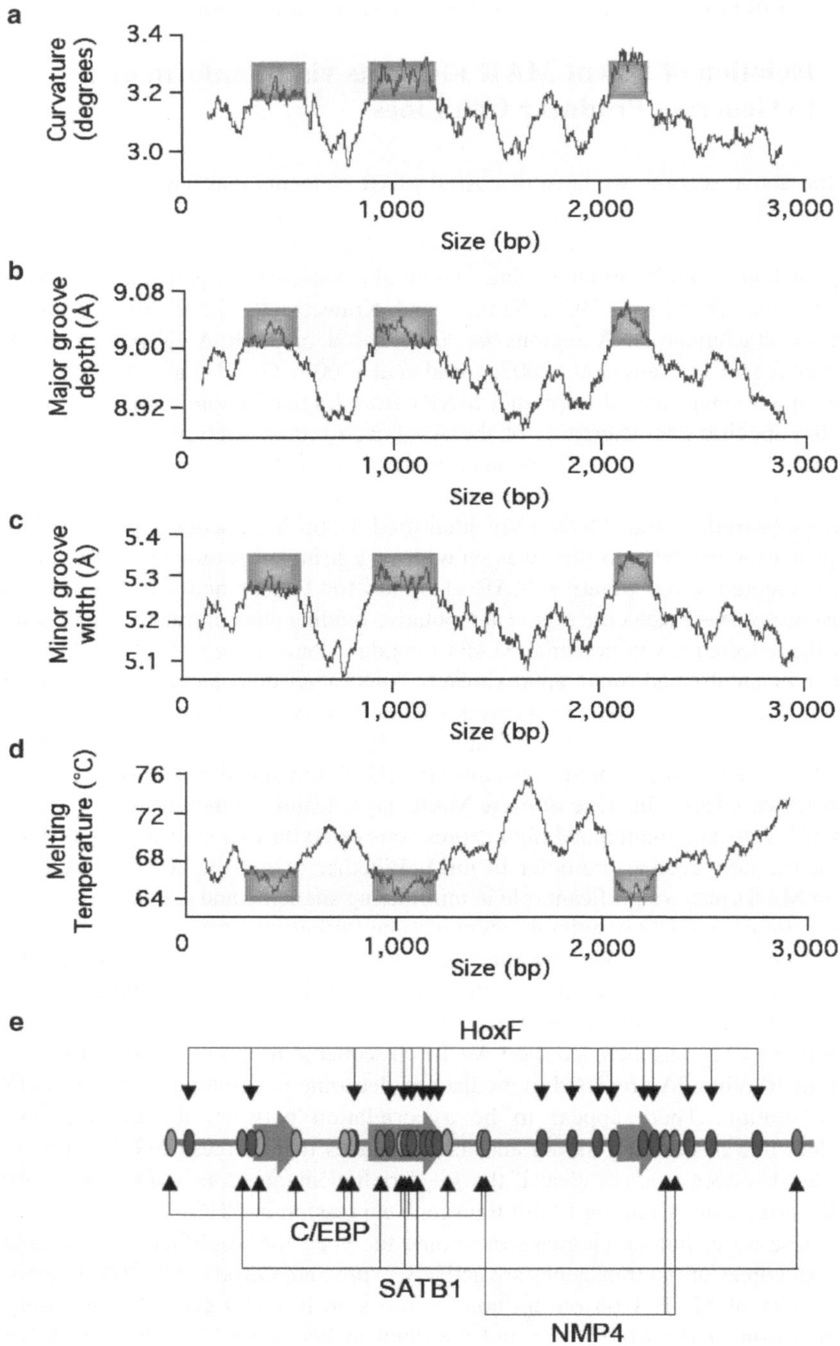

Fig. 7 Computational (SMARScan) analysis of chicken lysozyme MAR. (**a**) Double helix bending angle, (**b**) major groove depth, (**c**) minor groove width, (**d**) DNA melting temperature, and (**e**) schematic diagram of chicken lysozyme MAR with putative binding sites for transcription factors C/EBP, Hox F, NMP4 and SATB1 (marked by *colored ellipses*) (from the authors own work, Girod et al., 2007)

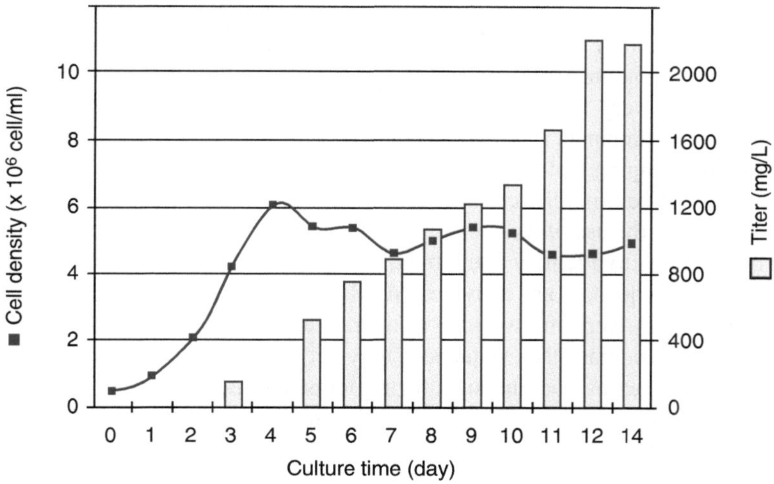

Fig. 8 Cell densities and IgG titers are from 1-liter bench-top bioreactors seeded at a target seed density of 0.5×10^6 cells/ml with a CHO cell clone producing an immunoglobulin gamma (IgG) under the control of a human MAR element. Cell line identification was performed without transgene amplification during a 15 week period, and titers and cell densities were determined before process or media optimization (from the authors own work, Varghese et al., 2008)

the levels achieved with the chicken lysozyme MAR elements, which peaked at 30 p/c/d. Approximately one clone in 30 shows a productivity of 30 p/c/d or higher with the human MAR, whereas isolation and screening of more than 300 clones was required using the chicken lysozyme MAR. Clones secreting large amounts of immunoglobulin were adapted to growth in suspension in serum-free synthetic medium, and they maintained high and stable expression without selection pressure as long as tested, during several months.

These human MARs, and new sequences derived from these elements, are currently being used to generate cell lines for the commercial production of pharmaceuticals as well as for diagnostic kits. An example of the productivities that are routinely obtained is illustrated in Fig. 8. The main benefits of the incorporation of such elements in expression vectors are reduced time, efforts and costs for the generation, screening and characterization of cell lines, often coupled to a gain in productivity, because stable and very high producer clones can be obtained from the screening of few cell lines.

9 Conclusions

MAR elements have been linked to a bewildering array of activities, including the formation of higher order chromosomal loops and their positioning to sub-nuclear compartments enriched in proteins mediating DNA transcription and RNA maturation, the recruitment of proteins mediating chromatin modifications that decrease silencing effects, the reduction of variegation effects that limits transgene expression, and increased transgene integration into the cell genome. All of these effects are likely to contribute to very high expression levels of adjacent genes, yielding elevated production of recombinant proteins by cells such as the CHO or HEK293 lines. Specific productivities ranging from 20 to 100 picogram p/c/d have been reported during the development of commercial cell lines, and titers above 5 g/l have been achieved.

In addition to the use of these elements to generate producer cell lines, progresses have been made to identify the fundamental constituents of MARs, despite a wide variety of sequences and activities. MARs appear to act as scaffolds that combine DNA and protein elements working cooperatively to control chromatin structure. For instance, particular DNA sequences may act to position nucleosomes, whereas other sequences act as docking sites for proteins that mediate modifications of the histones and a gene expression-permissive chromatin structure. At present, however, a detailed molecular understanding of the contribution of these elements to the action of MARs on gene expression or DNA recombination is missing, which has precluded the assembly of totally synthetic MARs from individual optimized building blocks. Nevertheless, we speculate that these goals that will be found worthy of further research efforts, and that these efforts in turn will yield even simpler procedures to construct mammalian cell lines that produce high titers of recombinant proteins.

References

Allen G, Hall GJ, Michalowski S, Newman W, Spiker S, Weissinger A, Thompson W (1996) High-level transgene expression in plant cells: effects of a strong scaffold attachment region from tobacco. Plant Cell 8:899–913

Almeida R, Robin C (2005) RNA silencing and genome regulation. Trends Cell Biol 15:251–258

Andrulis IL, Duff C, Evans-Blackler S, Worton R, Siminovitch L (1983) Chromosomal alterations associated with overproduction of asparagine synthetase in albizziin-resistant Chinese hamster ovary cells. Mol Cell Biol 3:391–398

Baldassarre G, Fedele M, Battista S, Vecchione A, Klein-Szanto AJ, Santoro M, Waldmann TA, Azimi N, Croce CM, Fusco A (2001) Onset of natural killer cell lymphomas in transgenic mice carrying a truncated HMGI-C gene by the chronic stimulation of the IL-2 and IL-15 pathway. Proc Natl Acad Sci USA 98:7970–7975

Barnes LM, Dickson AJ (2006) Mammalian cell factories for efficient and stable protein expression. Curr Opin Biotech 17:381–386

Barnes LM, Bentley CM, Dickson AJ (2001) Characterization of the stability of recombinant protein production in the GS-NSO expression system. Biotechnol Bioeng 73:261–270

Barnes LM, Bentley CM, Dickson AJ (2003) Stability of protein production from recombinant mammalian cells. Biotechnol Bioeng 81:631–639

Barnes LM, Bentley CM, Dickson AJ (2004) Molecular definition of predictive indicators of stable protein expression in recombinant NSO myeloma cells. Biotechnol Bioeng 85:115–121

Barnes LM, Moy N, Dickson AJ (2006b) Phenotypic variation during cloning procedures: analysis of the growth behavior of clonal cell lines. Biotechnol Bioeng 94:530–537

Barnes LM, Bentley CM, Moy N, Dickson AJ (2007) Molecular analysis of successful cell line selection in transfected GS-NSO myeloma cells. Biotechnol Bioeng 96:337–348

Baur JA, Shay JW, Wright WE (2004) Spontaneous reactivation of a silent telomeric transgene in a human cell line. Chromosoma 112:240–246

Bode J, Kohwi Y, Dickinson L, Joh T, Klehr D, Mielke C, Kohwi-Shigematsu T (1992) Biological significance of unwinding capability of nuclear matrix-associating DNAs. Science 255:195–197

Bode J, Stengert-Iber M, Kay V, Schlake T, Dietz-Pfeilstetter A (1996) Scaffold/matrix-attached regions: topological switches with multiple regulatory functions. Crit Rev Eukaryot Gene Expr 6:115–138

Bode J, Goetze S, Heng H, Krawetz SA, Benham C (2003) From DNA structure to gene expression: mediators of nuclear compartmentalization and dynamics. Chromosome Res 11:435–445

Bode J, Winkelmann S, Gotze S, Spiker S, Tsutsui K, Bi C, A KP, Benham C (2006) Correlations between scaffold/matrix attachment region (S/MAR) binding activity and DNA duplex destabilization energy. J Mol Biol 358:597–613

Bonifer C, Vidal M, Grosveld F, Sippel AE (1990) Tissue specific and position independent expression of the complete gene domain for chicken lysozyme in transgenic mice. EMBO J 9:2843–2848

Bonifer C, Yannoutsos N, Kruger G, Grosveld F, Sippel AE (1994) Dissection of the locus control function located on the chicken lysozyme gene domain in transgenic mice. Nucl Acids Res 22:4202–4210

Boulikas T (1993) Nature of DNA sequences at the attachment regions of genes to the nuclear matrix. J Cell Biochem 52:14–22

Cai S, Han HJ, Kohwi-Shigematsu T (2003) Tissue-specific nuclear architecture and gene expression regulated by SATB1. Nat Genet 34:42–51

Cai S, Lee CC, Kohwi-Shigematsu T (2006) SATB1 packages densely looped, transcriptionally active chromatin for coordinated expression of cytokine genes. Nat Genet 38:1278–1288

Chung JH, Whiteley M, Felsenfeld G (1993) A 5′ element of the chicken beta-globin domain serves as an insulator in human erythroid cells and protects against position effect in Drosophila. Cell 74:505–514

Clark DJ, Felsenfeld G (1991) Formation of nucleosomes on positively supercoiled DNA. EMBO J 10:387–395

Cui T, Leng F (2007) Specific recognition of AT-rich DNA sequences by the mammalian high mobility group protein AT-hook 2: a SELEX study. Biochemistry 46:13059–13066

Derouazi M, Martinet D, Besuchet Schmutz N, Flaction R, Wicht M, Bertschinger M, Hacker DL, Beckmann JS, Wurm FM (2006) Genetic characterization of CHO production host DG44 and derivative recombinant cell lines. Biochem Biophys Res Commun 340:1069–1077

Dickinson LA, Joh T, Kohwi Y, Kohwi-Shigematsu T (1992) A tissue-specific MAR/SAR DNA-binding protein with unusual binding site recognition. Cell 70:631–645

Eissenberg JC (1989) Position effect variegation in *Drosophila:* Towards genetics of chromatin assembly. Bioessays 11:14–17

Eissenberg JC, Morris GD, Reuter G, Hartnett T (1992) The heterochromatin-associated protein HP-1 is an essential protein in Drosophila with dosage-dependent effects on position-effect variegation. Genetics 131:345–352

Esnault G, Majocchi S, Martinet D, Besuchet-Schmute N, Beckmann JS, Mermod N (2009) Transcription factor CTFA acts as a chromatin domain boundary that Shields human telomeric genes from silencing. Moll Cell Biol 29:2409–2418

Esteller M (2008) Epigenetics in cancer. N Engl J Med 358:1148–1159

Evans K, Ott S, Hansen A, Koentges G, Wernisch L (2007) A comparative study of S/MAR prediction tools. BMC Bioinformatics 8:71–100

Fann CH, Guirgis F, Chen G, Lao MS, Oiret JM (2000) Limitations to the amplification and stability of human tissue-type plasminogen activator expression by Chinese hamster ovary cells. Biotechnol Bioeng 69:204–12

Fedoriw AM, Engel NI, Bartolomei MS (2004) Genomic imprinting: antagonistic mechanisms in the germ line and early embryo. Cold Spring Harb Symp Quant Biol 69:39–45

Feinberg AP (2008) Epigenetics at the epicenter of modern medicine. JAMA 299:1345–1350

Felsenfeld G (1992) Chromatin as an essential part of the transcriptional mechanism. Nature 355:219–224

Felsenfeld G (1996) Chromatin unfolds. Cell 86:13–19

Filippova GN, Fagerlie S, Klenova EM, Myers C, Dehner Y, Goodwin G, Neiman PE, Collins SJ, Lobanenkov VV (1996) An exceptionally conserved transcriptional repressor, CTCF, employs different combinations of zinc fingers to bind diverged promoter sequences of avian and mammalian c-myc oncogenes. Mol Cell Biol 16:2802–2813

Flintoff WF, Livingston E, Duff C, Worton RG (1984) Moderate-level gene amplification in methotrexate-resistant Chinese hamster ovary cells is accompanied by chromosomal translations at or near the site of amplified DHFR gene. Mol Cell Biol 4:69–76

Folger KR, Wong EA, Wahl G, Capecchi MR (1982) Patterns of integration of DNA microinjected into cultured mammalian cells: evidence for homologous recombination between injected plasmid DNA molecules. Mol Cell Biol 2:1372–1387

Folger KR, Thomas K, Capecchi MR (1985) Nonreciprocal exchanges of information between DNA duplexes coinjected into mammalian cell nuclei. Mol Cell Biol 5:59–69

Forrester WC, Novak U, Gelinas R, Groudine M (1989) Molecular analysis of the human beta-globin locus activation region. Proc Natl Acad Sci USA 86:5439–5443

Fouremana P, Winfield JA, Peter J, Hahnb PJ (1998) Chromosome breakpoints near CpG islands in double minutes. Gene 218:121–128

Galande S, Purbey PK, Notani D, Kumar PP (2007) The third dimension of gene regulation: organization of dynamic chromatin loopscape by SATB1. Curr Opin Genet Dev 17:408–414

Galbete JL, Buceta M, Mermod N (2009) MAR elements regulate the probability of epigenetic switching between active and inactive gene expression. Mol Biosyst 5:143–150

Gaszner M, Felsenfeld G (2006) Insulators: exploiting transcriptional and epigenetic mechanisms. Nat Rev Genet 7:703–713

Gilmour DS, Pflugfelder G, Wang JC, Lis JT (1986) Topoisomerase I interacts with transcribed regions in Drosophila cells. Cell 44:401–407

Girod P-A and Mermod N (2003) Use of scaffold/matrix attachment regions for protein production. Gene Transfer and Expression in Mammalian cells. In S.C. Makrides (ed.), Elsevier Science pp. 359–379

Girod P-A, Zahn-Zabal M, Mermod N (2005) Use of the chicken lysozyme 5′ matrix attachment region to generate high producer CHO cell lines. Biotechnol Bioeng 91:1–11

Girod P-A, Nguyen DQ, Calabrese D, Puttini S, Grandjean M, Martinet D, Regamey A, Saugy D, Beckmann JS, Bucher P, Mermod N (2007) Genome-wide prediction of matrix attachment regions that increase gene expression in mammalian cells. Nat Methods 4:747–753

Grewal SIS, Moazed D (2003) Heterochromatin and epigenetic control of gene expression. Science 301:798–802

Gutierrez-Adan A, Pintado B (2000) Effect of flanking matrix attachment regions on the expression of microinjected transgenes during preimplantation development of mouse embryos. Transgenic Res 9:81–89

Han HJ, Russo J, Kohwi Y, Kohwi-Shigematsu T (2008) SATB1 reprograms gene expression to promote breast tumor growth and metastasis. Nature 452:187–193

Hart CM, Laemmli UK (1998) Facilitation of chromatin dynamics by SARs. Curr Opin Genet Dev 8:519–525

Henikoff S (1996) Dosage-dependent modification of position-effect variegation in Drosophila. Bioessays 18:401–409

Herrscher RF, Kaplan MH, Lelsz DL, Das C, Scheuermann R, Tucker PW (1995) The immunoglobulin heavy-chain matrix-associating regions are bound by Bright: a B cell-specific trans-activator that describes a new DNA-binding protein family. Genes Dev 9:3067–3082

Homberger HP (1989) Bent DNA is a structural feature of scaffold-attached regions in Drosophila melanogaster interphase nuclei. Chromosoma 98:99–104

Iarovaia OV, Shkumatov P, Razin SV (2004) Breakpoint cluster regions of the AML-1 and ETO genes contain MAR elements and are preferentially associated with the nuclear matrix in proliferating HEL cells. J Cell Sci 117:4583–4590

John S, Reeves RB, Lin JX, Child R, Leiden JM, Thompson CB, Leonard WJ (1995) Regulation of cell-type-specific interleukin-2 receptor alpha-chain gene expression: potential role of physical interactions between Elf-1, HMG-I(Y), and NF-kappa B family proteins. Mol Cell Biol 15:1786–1796

Johnson CN, Levy LS (2005) Matrix attachment regions as targets for retroviral integration. Virol J 2:68–77

Jun SC, Kim MS, Hong HJ, Lee GM (2006) Limitations to the development of humanized antibody producing Chinese hamster ovary cells using glutamine synthetase-mediated gene amplification. Biotechnol Prog 22:770–780

Kaffer CR, Srivastava M, Park KY, Ives E, Hsieh S, Batlle J, Grinberg A, Huang SP, Pfeifer K (2000) A transcriptional insulator at the imprinted H19/Igf2 locus. Genes Dev 14:1908–1919

Kalos M, Fournier REK (1995) Position-independent transgene expression mediated by boundary elements from the apolipoprotein B chromatin domain. Mol Cell Biol 15:198–207

Kamiya H, Fukunaga S, Ohyama T, Harashima H (2007) The location of the left-handedly curved DNA sequence affects exogenous DNA expression in vivo. Arch Biochem Biophys 461:7–12

Kim SJ, Lee GM (1999) Cytogenetic analysis of chimeric antibody-producing CHO cell sin the course of dihydrofolate reductase-mediated gene amplification and their stability in the absence of selective pressure. Biotechnol Bioeng 64:741–749

Kim SJ, Chun JR, Hong HJ, Lee GM (1998) Characterization of chimeric antibody producing CHO cells in the course of dihydrofolate reductase-mediated gene amplification and their stability in the absence of selective pressure. Biotechnol Bioeng 58:73–84

Kim NS, Kim SJ, Lee GM (1998) Clonal variation within dihydrofolate reductase-mediate gene amplified Chinese hamster ovary cells: stability in the absence of selective pressure. Biotech Bioeng 60:679–88

Kim NS, Byun TH, Lee GM (2001) Key determinants in the occurrence of clonal variation in humanized antibody expression of CHO cells during dihydrofolate reductase mediated gene amplification. Biotechnol Prog 17:69–75

Kim JM, Kim JS, Park DH, Kang HS, Yoon J, Baek K, Yoon Y (2004) Improved recombinant gene expression in CHO cells using matrix attachment regions. Biotechnol J 107:95–105

Kim JD, Yoon Y, Hwang HY, Park JS, SYu J Lee, Baek K, Yoon J (2005a) Efficient selection of stable Chinese hamster ovary (CHO) cell lines for expression of recombinant proteins by using human β-interferon SAR element. Biotechnol Prog 21:933–937

Kim TH, Barrera LO, Zheng M, Qu C, Singer MA, Richmond TA, Wu Y, Green RD, Ren B (2005b) A high-resolution map of active promoters in the human genome. Nature 436:876–880

Kim TH, Abdullaev ZK, Smith AD, Ching KA, Loukinov DI, Green RD, Zhang MQ, Lobanenkov VV, Ren B (2007) Analysis of the vertebrate insulator protein CTCF-binding sites in the human genome. Cell 128:1231–1245

Klehr D, Maass K, Bode J (1991) Scaffold-attached regions from the human interferon beta domain can be used to enhance the stable expression of genes under the control of various promoters? Biochemistry 30:1264–1270

Klenova EM, Nicolas RH, Paterson HF, Carne AF, Heath CM, Goodwin GH, Neiman PE, Lobanenkov VV (1993) CTCF, a conserved nuclear factor required for optimal transcriptional activity of the chicken c-myc gene, is an 11-Zn-finger protein differentially expressed in multiple forms. Mol Cell Biol 13:7612–7624

Kohwi-Shigematsu T, deBelle I, Dickinson LA, Galande S, Kohwi Y (1998) Identification of base-unpairing region-binding proteins and characterization of their in vivo binding sequences. Methods Cell Biol 53:323–354

Kramer JA, Krawetz SA (1995) Matrix-associated regions in haploid expressed domains. Mamm Genome 6:677–679

Kumar PP, Purbey PK, Ravi DS, Mitra D, Galande S (2005) Displacement of SATB1-bound histone deacetylase 1 corepressor by the human immunodeficiency virus type 1 transactivator induces expression of interleukin-2 and its receptor in T cells. Mol Cell Biol 25:1620–1633

Kumar PP, Purbey PK, Sinha CK, Notani D, Limaye A, Jayani RS, Galande S (2006) Phosphorylation of SATB1, a global gene regulator, acts as a molecular switch regulating its transcriptional activity *in vivo*. Mol Cell 22:231–243

Kumar PP, Bischof O, Purbey PK, Notani D, Urlaub H, Dejean A, Galande S (2007) Functional interaction between PML and SATB1 regulates chromatin-loop architecture and transcription of the MHC class I locus. Nat Cell Biol 9:45–56

Kurukuti S, Tiwari VK, Tavoosidana G, Pugacheva E, Murrell A, Zhao Z, Lobanenkov V, Reik W, Ohlsson R (2006) CTCF binding at the H19 imprinting control region mediates maternally inherited higher-order chromatin conformation to restrict enhancer access to Igf2. Proc Natl Acad Sci USA 103:10684–10689

Kwaks THJ, Otte AP (2006) Employing epigenetics to augment the expression of therapeutic proteins in mammalian cells. Trends Biotechnol 24:127–142

Labrador M, Corces VG (2002) Setting the boundaries of chromatin domains and nuclear organization. Cell 111:151–154

Landsman D, Bustin M (1993) A signature for the HMG-1 box DNA-binding proteins. Bioessays 15:539–546

Liebich I, Bode J, Reuter I, Wingender E (2002) Evaluation of sequence motifs found in scaffold/matrix-attached regions (S/MARs). Nucleic Acids Res 30:3433–3442

Ling JQ, Li T, Hu JF, Vu TH, Chen HL, Qiu XW, Cherry AM, Hoffman AR (2006) CTCF mediates interchromosomal colocalization between Igf2/H19 and Wsb1/Nf1. Science 312:269–272

Liu WM, Guerra-Vladusic FK, Kurakata S, Lupu R, Kohwi-Shigematsu T (1999) HMG-I(Y) recognizes base-unpairing regions of matrix attachment sequences and its increased expression is directly linked to metastatic breast cancer phenotype. Cancer Res 59:5695–5703

Maher JF, Nathans D (1996) Multivalent DNA-binding properties of the HMG-1 proteins. Proc Natl Acad Sci USA 93:6716–6720

Matzke M, Matzke A, Kooter J (2001) RNA: guiding gene silencing. Science 293:1080–1083

Melton DW, Brennand J, Ledbetter DH, Konecki DS, Chinault AC, Caskey CT (1982) Phenotypic reversion at the hprt locus as a consequence of gene amplification. In: Schimke RT (ed) Gene Amplification. Cold Spring Harbor Laboratory, New York, pp 59–65

Morris K (2008) RNA-mediated transcriptional gene silencing in human cells. Curr Top Microbiol Immunol 320:211–214

Nishikawa J, Amano M, Fukue Y, Tanaka S, Kishi H, Hirota Y, Yoda K, Ohyama T (2003) Left-handedly curved DNA regulates accessibility to cis-DNA elements in chromatin. Nucleic Acids Res 31:6651–6662

Nowak W, Gawlowska M, Jarmolowski A, Augustyniak J (2001) Effect of nuclear matrix attachment regions on transgene expression in tobacco plants. Acta Biochim Pol 48:637–646

Oh SJ, Jeong JS, Kim EH, Yi NR, Yi SI, Jang IC, Kim YS, Suh SC, Nahm BH, Kim JK (2005) Matrix attachment region from the chicken lysozyme locus reduces variability in transgene expression and confers copy number-dependence in transgenic rice plants. Plant Cell Rep 24:145–154

Ostermeier GC, Liu Z et al (2003) Nuclear matrix association of the human beta-globin locus utilizing a novel approach to quantitative real-time PCR. Nucleic Acids Res 31(12):3257–3266

Pankiewicz R, Karlen Y, Imhof M, Mermod N (2005) Reversal of the silencing of tetracycline-controlled genes requires the coordinate action of distinctly-acting transcription factors. J Gene Med 7:117–132

Park F, Kay MA (2001) Modified HIV-1 based lentiviral vectors have an effect on viral transduction efficiency and gene expression in vitro and in vivo. Mol Ther 4:164–173

Phi-Van L, Strätling WH (1996) Dissection of the ability of the chicken lysozyme gene 50 matrix attachment region to stimulate transgene expression and to dampen position effects. Biochemistry 35:10735–10742

Recillas-Targa F, Pikaart MJ, Burgess-Beusse B, Bell AC, Litt MD, West AG, Gaszner M, Felsenfeld G (2002) Position-effect protection and enhancer blocking by the chicken β-globin insulator are separable activities. Proc Natl Acad Sci USA 99:6883–6888

Reeves R, Edberg DD, Li Y (2001) Architectural transcription factor HMGI(Y) promotes tumor progression and mesenchymal transition of human epithelial cells. Mol Cell Biol 21:575–594
Rincon-Arano H, Furlan-Magaril M, Recillas-Targa F (2007) Protection against telomeric position effects by the chicken cHS4 beta-globin insulator. Proc Natl Acad Sci USA 104:14044–14049
Robertson G, Garrick D, Wu W, Kearns M, Martin D, Whitelaw E (1995) Position-dependent variegation of globin transgene expression in mice. Proc Natl Acad Sci USA 92:5371–5375
Robins DM, Ripley S, Henderson AS, Axel R (1981) Transforming DNA integrates into the host chromosome. Cell 23:29–39
Romig H, Fackelmayer FO, Renz A, Ramsperger U, Richter A (1992) Characterization of SAF-A, a novel nuclear DNA binding protein from HeLa cells with high affinity for nuclear matrix/scaffold attachment DNA elements. EMBO J 11:3431–3440
Saitoh Y, Laemmli UK (1993) From the chromosomal loops and the scaffold to the classic bands of metaphase chromosomes. Cold Spring Harb Symp Quant Biol 58:755–765
Saitoh Y, Laemmli UK (1994) Metaphase chromosome structure: bands arise from a differential folding path of the highly AT-rich scaffold. Cell 76:609–622
Schubeler D, Mielke C, Maass K, Bode J (1996) Scaffold/matrix-attached regions act upon transcription in a context-dependent manner. Biochemistry 35:11160–11169
Selker EU (1999) Gene silencing: repeats that count. Cell 97:157–160
Sgarra R, Rustighi A, Tessari MA, Di Bernardo J, Altamura S, Fusco A, Manfioletti G, Giancotti V (2004) Nuclear phosphoproteins HMGA and their relationship with chromatin structure and cancer. FEBS Lett 574:1–8
Sjakste NR, Sjakste TG (2001) Use of scaffold/matrix-attachment regions for protein production. Mol Biol 35:627–635
Solomon MJ, Strauss F, Varshavsky A (1986) A mammalian high mobility group protein recognizes any stretch of six A.T base pairs in duplex DNA. Proc Natl Acad Sci USA 83:1276–1280
Stief A, Winter DM, Strätling WH, Sippel AE (1989) A nuclear DNA attachment element mediates elevated and position-independent gene activity. Nature 341:343–345
Strutzenberger K, Borth N, Kunert R, Steinfellner W, Katinger H (1999) Changes during subclone development, ageing of human antibody- producing recombinant CHO cells. J Biotechnol 69:215–226
Sumida N, Nishikawa J, Kishi H, Amano M, Furuya T, Sonobe H, Ohyama T (2006) A designed curved DNA segment that is a remarkable activator of eukaryotic transcription. FEBS J 273:5691–5702
Thanos D, Maniatis T (1992) The high mobility group protein HMG I(Y) is required for NF-kappa B-dependent virus induction of the human IFN-beta gene. Cell 71:777–789
Thomas KR, Folger KR, Capecchi MR (1986) High frequency targeting of genes to specific sites in the mammalian genome. Cell 44:419–428
Thompson EM, Christians E, Stinnakre MG, Renard JP (1994) Scaffold attachment regions stimulate HSP70.1 expression in mouse preimplantation embryos but not in differentiated tissues. Mol Cell Biol 14:4694–4703
Thorvaldsen JL, Duran KL, Bartolomei MS (1998) Deletion of the H19 differentially methylated domain results in loss of imprinted expression of H19 and Igf2. Genes Dev 12:3693–3702
Torrungruang K, Alvarez M, Shah R, Onyia JE, Rhodes SJ, Bidwell JP (2002) DNA binding and gene activation properties of the Nmp4 nuclear matrix transcription factors. J Biol Chem 277:16153–16159
Varghese J, Alves W, Brill BJ, Wallace M, Calabrese D, Regamey A, Girod PA. (2008) Rapid development of high-performance, stable mammalian cell lines for improved clinical development. Bioprocess J 7:30–36
Volfson D et al (2006) Origins of extrinsic variability in eukaryotic gene expression. Nature 439:861–864
Wakimoto BT (1998) Beyond the nucleosome: epigenetic aspects of position-effect variegation in Drosophila. Cell 93:321–324
Wang T, Xue L, Hou W, Yang B, Chai Y, Ji X, Wang Y (2007) Increased expression of transgene in stably transformed cells of *Dunaliella salina* by matrix attachment regions. Appl Microbiol Biotechnol 76:651–657

West A, Gaszner M, Felsenfeld G (2002) Insulators: many functions, many mechanisms. Genes Dev 16:271–288
Wilson C, Bellen HJ, Gehring WJ (1990) Position effects on eukaryotic gene expression. Annu Rev Cell Biol 6:679–714
Woodcock CL, Dimitrov S (2001) Higher-order structure of chromatin and chromosomes. Curr Opin Genet Dev 11:130–5
Xu Y, Davidson L, Alt FW, Baltimore D (1996) Deletion of the Ig kappa light chain intronic enhancer/matrix attachment region impairs but does not abolish V kappa J kappa rearrangement. Immunity 4:377–385
Yamasaki K, Akiba T, Yamasak T, Harata K (2007) Structural basis for recognition of the matrix attachment region of DNA by transcription factor SATB1. Nucleic Acids Res 35:5073–5084
Yasui D, Miyano M, Cai S, Varga-Weisz P, Kohwi-Shigematsu T (2002) SATB1 targets chromatin remodelling to regulate genes over long distances. Nature 419:641–645
Yoshikawa T, Nakanishi F, Ogura Y, Oi D, Omasa T, Katakura Y, Kishimoto M, Suga K (2000) Amplified gene location in chromosomal DNA affected recombinant protein production and stability of amplified genes. Biotechnol Progr 16:710–715
Zahn-Zabal M, Kobr M, Girod PA, Imhof M, Chatellard P, de Jesus M, Wurm F, Mermod N (2001) Development of stable cell lines for production or regulated expression using matrix attachment regions. J Biotechnol 87:29–42
Zhao K, Kas E, Gonzalez E, Laemmli UK (1993) SAR-dependent mobilization of histone H1 by HMG-I/Y in vitro: HMG-I/Y is enriched in H1-depleted chromatin. Embo J 12:3237–3247

Expression Engineering – The IE2 Promoter/Enhancer from Mouse CMV

Markus O. Imhof, Philippe Chatellard, Michel Kobr, Renata Pankiewicz, Valérie Duverger, Léonard Bagnoud, Christophe Sauvage, and Christine Mossu

Abstract Cell engineering aims at changing the gene expression program of a specific host cell line for instance by molecular techniques. A widely used strategy is the overexpression of artificially introduced foreign genes. This is achieved by combining the DNA encoding the respective protein of interest with constitutive or regulated promoter sequences, and transfecting the host cell with such recombinant vectors. This approach is taken in a majority of engineering strategies, and a still growing toolbox of different expression governing sequences is becoming available. Some fine examples for such sequences, acting on different aspects of cell engineering, are reviewed throughout this book. Here we introduce the mouse CMV IE2 promoter/enhancer sequence as a new and extremely powerful tool for overexpression of genes of interest. This promoter/enhancer sequence withstands the comparison with established promoter/enhancer combinations commonly used for cell engineering, and often exceeds their performance. We highlight the usefulness of mouse CMV IE2 sequences for the expression of recombinant monoclonal antibodies in connection with the 'related' mouse CMV IE1 promoter/enhancer region. This strategy allows us to achieve high expression levels for the production of therapeutic proteins in serum-free mammalian cell culture.

1 Introduction

One of the main aspects in mammalian cell line engineering is the introduction and expression of heterologous genes in these host cells. Typically, this involves exposure of the host cells to viral or non-viral vector DNA, and uptake of these vectors mediated by diverse agents or mechanisms. Techniques for transient and stable gene transfections are broadly available and have been described in excellent

M.O. Imhof (✉), P. Chatellard, M. Kobr, R. Pankiewicz, V. Duverger, L. Bagnoud, C. Sauvage, and C. Mossu
Merck Serono Research, Protein and Cell Sciences, Merck Serono Biotechnology Center, 1809 Fenil-sur-Corsier, Switzerland
e-mail: Markus.Imhof@merckserono.net

M. Al-Rubeai (ed.), *Cell Line Development*, Cell Engineering 6,
DOI 10.1007/978-90-481-2245-5_2, © Springer Science+Business Media B.V. 2009

laboratory manuals (Kingston, 1997). This includes the description and use of selection genes for the isolation of transfectants that have stably introduced the foreign DNA into their genome.

In this book chapter we will mainly focus on the constitutive expression of transgenes mediated by the strong immediate early 2 (IE2) promoter and enhancer region of the mouse cytomegalovirus (mCMV). We use plasmidic DNA vectors and Chinese hamster ovary cells (CHO), cultivated in serum-free conditions, in the examples of cell line engineering described in the following. The data reported exemplify the usefulness of these promoter and enhancer sequences for expression of various transgenes. Their protein products are either accumulated intra-cellularly for an experimental readout, or released into the culture medium by active secretion and processing via the endoplasmic reticulum and Golgi apparatus. This latter pathway is extremely important for the efficient production of glycosylated recombinant therapeutic proteins in mammalian cell culture manufacturing.

Before focusing on the application side, it is of interest to review the molecular basis for transgene expression. A better understanding of these mechanisms will lead to constant improvements of cell engineering strategies and technologies.

2 Enhancer and Promoter Regions for Gene Expression

Expression of eukaryotic genes depends on a multi-step process of gene activation, transcription, mRNA processing and transport, translation, and finally targeting of the protein into different compartments, such as e.g. the secretory pathway (Kaufman, 2004). All these steps are subject to regulation allowing gene-specific fine-tuning. A prerequisite for all the following events is gene activation and the initiation of transcription. For nuclear genes this step is regulated by the chromatin environment and specific transcription factors promoting the productive association of multi-protein complexes, and recruitment of RNA polymerase to specific DNA regions near the transcription start site (Kim et al., 1997). RNA polymerase II is responsible for the transcription of protein encoding genes, and strategies for cell engineering for protein expression are almost exclusively based on systems taking advantage of RNA polymerase II-regulated expression cassettes.

The widely accepted model for gene activation is based on the interaction of specific DNA elements with DNA-binding proteins, and further association with multi-protein complexes comprising RNA polymerase II. Directional positioning of discrete promoter DNA elements is responsible for basal levels of transcription and recruitment of RNA polymerase II for initiation of transcription. In a classical model this includes at least two important DNA elements, a TATA box and, spaced by about 25 nucleotides, the downstream initiator region. This region, important for directional transcription initiation, is called the minimal or core promoter region, however, in many genes either of the elements is missing but complemented by other sequences such as the downstream promoter element DPE (Kim et al., 2005; Smale and Kadonaga, 2003). Promoters extend further to comprise additional upstream elements important for transcription. These proximal promoter elements assure basal transcription by ubiquitously expressed transcription factors binding to them, such as SP1 or CTF/NF1 (Latchman, 1991). In general, actively transcribed promoters also differ in their chromatin state as nucleosomes around the transcription start site (TSS) slide to a more downstream position (Schones et al., 2008) and often adopt specific post-translational modifications (Mellor, 2005).

Most promoters are further influenced by distal enhancer sequences binding to gene-specific transcription factors. These enhancer sequences vary in their positioning towards the core promoter. They are able to activate or increase transcription over large distances and in both orientations, and they may even be positioned downstream of the transcription initiation site. Enhancer-binding transcription factors and mechanistic aspects for their functioning in the context of gene activation and interaction with promoters have been reviewed (Lemon and Tjian, 2000; Szutorisz et al., 2005).

In the last decade or so the relatively linear view of the interplay between specific DNA sequences and transcription factors got complemented by new findings on chromatin as a major player in regulation of gene expression. The analysis of old paradigms on the activity of genes with respect to their localization in interphase nuclei got a renaissance as full genome sequences became available and new genetic, biochemical, and bioinformatic tools allowed to elaborate chromatin composition on the genomic level (Goetze et al., 2007). The main outcome is the precise characterization of DNA and histone modifications with respect to active and inactive genes and regulatory regions (Boyle et al., 2008; Kouzarides, 2007; Trojer and Reinberg, 2007). Some of this information is helpful to review in order to set the context concerning cell engineering aspects.

The nucleus is highly compartmentalized with chromosomes occupying specific territories, and the interchromatin space between these territories. The interchromatin space contains the nucleoplasm and specialized nuclear bodies required for transcription, splicing, replication and repair (Cremer and Cremer, 2001; Handwerger and Gall, 2006). On the other hand, genome architecture is also rather dynamic. Repositioning of large chromosome domains to the interchromatin space is often correlated with gene expression (Lanctôt et al., 2007; Schneider and Grosschedl, 2007). Gene-rich regions tend to cluster in the periphery of chromosome territories. When activated, these clusters loop out as decondensed chromatin reaching into the interchromatin compartment (Heard and Bickmore, 2007). In this line evidence is accumulating to support a model where genes move towards so-called transcription factories (Bartlett et al., 2006). Transcription factories contain RNA polymerases, and they are thought to exist as pre-assembled structures associated with the insoluble nuclear fraction. They are ready to functionally interact with genes clustered in active chromatin hubs (Mitchell and Fraser, 2008). Recent data support this model and reveal an essential role for enhancer and promoter regions for triggering the relocation process of genes towards transcription factories. For example, tandem arrays of glucocorticoid-inducible MMTV promoters upstream of a reporter gene were shown microscopically to localize next to discrete RNA polymerase II transcription sites. Two chromatin compartments were observed with these arrays: a relatively condensed region near the promoter and glucocorticoid receptor binding sites, and a decondensed region involved in RNA polymerase II transcription and surrounding the transcription factory. This latter area displayed histone modifications that are characteristic for 'recently transcribed' genes (Müller et al., 2007). The data are thus consistent with a model for immobilized RNA polymerases that reel chromatin through the site and extrude it in a decondensed form.

At least three other examples are worth noting as they support the model of transcribed genes moving towards transcription factories. This is of relevance for cell engineering, as the transcription factors binding to promoter and enhancer regions are thought to orchestrate the relocation of the linked genes and hence effect their transcription. First, in erythroid, but not in non-erythroid cells, the mouse α-globin genes (α1 and α2) relocate to a nearby active chromatin hub that is associated with a transcription factory (Zhou et al., 2006). This active hub contains clustered housekeeping genes that are transcribed in both cell lineages. Erythroid-specific transcription factors (GATA-1 and NF-E2) are thought to bind to the α-globin promoters and the major upstream regulatory elements mediating cell-specific expression of these genes. On the other hand, the theta-globin gene, located between the α-genes and their main upstream regulatory elements, remains inactive and excluded from the transcription factory in both cell types. Similarly, the transcription factors GATA-1 and FOG-1 were shown to be essential for interaction of the distant LCR (locus control region) with the promoter for the β-globin genes, and relocation of the gene locus towards transcription factories in erythroid cells (Ragoczy et al., 2006; Vakoc et al., 2005). Second, co-transfection and episomal replication of plasmids containing different marker genes revealed that the relocation to the same or distinct transcription factories is related to the identity or difference of promoter/enhancer regions (Xu and Cook, 2008). And third, similar results regarding relocation of homologous sequences were observed in a study with transiently transfected plasmids, however, localization to the same transcription factory further depended on the homology of the transcribed region (Binnie et al., 2006). All these data point towards a role for enhancers and promoters to target the linked genes to a transcription factory, a process orchestrated by transcription factor occupancy and transcription factor activity in a cell-specific mode. This is in line with earlier data describing enhancers to provide anti-silencing activity by preventing genes to relocate to inactive chromatin (Francastel et al., 1999). Interestingly, a similar relocation has been shown for the IFN-β enhancer upon viral challenge of HeLa cells, but in this case the enhancer interacted with remote chromosomal regions thought to constitute low affinity storage sites for limiting (NF-κB) transcription factors (Apostolou and Thanos, 2008).

All these findings are important for defining successful cell engineering strategies. They highlight the need for functional enhancer and promoter elements for expressing genes of interest from artificial vectors. The assembly of such expression cassettes has been thoroughly reviewed (Makrides, 1999), and the optimal combination of expression elements for different host cell lines has been addressed (Xia et al., 2006; Xu et al., 2001).

Here we introduce the use of the mouse CMV IE2 promoter/enhancer, a sequence only recently discovered for applications in biotechnology (Chatellard et al.,

2007; Kobr et al., 2008). We will highlight some characteristics of these sequences and report their use for expressing recombinant proteins in CHO cells.

3 The Human CMV Major Immediate Early Promoter Region

The major immediate early (MIE) region from the human cytomegalovirus (CMV) is one of the strongest enhancer/promoter regions used for heterologous gene expression, either in academic research or biopharmaceutical production. This region is responsible for the very early events of productive viral replication in differentiated human epithelial, endothelial, neuronal, glial, muscle, fibroblast, and macrophage cells (Meier and Stinski, 1996). Transcription of the MIE region is independent of de novo synthesis of viral proteins, i.e. host cells provide all factors necessary for immediate early transcription and replication steps, that are then followed by early and late phases of replication. The regulation of human CMV transcription during the immediate early phase has been reviewed (Meier and Stinski, 1996; Stamminger and Fleckenstein, 1990).

CMV belongs to the family of herpes viruses, and it is a widespread pathogen. Regionally up to 100% of adults are asymptomatically life-long infected with this virus in its latent form. Sporadic reactivation of the virus leads to an immune response keeping the virus under control, but the response is insufficient to eradicate it. CMV becomes an opportunistic pathogen in immunosuppressed individuals leading to disease involving the retina, lung, gastrointestinal tract, liver, kidney or nervous system. It is also the major infective cause for birth defects such as mental retardation or hearing loss (Powers and Früh, 2008).

The IE1 promoter/enhancer region from human CMV, responsible for the main transcript expressed from the MIE region, was dissected early on regarding its structural composition. It is harboring at least four distinct types of imperfect repeat sequences, interspersed by non-repetitive DNA (Boshart et al., 1985). These repeats were later identified to bind to specific transcription factors that confer constitutive and differentiation state-dependent transcriptional activation (Meier and Stinski, 1996). This enhancer region, in combination with its own or heterologous promoters, is used in numerous commercial expression vectors, and both ubiquitous (Sawicki et al., 1998) and cell-specific (Gruh et al., 2008; Liu et al., 1997) enhancement of transgene expression has been described.

The so-called unique and modulator regions, both upstream of the enhancer area that extends to approximately −580, are dispensable for promoter/enhancer activity (Meier and Stinski, 2006). Interestingly, this unique region contains a promoter for an early gene, UL127, transcribed in opposite direction. It was recently shown that transcription factors binding to the unique region repress UL127 expression and act as a boundary between this gene and the MIE enhancer (Lashmit et al., 2004; Lee et al., 2007).

CMV MIE promoters and enhancers vary between different species with respect to arrangement and composition of regulatory elements (Meier and Stinski, 2006).

Nevertheless, they are all characterized by a high density and redundancy of transcription factor binding sites. This is probably key to their strength for enhancing transcription in many different biological models, vector conformations, and cellular environments.

4 The Mouse CMV Major Immediate Early Promoter Region

Related to human CMV, the mouse CMV was extensively studied as a model for virus reactivation from latency in normal and transgenic mice (Hummel and Abecassis, 2002). Its MIE promoter region was first characterized with respect to the IE1/IE3 transcripts that are generated by differential splicing of the same mRNA expressed from the same promoter (Messerle et al., 1992). An enhancer for this promoter was described even earlier, containing different repetitive elements similar to the hCMV MIE enhancer (Dorsch-Häsler et al., 1985). The core IE1 enhancer region was mapped to the region between –500 and –150 relative to the IE1/3 transcription start site, however, sequences extending further upstream (from –1330 to –490) were reported to exhibit similar enhancer activity (Dorsch-Häsler et al., 1985). Altogether, the core IE1 enhancer met all criteria of a bona fide enhancer such as orientation, position and distance independence, and it is capable to activate heterologous promoters. Furthermore, activity was retained in numerous cell lines. This led to its use for expression of recombinant proteins in various cell lines, and in combination with other vector elements (Kim et al., 2002). Differences in expression levels of linked transgenes were reported in comparison to the rat (Sanford and Burns, 1996) and hCMV promoter/enhancer (Addison et al., 1997; Xia et al., 2006).

Most notably, it was realized that the mouse MIER is also responsible for the expression of the IE2 gene. This gene is transcribed in opposite direction and its genomic organization was characterized (Messerle et al., 1991). The mouse IE2 gene, however, seems to be dispensable for viral replication, and no definite function has been yet identified (Cardin et al., 1995). This unique arrangement of the IE1/3 and IE2 genes, expressed in opposite orientations, leaves an intergenic region of 1,374 bp between the two transcription start sites. This region contains the two promoters adjacent to their respective transcription start sites, and a long sequence with at least two separable enhancers (Chatellard et al., 2007; Dorsch-Häsler et al., 1985). Initially, the sequences remote to the IE1/3 promoter were reported to have repressive activities (Kim and Risser, 1993), however the fragment used in those experiments also contained the IE2 promoter and start site. It is conceivable that promoter competition was responsible for the reduced transcription by the IE1/3 promoter in those experiments. Similar data suggesting repressive effects were obtained in stably transfected cells, but transcription levels from IE1/3 were restored when the same vector was used without the core promoter region of IE2 (Chatellard et al., 2007). In the converse experiment with or without the IE1/3 promoter, a similar effect on IE2 transcription was observed. Therefore, we currently believe that this intergenic region contains two separable and strong enhancer regions that work well alone or in combination, but with modest additive effect when linked to a single promoter.

5 Dissection of the Bi-directional Mouse CMV MIE Region

As mentioned above, the mouse CMV MIE region is special in its architecture of expressing two immediate early genes in opposite direction (Messerle et al., 1991). It was shown that both promoters are transcribed in an independent but asynchronous way (Grzimek et al., 2001). More recently, a switching model was proposed based on the stochastic activation of either promoter by unknown signals. Models for either synchronized expression of both genes or high frequency oscillation between activation of either promoter were formally excluded in the context of viral latency (Simon et al., 2007). This means that using the whole mCMV MIER for expressing two different recombinant proteins may be inadequate: it could be that for each copy only one of the expression cassettes is actively transcribed. On the other hand, stable transfection of cells generally leads to multi-copy integration. The stochastic activity of either promoter is therefore likely to be complemented by the activity of the other promoter being active on a different gene copy. Alternatively, in the context of cell engineering, both promoters may be active simultaneously. In any case, we decided to exploit the whole mCMV MIER for expressing recombinant proteins in CHO cells. First we performed an analysis for putative transcription factor binding sites in the intergenic region. In analogy to the human CMV we found a plethora of repetitive sequences that contain (near-) consensus sequences for transcription factor binding sites. The results of this analysis are reported in the following.

5.1 Putative Binding Sites for Cellular Transcription Factors

The mCMV intergenic region contains several predicted binding sites for the transcription factors NF-κB and AP1, a situation very similar to the simian and human MIE enhancers (Meier and Stinski, 2006; Sanford and Burns, 1996). Figure 1a is an attempt to summarize the in silico prediction for binding sites for these and other transcription factors (TF) with more than 90% scoring when analyzed by the public software Consite, based on the JASPAR database (Sandelin et al., 2004). The analysis is complemented by the predictions as made by TFSEARCH, a bioinformatic tool based on the Transfac matrix tables (Heinemeyer et al., 1998). Further useful information on TFs is compiled in a comprehensive review (Faisst and Meyer, 1992). Sequence coordinates are shown with respect to the IE2 and IE1 transcription start sites. The minimal enhancer regions, as roughly defined for IE1 (Dorsch-Häsler et al., 1985) and IE2 (Chatellard et al., 2007), are shown as closed boxes. Over all, the intergenic region contains numerous imperfect repeats explaining

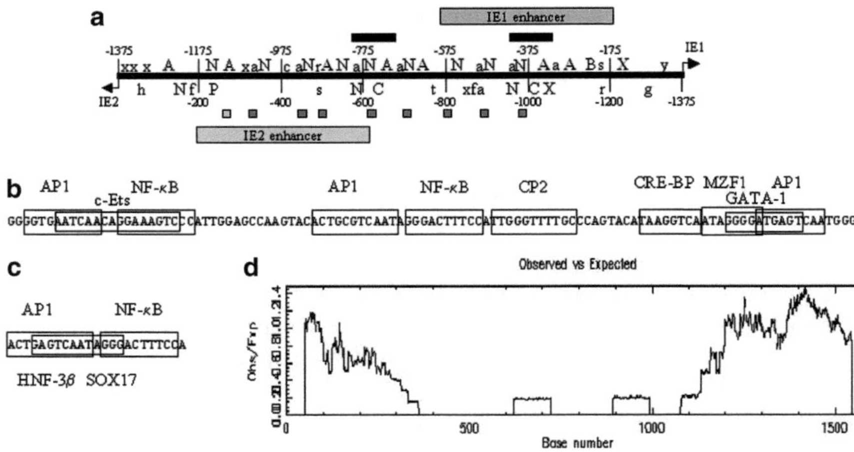

Fig. 1 Putative transcription factor binding sites located in the mouse IE2/IE1 intergenic region. (**a**) Arrangement of putative transcription factor binding sites with 100% (bold), 95% (higher case), and 90% (lower case) scoring, respectively, according to JASPAR and Transfac matrices (for references and information on transcription factors see text). A: AP1, N: NF-κB, h: HNF-3β, S: Sry/Sox-5/17, r: ROR-α1, y: YY1, B: CREB, C: CP2/NF-Y, f: HSF2, P: C/EBP, X: CdxA, and g: Sp1. The IE2 and IE1 core enhancer regions are depicted, as well as the positions for the long (black bar) and short repeats (dark gray squares) described in the following. (**b**) Long repeat sequence and putative transcription factor binding sites on it (see text for description of transcription factors). (**c**) Small sequence motif repeated nine times containing nearby binding sites for AP1 and NF-κB, overlapping with lower scoring sites for HNF-3β and the High Mobility Group protein Sox-17. (**d**) Density (observed versus expected) of CpG dinucleotides over the intergenic region, the core promoter area, and the first exons of both transcription units. The IE2 promoter is on the *left*, the IE1 promoter on the *right*

the frequent presence of the same binding sites. The longest repeat sequence of 109 bp is occurring twice, depicted as a closed black bar in Fig. 1a and drawn in detail in Fig. 1b. This sequence contains an AP1 site at each end and in the center of the repeat, two NF-κB sites, one CREB1 (cyclic AMP response element binding protein), and one Ets binding site. There are also sites for the ubiquitously expressed CP2 (NF-Y) transcription factor, the differentiation-specific GATA-1, and the zink finger transcription factor MZF1. Some of these sites overlap with each other. Further putative binding sites are not shown due to lower consensus homology scores.

Another repeat sequence (23 bp) is spread eight to nine times over the intergenic region (gray squares in Fig. 1a). The sequence contains a putative binding site each for AP1 and NF-κB, and overlapping binding sites for the differentiation-specific factor HNF-3β and SOX-17, both with a slightly lower homology score of 80% (Fig. 1c). SOX-17 is a member of a transcription factor family that induces DNA bending upon binding to the minor groove. SOX family members are important for development and differentiation, and they were shown to interact with protein partners such as other transcription factors to stabilize their interaction with DNA (Wilson and Koopman, 2002).

Of course, the relevance of all these putative binding sites is questionable. A detailed functional study has not been performed to date. Site occupancy and function (activation versus repression) will critically depend on the pool of available transcription factors in a specific cell line. From a non-saturating high throughput mRNA sequencing effort we know that, among other TFs, the following ones are expressed in the CHO cells that we use as a host for our engineering activities: AP1 site–binding JunB and JunD proteins, relA proteins binding to NF-κB sites, GATA, C/EBP, Ets, and different SOX family members (M. Ibberson and C. Power, personal communication). It is therefore likely, that the majority of sites described in Figs. 1b and 1c are occupied in CHO cells.

The repeats described above are just examples to illustrate the composition of the enhancer regions that include many other repeat and non-repeat sequences with potential binding sites for transcription factors. Among them the factors with high consensus homology scores are noted in Fig. 1a. Some of these factors may also cooperate with each other by integrating responses to cellular and environmental conditions. For instance, AP1-binding sites often overlap with sequences responsive to oxidative stress (Venugopal and Jaiswal, 1998), and cooperate with tonicity-response sequences that are bound by NF-κB-family members (Irarrazabal et al., 2008). Interestingly, preliminary results from our laboratory suggest that extracellular salt concentrations influence expression from mCMV constructs in specific culture conditions. Moreover, DNA-binding and transcriptional activity of both AP1 and NF-κB were described to be sensitive to cellular redox status (Ando et al., 2008). In summary, the unusual clustering of putative transcription factor binding sites, extending over the whole intergenic region, indicates that the mouse CMV MIE region is prone to be activated by various cellular factors and signals, possibly resulting in mediating ubiquitous and high transcriptional activity. On the other hand, some of these sequences are known targets for transcriptional repressors. This may lead to cell-specific differences on mCMV MIER gene expression or variable response to signaling events.

Mouse CMV reactivation from latency was shown to correlate with IE gene expression through activation of the transcription factors NF-κB and AP1 by tumor necrosis factor (TNF) (Hummel and Abecassis, 2002; Simon et al., 2005). However, involvement of NF-κB in MIE gene expression and replication of CMV is controversial. Mouse CMV replication is observed at elevated levels in fibroblasts deficient for p65 of NF-κB (Benedict et al., 2004), and mutations in the NF-κB binding sites in the human CMV MIE did not affect gene expression or replication (Gustems et al., 2006). This underlines the observation noted above that the MIE enhancer is a robust region for promoting transcription, most likely due to high transcription factor binding site redundancy, and interaction between transcription factors integrating various cell differentiation states and signaling events, resulting in high expression of naturally or artificially linked genes.

Interestingly, the artificial creation of transcription factor binding site redundancy by multimerisation of NF-κB binding sites has recently been shown to be a strategy to overcome the need of chromatin modification for transcriptional activation. Binding site redundancy resulted in nucleosome sliding near the transcription start site in the absence of histone acetylation, a process shown to be dependent on the recruitment of specific histone modifying complexes with the native promoter region (Koutroubas et al., 2008). Hence, the high redundancy of transcription factor binding sites in the mouse CMV IE1 and IE2 promoter/enhancer regions supports the idea that these promoters are good candidates for driving expression of heterologous genes in cell engineering strategies.

5.2 Enrichment of CpG Content Around the IE2 and IE1 Transcription Start Sites

Another interesting aspect of the intergenic region is that it is essentially devoid of CpG dinucleotides over the whole enhancer regions. However, CpGs are found around and downstream of the respective IE1 and IE2 transcription start sites (TSS). CpG sequences are generally underrepresented in vertebrate genomes as they are prone to cytosine methylation and spontaneous deamination to TpG. However, it has been observed that the frequency of CpGs is elevated near core promoter regions, especially of constitutively expressed genes (Kim et al., 2005; Saxonov et al., 2006). Such CpG-rich regions were called CpG islands (Gardiner-Garden and Frommer, 1987), and promoter-associated CpGs typically remain unmethylated. Methylation interferes with transcription possibly because transcription factors are impaired from binding their CpG-target sites (Rozenberg et al., 2008), or by methylation-specific binding of repressor proteins (Bird and Wolffe, 1999). Genome-wide analysis revealed that there is a high coincidence of CpG islands and active promoters within the first 10 kb from the border to lamin-associated inactive chromatin (Guelen et al., 2008).

For both the IE1 and IE2 promoters the stringent criteria for CpG islands are not quite met (data not shown), but it is remarkable how the CpGs are clustered around both TSS. Figure 1d shows this clustering using cpgplot of the EMBOSS software for CpG analysis (Rice et al., 2000). Note that we linked the respective promoters with the 5′ untranslated regions as used in our expression vectors (see below) and performed the CpG analysis on the sequences including the first exons. This artificial construct results in the IE1 region reaching the CpG island criteria (at least 200 bp long, >60% observed over expected CpG, >50% GC content).

Although we could not specifically localize binding sites for the transcription factors cited by Rozenberg et al. in the two core promoters (by applying the criteria of 90% scoring according to JASPAR), we tested whether in vitro CpG methylation using the SssI methyltransferase would interfere with expression from an IE1 expression vector. Transient transfection experiments revealed efficient silencing by in vitro methylation when compared to the unmethylated vector. The same plasmid did not lead to the recovery of stable transfectants with the IE1 promoter responsible for the polycistronic expression of the gene of interest and the selection marker (linked by an IRES). This suggests that CpG methylation strongly interfered with selection marker expression. Conversely, a population of stable CHO transfectants did not increase IE1-dependent expression of the gene of interest when the culture was treated with different concentrations of the DNA methyltransferase blocker RG108 (M. Kobr, personal communication). Although indirect, these data suggest that the CpGs in the IE1 promoter are not methylated in these cells. This is reminiscent of the unmethylated status of CpGs in active promoters (Rozenberg et al., 2008).

In summary, it is interesting to note that the mCMV IE promoters / enhancers are similar to cellular genes in their architecture. Sequence analysis suggests that the IE promoters of mouse CMV have all the hallmarks of highly active mammalian promoters, namely a CpG-enriched core region with a TATA-box and TSS (although the initiator sequence poorly matches the consensus), a DPE homologue (Smale and Kadonaga, 2003), and a strong upstream enhancer. Therefore these promoters, along with their enhancers, are of interest for expressing recombinant proteins in cell engineering projects. In the following we review some of our recent experiments with focus on the use of the new IE2 promoter / enhancer combination along with the previously described IE1 sequences.

6 Vector Architectures Using IE2 and IE1 Promoter/ Enhancers for Expression of Heterologous Genes

In previous work we reported the use of the bi-directional mouse CMV MIE region for the expression of recombinant proteins in transient and stable transfections of CHO cells (Chatellard et al., 2007). We further applied these promoter/enhancers for cell engineering and secretion of a monomeric recombinant protein of therapeutic value. On a molecular level we identified the minimal IE2 enhancer as an independent expression promoting element fulfilling all the criteria of a bona fide enhancer. We even showed that the IE2 enhancer activated both the IE2 and IE1 promoters in a construct bearing a deletion of the IE1 enhancer region. The consequence of those experiments is that the MIE region conceptually contains two independent enhancers that act independently on their respective promoters. This is somewhat in contradiction to recent work of Simon et al. (2007) who nicely demonstrated that the 'combined' enhancer region stochastically switches transcription from either promoter to the other during TNF-induced activation of gene expression. Their experiments also excluded high frequency switching of transcription between those promoters. As mentioned above, these data do not necessarily favor the use of the whole MIE region along with the two promoters for expression of synthetically linked transgenes. One possible implication is that only one of the promoters is activated at a time, and that this would result in the production of only one protein when the bi-directional promoter region is linked to two distinct genes of interest. However, it is also possible that unknown factors, restricted to the viral environment, are responsible for the reported stochastic switching behavior of the full enhancer region. Taken out of this viral context, the two promoters could act independently and express linked genes simultaneously.

To test the applicability of the bi-directional intergenic region for cell engineering purposes, we explored these promoters in different architectures. Figure 2 shows

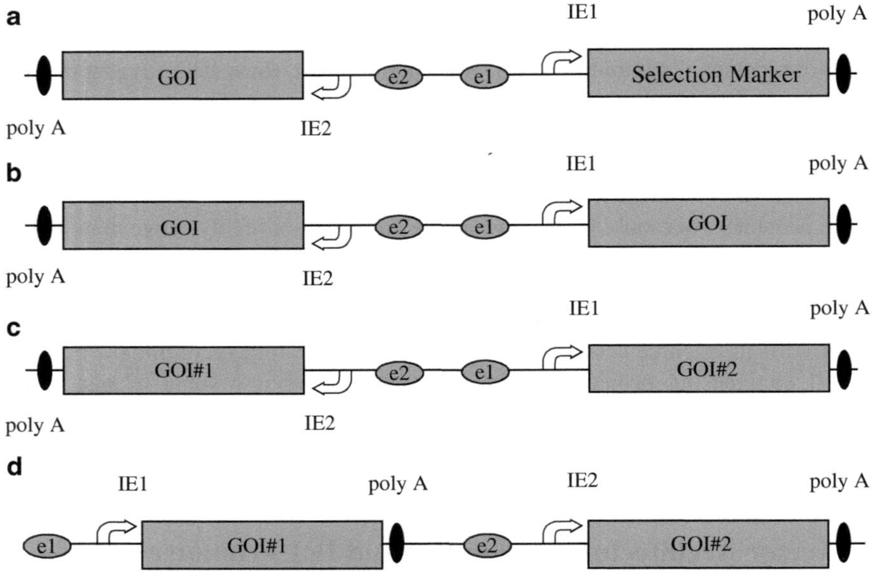

Fig. 2 Conceptual design of expression vectors containing bi-directional and tandem arrangements of the mCMV MIE promoters/enhancers. (**a**) Expression of a gene of interest (GOI) and a selection marker. (**b**) Expression of the same GOI from both promoters. (**c**) Expression of two different GOIs from bi-directional promoter arrangement. (**d**) Expression of two different GOIs from tandem promoter/enhancer arrangement

a non-exhaustive selection of combinations how the IE1 and IE2 promoters can be linked to different genes of interest in bi-directional and tandem orientation. Expression from some of these constructs has been reported previously for monomeric proteins (Chatellard et al., 2007). Conceptually it is possible to express a selection marker from one promoter and a gene of interest (GOI) from the other one (a), or the same gene from both promoters (b), or two different GOIs encoding e.g. the subunits of heteromeric proteins (e.g. peptide hormones, monoclonal antibodies) from either promoter (c and below). Of course even more genes can be co-expressed by using polycystronic arrangements, or by co-transfection of several distinct expression vectors, or by use of other techniques such as e.g. 2A elements (de Felipe et al., 2006).

Alternatively, the bi-directional architecture can be sacrificed by aligning the transcription units in tandem (Fig. 2d). We tested these new settings with different vectors all containing the IE1 promoter/enhancer (from –650) expressing a fluorescent protein (GFP) in upstream position, and the IE2 promoter expressing luciferase in downstream position (Fig. 3). Three variants of luciferase expression cassettes were made: the first (a in Fig. 3) containing both the IE2 and IE1 enhancers (from –1230 with respect to the IE2 TSS), the second (b) containing only the IE2 enhancer

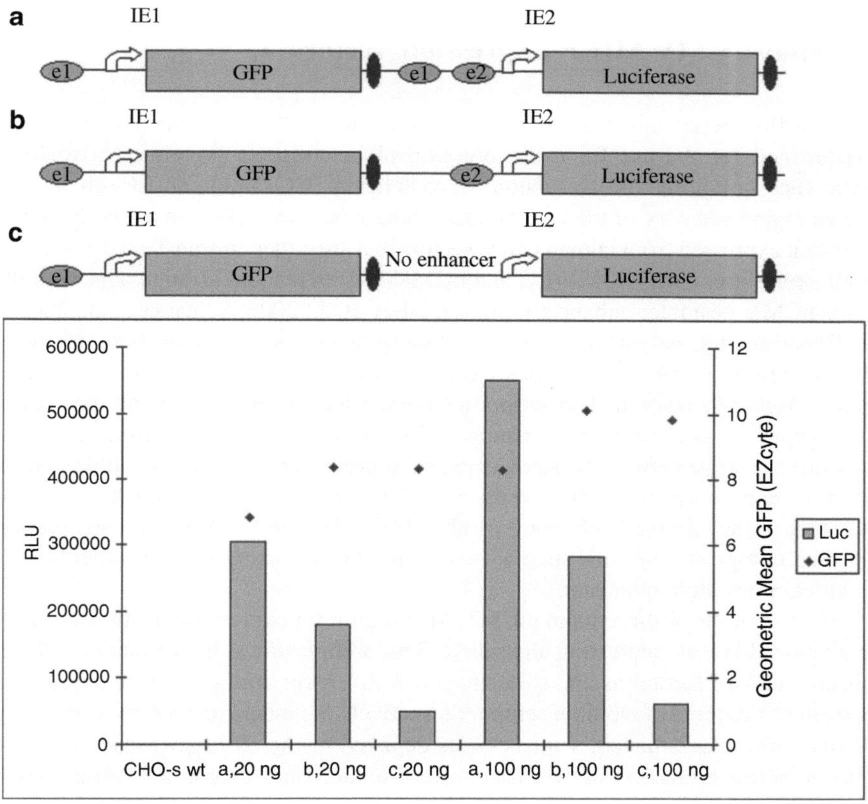

Fig. 3 Functional testing of vectors with tandem arrangement of IE1-GFP (from –650) and IE2 promoter-luciferase expression cassettes. On top the drawings representing three different vectors that vary in their respective enhancer regions for expression from IE2: (**a**) from –1230 containing both enhancer regions, (**b**) from –607 containing the IE2 enhancer, and (**c**) from –208 containing no enhancer. The graph on the bottom shows the results from all three expression vectors used at two different DNA concentrations in transient transfections of CHO-S cells. Luciferase expression data (*bars*) were normalized for the number of transfected cells (as measured by flow cytometry). Mean expression of GFP for each population of transfected cells is shown by *diamonds*

(from –607), and the third (c) containing no enhancer (from –208). As shown in the graph of Fig. 3, both marker genes are expressed efficiently in transient transfection assays. Progressive deletion of enhancer sequences resulted in diminished luciferase expression, but also elevated levels of fluorescent protein expression when the IE1 enhancer is removed from the luciferase cassette. In summary the data suggest that tandem arrangement of the IE1 and IE2 promoter/enhancer pairs allows for expression of two distinct recombinant proteins. This prompted us to consider bi-directional and tandem arrangement of expression units for the production of monoclonal antibodies in engineered CHO cells.

7 Expression of Recombinant Antibodies Using Mouse CMV MIEP Expression Vectors

The bi-directional and tandem arrangement architectures presented above are predestined for the use for the expression of heteromeric proteins. Monoclonal antibodies, composed of two subunits of each heavy (HC) and light (LC) chains, are the emerging category of therapeutic recombinant proteins. The two antibody chains are often expressed from human CMV vectors that are either co-transfected or contain both expression units (hCMV-LC and hCMV-HC) in tandem arrangement with the same hCMV promoter/enhancer regions (Kalwy et al., 2006; Schlatter et al., 2005).

Experimental analysis of vectors expressing two marker proteins, both driven by the same promoters/enhancers, has shown that the expression of the first (upstream) gene is typically favored. Transcription from the second expression unit may suffer from promoter occlusion or interference (Eszterhas et al., 2002). Furthermore, the repetitive arrangement of the same expression promoting sequences could result in instable expression through repeat-induced silencing in stably transfected cells (Garrick et al., 1998; McBurney et al., 2002). Therefore long term expression may be compromised with such a vector architecture using tandem repetition of identical promoter/enhancers.

The use of the bi-directional mCMV MIE region for expressing the two antibody chains would be an interesting alternative. This architecture is in agreement with the observations of Eszterhas and colleagues that divergent arrangement of expression cassettes is the most favorable setup. Alternatively, a tandem arrangement by using distinct promoter/enhancer sequences, as depicted in Fig. 2d, may reduce negative effects such as promoter interference or repeat-induced silencing in stably transfected cells. Therefore we explored these two arrangements for expressing the heavy and light chains of antibodies. Figure 4a shows the two expression constructs used to this end. Note that we previously observed that the shown bi-directional expression vector leads to superior monoclonal antibody expression compared to the same vector having the LC and HC interchanged with respect to the mCMV promoters. Furthermore, comparison of the bi-directional mCMV expression vector shown in Fig. 4a to a vector with tandem arrangement of hCMV-LC and hCMV-HC showed increased and more stable expression with the bi-directional mCMV vector in stably transfected CHO cells (R. Pankiewicz, unpublished observation). Note that here we use glutamine synthetase (labeled Q-Sy) as a selection marker for stable transfectants, expressed from the HC mRNA by inclusion of an IRES sequence (Mountford and Smith, 1995). All antibody chain encoding cDNAs are preceded by an intron (not shown). In both vectors the expression cassette is flanked by two insulator sequences to protect against gene silencing (Kobr et al., 2008).

In order to analyze the activity of the described expression vectors CHO cells were stably transfected using linearized vectors and DMRIE-C as transfection agent. DNA uptake and transport to the nucleus (Wong et al., 2007) is followed by random integration into the host cell genome. Expression depends on the global activity of the genomic integration region (Gierman et al., 2007), with the local

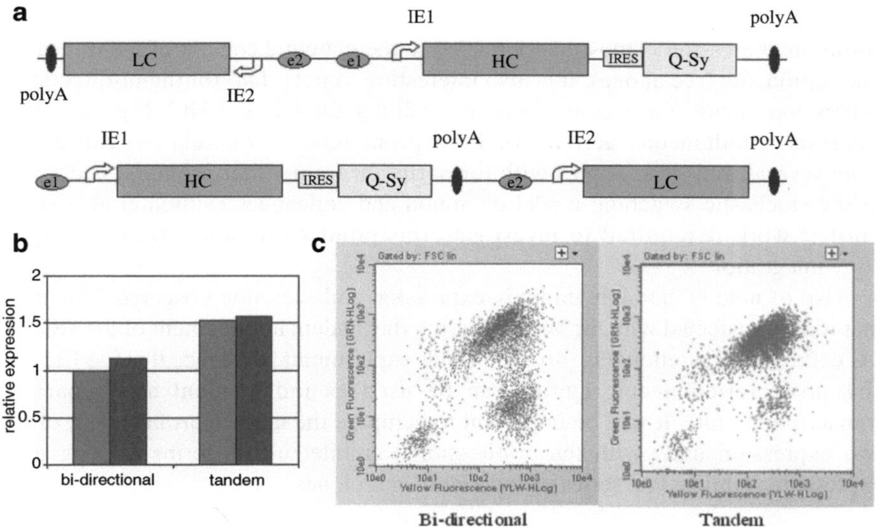

Fig. 4 Stable transfection of CHO cells with antibody expression vectors using IE1 and IE2 promoter/enhancers in bi-directional and tandem arrangements. (**a**) Schematic view of the two expression vectors. Note that all mRNAs contain hCMV intron A upstream of the antibody encoding sequences and the shown regions are flanked by insulators (not shown in the graph); ovals: polyadenylation sites. (**b**) Relative expression level measured by antibody titre secreted into the medium, normalized for cell number. Analysis was performed on the stably transfected population after selection. Data shown were measured at one week interval for each population. (**c**) Expression analysis of stably transfected cells by intracellular staining of LC (y-axis) and HC (x-axis). Cells were fixed, stained, and analyzed by flow cytometry. Population data correspond to the first measure in (**b**)

properties of the integration site affecting the stability of expression (Migliaccio et al., 2000; Pikaart et al., 1998). To isolate stable transfectants, selection was applied using glutamine-free medium containing methionine sulfoximine to increase selective pressure by inhibiting glutamine synthetase activity. Stably transfected pools of cells were used for measuring secreted antibody titres in conditioned serum-free media (Fig. 4b), and for monitoring expression of the individual antibody chains by cell fixation, permeabilisation, and intracellular labeling with fluorescent anti-LC and anti-HC antibodies (Fig. 4c).

Absolute expression levels were assessed for cell populations transfected with either expression vector by measuring the secreted antibody titre, normalized by cell number (Fig. 4b). From these data it is clear that the mouse CMV IE1 and IE2 promoter/enhancer regions prove their usefulness for expressing recombinant proteins such as for heteromeric monoclonal antibodies. Figure 4c shows the intracellular labeling of expressed HCs and LCs in populations of stably transfected cells, as assessed by flow cytometry on the level of individual cells. These graphs show that the levels of expression of either chain are highly variable for both constructs, but the majority of cells (about 90%) is producing both chains. The variability in expression is most likely related to the genomic context of the transgene integration site (see above). It is also interesting to note that, for the bi-directional expression vector, most cells shows co-staining for LC and HC (Fig. 4c). This suggests simultaneous activity of both promoters, or cumulative expression from several transgene copies with transcription from either promoter according to the stochastic switching model of Simon and colleagues (Simon et al., 2007). Further work is required to investigate this point with clones bearing single-copy integration.

Also of note is the high antibody expression and secretion observed from cells that were transfected with the vector bearing the tandem arrangement of the HC and LC expression cassettes (Fig. 4b). This adds experimental evidence that the IE2 and IE1 promoter/enhancer regions can be used as independent and separable transcription units. It will be important to compare the tandem arrangement of the two expression units with the bi-directional architecture in terms of long term expression stability in a series of individual cell clones.

8 Conclusions

The data presented here and in previous publications (Chatellard et al., 2007; Kobr et al., 2008) demonstrate that the mouse CMV IE2 promoter and enhancer regions are useful for cell engineering applications in biotechnology. High levels of recombinant protein production, exemplified by the expression of a monoclonal antibody, are achieved using different vector architectures. This is promising for future use of mouse CMV MIE promoters as alternative to other expression elements providing elevated and stable protein production. Alternatively, these promoters could also be used for the production of inhibitory RNA. Cell engineering could hence be extended to constitutive downregulation of endogenous genes. Altogether these sequences offer a variety of applications in research and biotechnology.

Acknowledgements We would like to thank the members of the Merck Serono Biotechnology Center Cell Sciences group for their constant help with experiments and support for generating the reported data. Our special thanks go to the Process Development Analytics group. Christine Power and Mark Ibberson provided essential information on expressed CHO cell sequences.

References

Addison CL, Hitt M, Kunsken D, Graham, FL (1997). Comparison of the human versus murine cytomegalovirus immediate early gene promoters for transgene expression by adenoviral vectors. J Gen Vir 78:1653–1661

Ando K, Hirao S, Kabe Y, Ogura Y, Sato I, Yamaguchi Y, Wada T, Handa H (2008) A new APE1/Ref-1-dependent pathway leading to reduction of NF-κB and AP-1, and activation of their DNA-binding activity. Nucl Acids Res 36:4327–4336

Apostolou E, Thanos D (2008) Virus infection induces NF-κB-dependent interchromosomal associations mediating monoallelic IFN-β gene expression. Cell 134:85–96

Bartlett J, Blagojevic J, Carter D, Eskiw C, Fromaget M, Job C, Shamsher M, Faro-Trindale I, Xu M, Cook PJ (2006) Specialized transcription factories. Biochem Soc Symp 73:67–75

Benedict CA, Angulo A, Patterson G, Ha S, Huang H, Messerle M, Ware CF, Ghazal P (2004) Neutrality of the canonical NF-κB-dependent pathway for human and murine cytomegalovirus transcription and replication in vitro. J Virol 78:741–750

Binnie A, Castelo-Branco P, Monks J, Proudfoot NJ (2006) Homologous gene sequences mediate transcription-domain formation. J Cell Sci 119:3876–3887

Bird AP, Wolffe AP (1999) Methylation-induced repression – belts, braces, and chromatin. Cell 99:451–454

Boshart M, Weber F, Jahn G, Dorsch-Häsler K, Fleckenstein B, Schaffner W (1985) A very strong enhancer is located upstream of an immediate early gene of human cytomegalovirus. Cell 41:521–530

Boyle AP, Davis S, Shulha HP, Meltzer P, Margulies EH, Weng Z, Furey TS, Crawford GE (2008) High-resolution mapping and characterization of open chromatin across the genome. Cell 132:311–322

Cardin RD, Abens GB, Stoddart CA, Mocarski ES (1995) Murine cytomegalovirus IE2, an activator of gene expression, is dispensable for growth and latency in mice. Virology 209:236–241

Chatellard P, Pankiewicz R, Meier E, Durrer L, Sauvage C, Imhof MO (2007) The IE2 promoter/enhancer region from mouse CMV provides high levels of therapeutic protein expression in mammalian cells. Biotech Bioeng 96:106–117

Cremer T, Cremer C (2001) Chromosome territories, nuclear architecture and gene regulation in mammalian cells. Nat Rev Genetics 2:292–301

de Felipe P, Luke GA, Hughes LE, Gani D, Halpin C, Ryan MD (2006) E unum pluribus: multiple proteins from a self-processing polyprotein. Trends Biotech 24:68–75

Dorsch-Häsler K, Keil GM, Weber F, Jasin M, Schaffner W, Koszinowski UH (1985) A long and complex enhancer activates transcription of the gene coding for the highly abundant immediate early mRNA in murine cytomegalovirus. Proc Natl Acad Sci USA 82:8325–8329

Eszterhas SK, Bouhassira EE, Martin DIK, Fiering S (2002) Transcriptional interference by independently regulated genes occurs in any relative arrangement of the genes and is influenced by chromosomal integration position. Mol Cell Biol 22:469–479

Faisst S, Meyer S (1992) Compilation of vertebrate-encoded transcription factors. Nucl Acids Res 20:3–26

Francastel C, Walters MC, Groudine M, Martin DIK (1999) A functional enhancer suppresses silencing of a transgene and prevents its localization close to centromeric heterochromatin. Cell 99:259–269

Gardiner-Garden M, Frommer M (1987) CpG islands in vertebrate genomes. J Mol Biol 196:261–282

Garrick D, Fiering S, Martin DIK, Whitelaw E (1998) Repeat-induced gene silencing in mammals. Nature Gen 18:56–59

Gierman HJ, Indemans MHG, Koster J, Goetze S, Seppen J, Geerts D, van Driel R, Versteeg R (2007) Domain-wide regulation of gene expression in the human genome. Genome Res 17:1286–1295

Goetze S, Mateos-Langerak J, van Driel R (2007) Three-dimensional genome organisation in interphase and its relation to genome function. Sem Cell Dev Biol 18:707–714

Gruh I, Wunderlich S, Winkler M, Schwanke K, Heinke J, Blömer U, Ruhparwar A, Rohde B, Li R-K, Haverich A, Martin U (2008) Human CMV immediate-early enhancer: a useful tool to enhance cell-type-specific expression from lentiviral vectors. J Gene Med 10:21–32

Grzimek NKA, Dreis D, Schmalz S, Reddehase MJ (2001) Random, asynchronous, and asymmetric transcriptional activity of enhancer-flanking major immediate-early genes ie1/3 and ie2 during murine cytomegalovirus latency in the lungs. J Virol 75:2692–2705

Guelen L, Pagie L, Brasset E, Meuleman W, Faza MB, Talhout W, Eussen BH, de Klein A, Wessels L, de Laat W, van Steensel B (2008) Domain organization of human chromosomes revealed by mapping of nuclear lamina interactions. Nature 453:948–951

Gustems M, Borst E, Benedict CA, Perez C, Messerle M, Ghazal P, Angulo A (2006) Regulation of the transcription and replication cycle of human cytomegalovirus is insensitive to genetic elimination of the cognate NF-κB binding sites in the enhancer. J Virol 80:9899–9904

Handwerger KE, Gall JG (2006) Subnuclear organelles: new insights into form and function. Trends Cell Biol 16:19–26

Heard E, Bickmore W (2007) The ins and outs of gene regulation and chromosome territory organisation. Curr Opin Cell Biol 19:311–316

Heinemeyer T, Wingender E, Reuter I, Hermjakob H, Kel A, Kel O, Ignatieva E, Ananko E, Podkolodnaya O, Kolpakov F et al (1998) Databases on transcriptional regulation: TRANSFAC, TRRD, and COMPEL. Nucl Acids Res 26:364–370

Hummel M, Abecassis MM (2002) A model for reactivation of CMV from latency. J Clin Virol 25:S123–S136

Irarrazabal CE, Williams CK, Ely MA, Birrer MJ, Garcia-Perez A, Burg MB, Ferraris JD (2008) Activator protein-1 contributes to high NaCl-induced increase in tonicity-responsive enhancer/osmotic response element-binding protein transactivating activity. J Biol Chem 283:2554–2563

Kalwy S, Rance J, Young R (2006) Toward more efficient protein expression. Mol Biotechnol 34:151–156

Kaufman RJ (2004) Regulation of mRNA translation by protein folding in the endoplasmic reticulum. Trends Biochem Sci 29:152–158

Kim Y-S, Risser R (1993) TAR-independent transactivation of the murine cytomegalovirus major immediate-early promoter by the Tat protein. J Virol 67:239–248

Kim TK, Lagrange T, Wang YH, Griffith JD, Reinberg D, Ebright RH (1997) Trajectory of DNA in the RNA polymerase II transcription preinitiation complex. Proc Natl Acad Sci USA 94:12268–12273

Kim S-Y, Lee J-H, Shin H-S, Kang H-J, Kim Y-S (2002) The human elongation factor 1 alpha (EF-1α) first intron highly enhances expression of foreign genes from the murine cytomegalovirus promoter. J Biotech 93:183–187

Kim TH, Barrera LO, Zheng M, Qu C, Singer MA, Richmond TA, Wu Y, Green RD, Ren B (2005) A high-resolution map of active promoters in the human genome. Nature 436:876–880

Kingston RE (1997) Introduction of DNA into mammalian cells. In: Ausubel FM, Brent R, Kingston RE, Moore DD, Seidman JG, Smith JA, Struhl K (eds) Current protocols in molecular biology. Wiley, Chapter 9, Boston.

Kobr M, Chatellard P, Imhof MO (2008) Expression vector engineering for cell line development – new roles for 'old' sequences. BioProcessing J 7:16–20

Koutroubas G, Merika M, Thanos D (2008) Bypassing the requirements for epigenetic modifications in gene transcritpion by increasing enhancer strength. Mol Cell Biol 29:926–938

Kouzarides T (2007) Chromatin modifications and their function. Cell 128:693–705

Lanctôt C, Cheutin T, Cremer M, Cavalli G, Cremer T (2007) Dynamic genome architecture in the nuclear space: regulation of expression in three dimensions. Nat Rev Genetics 8:104–115

Lashmit PE, Lundquist CA, Meier JL, Stinski MF (2004) A cellular repressor inhibits human cytomegalovirus transcription from the UL127 promoter. J Virol 78:5113–5123

Latchman DS (1991) Eukaryotic transcription factors. London, Academic Press

Lee J, Klase Z, Gao X, Caldwell JS, Stinski MF, Kashanchi F, Chao S-H (2007) Cellular homeoproteins, SATB1 and CDP, bind to the unique region between the human cytomegalovirus UL127 and major immediate-early genes. Virology 366:117–125

Lemon B, Tjian R (2000) Orchestrated response: a symphony of transcription factors for gene control. Genes Dev 14:2551–2569

Liu Z, Cashion LM, Twu JJ (1997) A systematic comparison of relative promoter/enhancer activities in mammalian cell lines. Anal Biochem 246:150–152

Makrides SC (1999) Components for vectors for gene transfer and expression in mammalian cells. Prot Express Purif 17:183–202

McBurney MW, Mai T, Yang X, Jardine K (2002) Evidence for repeat-induced gene silencing in cultured mammalian cells: inactivation of tandem repeats of transfected genes. Exp Cell Res 274:1–8

Meier JL, Stinski MF (1996) Regulation of human cytomegalovirus immediate-early gene expression. Intervirology 39:331–342
Meier JL, Stinski MF (2006) Major immediate-early enhancer and its gene products. In: Reddehase MJ, Reddehase MJ (eds) Cytomegaloviruses: molecular biology and immunology. Caister Academic Press, Wymondham, pp 151–166
Mellor J (2005) The dynamics of chromatin remodeling at promoters. Mol Cell 19:147–157
Messerle M, Keil GM, Koszinowski UH (1991) Structure and expression of murine cytomegalovirus immediate-early gene 2. J Virol 65:1638–1643
Messerle M, Bühler B, Keil GM, Koszinowski UH (1992) Structural organization, expression, and functional characterization of the murine cytomegalovirus immediate-early gene 3. J Virol 66:27–36
Migliaccio AR, Bengra C, Ling J, Pi W, Li C, Zeng S, Keskintepe M, Whitney B, Sanchez M, Migliaccio G, Tuan D (2000) Stable and unstable transgene integration sites in the human genome: extinction of the green fluorescent protein transgene in K562 cells. Gene 256:197–214
Mitchell JA, Fraser P (2008) Transcription factories are nuclear subcompartments that remain in the absence of transcription. Genes Dev 22:20–25
Mountford PS, Smith AG (1995) Internal ribosome entry sites and dicistronic RNAs in mammalian transgenesis. Trends Genet 11:179–184
Müller WG, Rieder D, Karpova TS, John S, Trajanoski Z, McNally JG (2007) Organization of chromatin and histone modifications at a transcription site. J Cell Biol 177:957–967
Pikaart MJ, Recillas-Targa F, Felsenfeld G (1998) Loss of transcriptional activity of a transgene is accompanied by DNA methylation and histone deacetylation and is prevented by insulators. Genes Dev 12:2852–2862
Powers C, Früh K (2008) Rhesus CMV: an emerging animal model for human CMV. Med Microbiol Immunol 197:109–115
Ragoczy T, Bender MA, Telling A, Byron R, Groudine M (2006) The locus control region is required for association of the murine β-globin locus with engaged transcription factories during erythroid maturation. Genes Dev 20:1447–1457
Rice P, Longden I, Bleasby A (2000) EMBOSS: the European Molecular Biology Open Software Suite. Trends Genetics 16:276–277
Rozenberg JM, Shlyakhtenko A, Glass K, Rishi V, Myakishev MV, FitzGerald PC, Vinson C (2008) All and only CpG containing sequences are enriched in promoters abundantly bound by RNA polymerase II in multiple tissues. BMC Genomics 9:67–79
Sandelin A, Alkema W, Engström P, Wasserman WW, Lenhard B (2004) JASPAR: an open-access database for eukaryotic transcription factor binding profiles. Nucl Acids Res 32:D91–D94
Sanford GR, Burns WH (1996) Rat Cytomegalovirus has a unique immediate early gene enhancer. Virology 222:310–317
Sawicki JA, Morris RJ, Monks B, Sakai K, Miyazaki J (1998) A composite CMV-IE enhancer/beta-actin promoter is ubiquitously expressed in mouse cutaneous epithelium. Exp Cell Res 244:367–369
Saxonov S, Berg P, Brutlag DL (2006) A genome-wide analysis of CpG dinucleotides in the human genome distinguishes two different classes of promoters. Proc Natl Acad Sci USA 103:1412–1417
Schlatter S, Stansfield SH, Dinnis D, Racher AJ, Birch JR, James DC (2005) On the optimal ratio of heavy to light chain genes for efficient recombinant antibody production by CHO cells. Biotechnol Prog 21:122–133
Schneider R, Grosschedl R (2007) Dynamics and interplay of nuclear architecture, genome organization, and gene expression. Genes Dev 21:3027–3043
Schones DE, Cui K, Cuddapah S, Roh T-Y, Barski A, Wang Z, Wei G, Zhao K (2008) Dynamic regulation of nucleosome positioning in the human genome. Cell 132:887–898
Simon CO, Seckert CK, Dreis D, Reddehase MJ, Grzimek NKA (2005) Role for tumor necrosis factor alpha in murine cytomegalovirus transcriptional reactivation in latently infected lungs. J Virol 79:326–340

Simon CO, Kühnapfel B, Reddehase MJ, Grzimek NK (2007) Murine cytomegalovirus major immediate-early enhancer region operating as a genetic switch in bidirectional gene pair transcription. J Virol 81:7805–7810
Smale ST, Kadonaga JT (2003) The RNA polymerase II core promoter. Ann Rev Biochem 72:449–479
Stamminger T, Fleckenstein B (1990) Immediate-early transcription regulation of human cytomegalovirus. Curr Top Microbiol Immunol 154:3–19
Szutorisz H, Dillon N, Tora L (2005) The role of enhancers as centres for general transcription factor recruitment. Trends Biochem Sci 30:593–599
Trojer P, Reinberg D (2007) Facultative Heterochromatin: is there a distinctive molecular signature? Mol Cell 28:1–13
Vakoc CR, Letting DL, Gheldof N, Sawado T, Bender MA, Groudine M, Weiss MJ, Dekker J, Blobel GA (2005) Proximity among distant regulatory elements at the β-globin locus requires ATA-1 and FOG-1. Mol Cell 17:453–462
Venugopal R, Jaiswal AK (1998) Nrf2 and Nrf1 in association with Jun proteins regulate antioxidant response element-mediated expression and coordinated induction of genes encoding detoxifying enzymes. Oncogene 17:3145–3156
Wilson M, Koopman P (2002) Matching SOX: partner proteins and co-factors of the SOX family of transcriptional regulators. Curr Opin Genet Dev 12:441–446
Wong AW, Scales SJ, Reilly DE (2007) DNA internalized via caveolae requires microtubule-dependent, Rab7-independent transport to the late endocytic pathway for delivery to the nucleus. J Biol Chem 282:22953–22963
Xia W, Bringmann P, McClary J, Jones PP, Manzana W, Zhu Y, Wang S, Liu Y, Harvey S, Madlansacay MR et al (2006) High levels of protein expression using different mammalian CMV promoters in several cell lines. Prot Express Purif 45:115–124
Xu M, Cook PR (2008) Similar active genes cluster in specialized transcription factories. J Cell Biol 181:615–623
Xu Z-L, Mizuguchi H, Ishii-Watabe A, Uchida E, Mayumi T, Hayakawa T (2001) Optimization of transcriptional regulatory elements for constructing plasmid vectors. Gene 272:149–156
Zhou G-L, Xin L, Song W, Di L-J, Liu G, Wu X-S, Liu D-P, Liang C-C (2006) Active chromatin hub of the mouse α-globin locus forms in a transcription factory of clustered housekeeping genes. Mol Cell Biol 26:5096–5105

Defeating Randomness – Targeted Integration as a Boost for Biotechnology

L. Gama-Norton, P. Riemer, U. Sandhu, K. Nehlsen, R. Schucht, H. Hauser, and D. Wirth

Abstract Genetic modification of mammalian cells is a prerequisite for the production of recombinant proteins, virus like particles, and viruses. For most applications long-term stability of such modifications is needed. Apart from episomal approaches which have not yet been sufficiently explored, integration of transgenes into the chromosomal DNA of the host cell is regarded to be the method of choice. Due to strong influences of the chromosomal surroundings on the expression of transgenes, targeted integration has obvious advantages over classical random integration procedures. Site directed integration leads to predictable expression properties, circumvents screening, is fast and provides high safety and is therefore advantageous for the integration of transgenes. In this chapter recombinase mediated cassette exchange (RMCE) using heterologous recombinases is described as an efficient and reliable method to target (integrate and replace) transgenes by site directed engineering of defined chromosomal sites for recombinant protein and virus expression. This technique, however, requires that the site of interest has been tagged before by specific sequence motifs. Thus, the chapter highlights current and forthcoming methods to (find and) tag chromosomal sites in order to make RMCE possible. This includes screening of randomly integrated reporters and spontaneous or Zinc Finger Nuclease enhanced homologous recombination.

L. Gama-Norton, P. Riemer, U. Sandhu, K. Nehlsen, R. Schucht,
H. Hauser (✉), and D. Wirth
Helmholtz Centre for Infection Research, Department of Gene Regulation and Differentiation and Research Group Model Systems for Infection and Immunity,
Inhoffenstrasse 7, D-38124, Braunschweig, Germany
e-mail: Hansjoerg.hauser@helmholtz-hzi.de

L. Gama-Norton
Instituto de Biologia Experimental e Tecnológica, Universidade Nova de Lisboa (IBET/ITQB-UNL), Animal Cell Technology, Oeiras, Portugal

M. Al-Rubeai (ed.), *Cell Line Development*, Cell Engineering 6,
DOI 10.1007/978-90-481-2245-5_3, © Springer Science+Business Media B.V. 2009

1 Transgene Expression in Mammalian Cells – Limitations of the Classical Random Gene Integration Strategy

Technologies enabling expression of recombinant genes in mammalian cells belong to the tools of the trade in molecular biology. These technologies can serve a wide variety of purposes, ranging from basic research to elucidate gene functions, e.g., in transgenic mouse models, to biotechnological applications like manufacturing antibodies. Efficient production of proteins and virus particles using mammalian cell lines often relies on random and stable integration of an expression construct into the cell's chromosomal DNA. Once a construct is incorporated its expression levels will be determined by neighbouring genetic elements (Festenstein et al., 1996; Bell and Felsenfeld, 1999). For instance, enhancers will support expression while silencers and formation of heterochromatin will suppress it (Fig. 1). A so called "position effect" constitutes a major drawback of the random approach of integration; expression from many integration sites is silenced or severely reduced and stability of long term expression is often constricted. This results in an unpredictable expression pattern. Expression can also differ due to variable gene copy number. Importantly, the intuition that a higher copy number supports stronger expression is not necessarily valid. Conversely, single copy integration can as well provide the required expression strength and stability (Yarranton, 1990; Schucht et al., 2006; Coroadinha et al., 2006). If a toxic protein is to be produced, regulated transgene expression is required. In this case, genomic loci have to meet even higher demands. Here, they need not simply support efficiently high expression but should display low basal expression levels as well as strong inducibility. Consequently, extensive screenings are necessary to identify cell clones (i.e., integration sites) with suitable expression characteristics. This selection step usually requires several

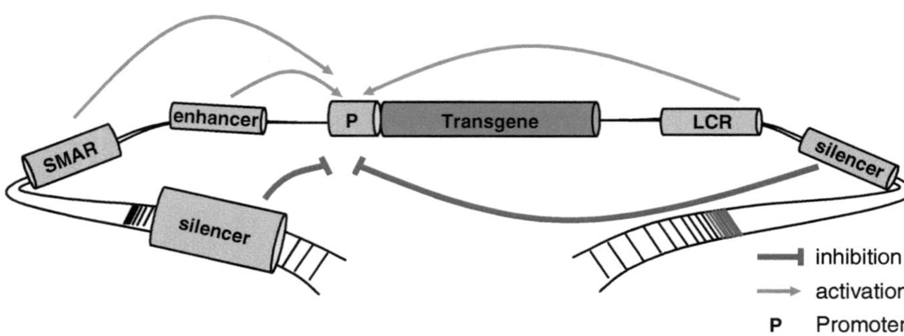

Fig. 1 Chromosomal position effects on transgene expression. Neighbouring genetic elements of an integrated transgene can support expression (e.g. enhancers), while others can suppress it (e.g. silencers). Expression of a transgene is also affected by the chromatin status that can be modulated by elements such as S/MARs and Locus control regions (LCRs). These so called "position effects" result in unpredictable expression patterns and could potentially lead to complete silencing of the genes of interest. The effective proximity of the elements can range from a few hundred up to several thousand base pairs.

weeks and it has to be repeated for each new expression construct, slowing down the generation of production cell lines to a significant degree. Thus, the ability to repeatedly reuse a single chromosomal locus that supports the desired expression characteristics is highly advantageous. To enable exploitation of a favourable chromosomal integration site, a primary genomic modification to first mark (tag) the site needs to be performed. This in turn, would create a genomic platform that eventually supports subsequent modifications of that particular site. The requirements needed to consecutively manipulate the genomic site can be met by the technology of Recombinase Mediated Cassette Exchange (RMCE). This technology permits the rapid exchange of different cassettes of choice in the desired genomic surrounding adding the dimension of flexibility and reusability needed for biotechnological purposes. In this respect, the next section describes the different methods to introduce

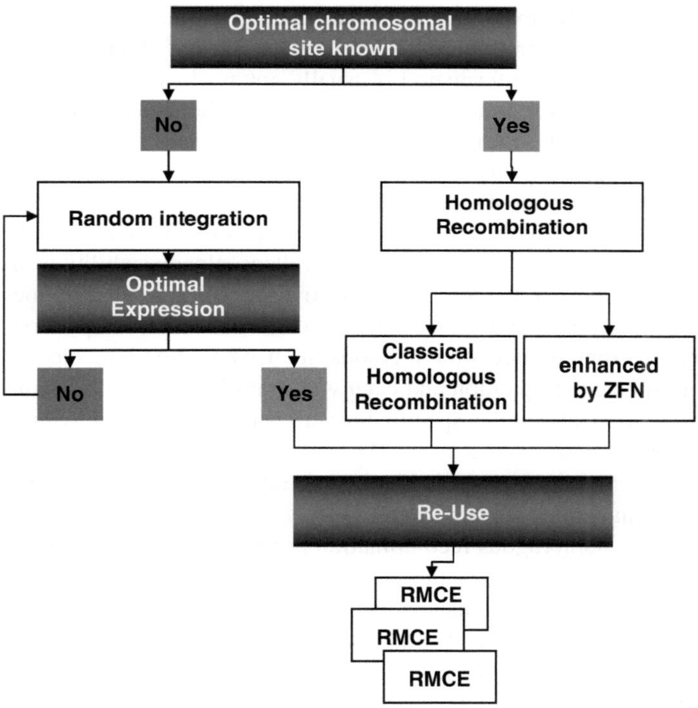

Fig. 2 How to create a platform for genetic engineering. In order to utilise a particular chromosomal site, two approaches can be considered: (**a**) functional screening of an unknown locus using random integration and (**b**) exploiting a characterised genomic site using homologous recombination. Random integration of recombinant genes into a host cell line influences productivity as well as clonal stability. Therefore, screening has to be performed to identify loci with desirable expression patterns. In contrast, homologous recombination permits the use of predefined chromosomal loci with known expression characteristics. The frequency of homologous recombination in mammalian cells can be enhanced by ZF-nucleases that stimulate the host cell's DNA repair mechanism via introducing site-specific double strand breaks. Once a particular chromosomal site has been identified that meets all the desired properties, RMCE allows efficient re-use of this locus.

a tag followed by the detailed description of the tools permitting the reuse of the tagged loci. Accordingly, Fig. 2 summarises the scope of this review.

2 Creating a Favorable Platform for Genetic Engineering – Primary Genomic Modification to Integrate a Tag

As mentioned in Section 1, tagging a genomic site that possesses the desired expression properties is a prerequisite to generate a platform for further engineering. To utilize a particular chromosomal site, two approaches can be considered for the primary integration of a tag: (a) functional screening of an unknown locus using random integration, (b) exploiting a characterised genomic site using homologous recombination. The second approach requires knowledge about the properties of chromosomal sites. This information is usually not available; in particularly not in cell lines that are used for protein production. On the other hand, random integration provides the possibility to screen for the desired properties amongst thousands of loci. This is why currently most endeavours follow the random approach.

The random integration of recombinant genes into a host cell line is characterised by unpredictable cis-effects which might either positively or negatively regulate gene expression (Fig. 1). Productivity as well as clonal stability consequently depends on the site(s) of integration. Exhaustive screening to identify loci supporting desirable (e.g., high and/or regulated) and stable expression patterns is hence mandatory. Current transfection methods such as lipofection, electroporation, calcium phosphate precipitation and even viral transduction predominantly lead to random integration, although a specific bias has been shown for individual methods (Schroeder et al., 2002; Wu et al., 2003; Mitchell et al., 2004). Thus, any of the above mentioned methods are suitable tools to integrate the tags randomly as long as single copy integrates are achieved.

Alternatively, homologous recombination permits the integration of tags to use predefined chromosomal loci with known expression characteristics. This strategy exploits the cell's own recombination apparatus to exchange a certain DNA sequence for the construct of interest which is flanked by ends homologous to the endogenous locus. Homologous recombination has been widely applied in murine embryonic stem cells to establish e.g., knockout mouse models. However, the frequency of homologous recombination in differentiated mammalian cells is very low (one in a million or even less) requiring a large number of clones to be screened for correct integration. In differentiated cells the high ratio of illegitimate recombination masks homologous recombination showing that this strategy is not suitable for routine use (Glaser et al., 2005). Although currently homologous recombination is not feasible for systematic integration of tags, recent developments in zinc finger-mediated DNA modifications let envision that this technology might gain importance in the near future.

The frequency of homologous recombination can be enhanced by stimulating the host cell's DNA repair mechanism via introducing site-specific double strand breaks. Making use of this, custom designed zinc finger (ZF) nucleases are emerging as an alternative approach towards site-directed integration (Porteus and Carroll, 2005). ZF-nucleases are modular artificial proteins that fuse specific ZF DNA-binding domains with an endonuclease activity domain, such as the *FokI* endonuclease (Kim et al., 1996). Thereby, these engineered molecules take advantage from synergistic effects of the DNA-binding site specificity of ZF and DNA cleavage domain of endonuclease moieties, respectively. Upon induction of double strand breaks at a specific genomic site the cell repair machinery is stimulated, leading to targeted DNA modifications which can be determined by an extrachromosomal donor template (Fig. 3).

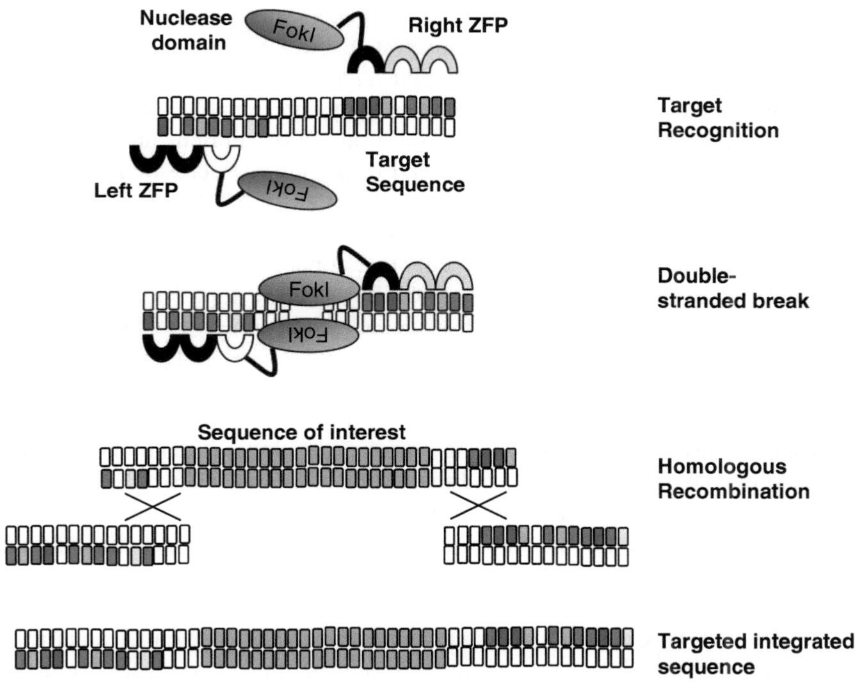

Fig. 3 DNA recognition and cleavage by ZF-nucleases for targeted integration. Each ZF motif (depicted in the figure as semi-circles) primarily binds to a triplet within the DNA substrate (Pavletich and Pabo, 1991). Due to their modular nature linked ZF units result in ZF-protein that recognises and binds longer DNA sequences (Liu et al., 1997). Each finger binds its cognate site independently of the neighbouring fingers (Pruett-Miller et al., 2008). Moreover, the DNA sequence specificity can be substantially increased by combining several ZF motifs (Porteus and Carroll, 2005). Due to the modular structure and modular binding properties of ZFs, they represent an attractive framework for designing ZF-nucleases with tailor-made sequence specificities. Fusion of a ZF motif with a non-specific DNA cleavage domain (such as that of the *FokI* endonuclease) results in specific double strand break 9 to 13 bp downstream of the recognition site (Kim et al., 1996). For high efficiency in double strand DNA recognition and cleavage, binding of two distinct ZF-nuclease monomers at their cognate sites of both DNA strands is required (Mani et al., 2005). Thereby, the specificity of a DNA cleavage by a pair of ZF-nucleases (composed of 3 ZF, as depicted in the figure) is guaranteed by the recognition of an 18 bp particular sequence. Site specific cleavage of DNA will induce DNA repair mediated by the homology-directed repair machinery. This can lead to the integration of heterologous DNA sequences in the site of interest.

Several studies indicate that ZF-nucleases are indeed powerful tools for making directed modifications in experimental organisms for functional studies and for creating models for human genetic diseases (Bibikova et al., 2002; Porteus and Baltimore, 2003; Alwin et al., 2005; Lloyd et al., 2005; Urnov et al., 2005; Wright et al., 2005; Beumer et al., 2006; Doyon et al., 2008; Meng et al., 2008). As an example for their applicability, Urnov et al. (2005) have demonstrated that ZF-nuclease composed of four ZF motifs could correct the gene encoding human interleukin 2 receptor gamma (*IL2R*γ) which underlies X-linked Severe Combined Immunodeficiency Disorder (SCID). Moreover, Moehle et al. (2007) have shown that ZF-nuclease can drive site-specific addition of 8 kb DNA stretches at a frequency of 5–15% into a pre-determined locus in the human genome. This suggests the feasibility to exploit this technology to integrate expression cassettes.

However, one current limitation of the application of ZF-nuclease for site-directed genome modifications is their cytotoxic activity, presumably associated with illegitimate cleavage of the DNA (Bibikova et al., 2002; Porteus and Baltimore, 2003). Attempts to increase recognition and cleavage specificity of ZF-nuclease have been shown to reduce cytotoxicity (Urnov et al., 2005; Pruett-Miller et al., 2008). Since stable expression of ZF-nuclease can have deleterious effects on the integrity of the cellular genome, protocols that support transient expression are currently being explored (Urnov et al., 2005). In this respect, potential alternatives could include either the delivery of ZF-nucleases via integrase-defective lentiviral vectors (Lombardo et al., 2007) or inducible gene expression systems.

The generation of many new ZF-nuclease molecules with tailor-made sequence specificities is an attractive and desirable approach for various applications. ZF-nucleases have the potential to bring obvious advantages in both, medical and industrial fields. Widespread testing and application of the engineered ZF technology will depend upon the availability of information to the academic scientific community. With respect to the highly patented field it is to be hoped that research and development of engineered ZF technology will be continued. For this purpose, the Zinc Finger Consortium was established (http://www.zincfingers.org). The development of ZF-nucleases as broadly applicable and readily accessible molecular tools for performing targeted genetic alterations would be thus enormously useful for biological research and molecular therapeutics.

3 Exploiting the Tagged Loci–Reusability via Recombinase Mediated Cassette Exchange (RMCE)

Once a good chromosomal integration site has been identified and tagged, repeated extensive screenings can be circumvented by the ability to constantly reuse the desired locus. These loci can hence be used to specifically integrate any DNA sequence of interest providing predictable expression levels. Homologous recombination in principle can permit this, but the above mentioned drawbacks limit the feasibility of this approach. Efficient and flexible ZF-nuclease based technologies are still being developed and have not yet found routine use for biotechnological applications

Currently, the technology fulfilling at best the requirements that allow the recognition of a specific genomic site are based on the use of site-specific recombinases. The most advanced application of these site-specific recombinases is RMCE. RMCE enables to repeatedly and rapidly modify a predefined chromosomal locus. Thereby, targeted integration of transgenes in any given locus supporting the desired expression level can now be routinely and efficiently performed. Site-specific recombinases recognize distinct sequence-specific motifs termed as recognition targets and catalyze efficient DNA recombination of the respective DNA regions (Branda and Dymecki, 2004; Schnütgen et al., 2006). The basic principle of their use involves a tagging and screening step, in which suitable chromosomal loci are identified and marked by the integration of specific recognition sequences for the recombinase enzyme. These sites can hence later be used for specifically integrating, i.e., targeting any expression construct of interest into the predefined locus providing predictable expression levels (Fig. 4). On the whole, screening for well expressing chromosomal loci has to be performed just once, reducing both time and effort to establish producer cell lines.

3.1 Site-Specific Recombinases as Tools for Targeting a Previously Tagged Locus

At present, the three main site-specific recombinase (SSR) systems used in mammalian cultured cells as well as in mice are Cre/*LoxP* (Sternberg et al., 1986; Sauer and Henderson, 1988; Lakso et al., 1992), Flp/*FRT* (Vetter et al., 1983; Andrews et al., 1985; O'Gorman et al., 1991; Dymecki, Dymecki, 1996) and C31/*att* (Groth et al., 2000; Belteki et al., 2003).

Both, Cre ('Cyclization recombination') from bacteriophage P1 and Flp (named because of its ability to "flip" DNA) from the 2 μ plasmid in *Saccharomyces cerevisiae* are tyrosine recombinases that catalyze a reciprocal and conservative DNA rearrangement between specific recognition target (RT)-sites, *LoxP* and *FRT*, respectively (Branda and Dymecki, 2004). This results in either excision, insertion, inversion or translocation of specific DNA sequences depending on the relative position and orientation of the target sites (Branda and Dymecki, 2004). The excision reaction between two directly repeated identical recognition targets is always more efficient as compared to the integration reaction (Baer and Bode, 2001). Each 34 bp *LoxP* site comprises two inverted palindromic 13 bp repeats flanking an 8 bp asymmetric spacer sequence (Fig. 5). The original 48 bp *FRT* site in addition contains a third 13 bp direct repeat upstream of the two inverted 13 bp repeats flanking the 8 bp core sequence. While this upstream part is required for the later described RMCE procedure, it is dispensable for recombination if simple excision is concerned. It is the spacer region which acts as the site for strand cleavage, exchange and ligation whereas each inverted repeat is a recombinase monomer binding site.

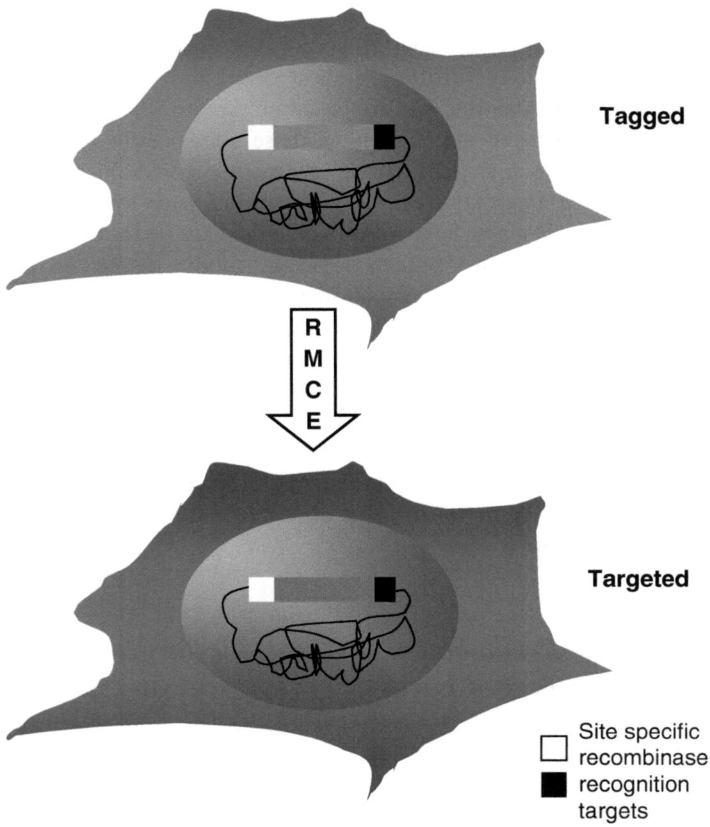

Fig. 4 Recombinase mediated cassette exchange. A suitable genomic site tagged by recognition sites of site-specific recombinases (upper part) is used to integrate any expression cassette of choice via RMCE (lower part).

The asymmetry of the core sequence imparts directionality such that directly oriented target sites would result in an excision of the intervening DNA and inversely oriented target sites would lead to an inversion of the intervening sequence (Fig. 5).

Several modifications produced enzyme variants with higher recombination activity, most notably the thermostable form of Flp, "Flpe" by Buchholz et al. (1998) (to be also noted iCre: Shimshek et al., 2002; Flpo and ΦC31o: Raymond and Soriano, 2007). Regulation of recombinase activity can be achieved by fusion of site-specific recombinases to steroid receptor ligand binding domains. In the inactive state, the fusion will be retained in the cytoplasm while induction by the ligand triggers nuclear localization (Cre: Feil et al., 1996, 1997; Danielian et al., 1998; Schwenk et al., 1998; Flpe: Hunter et al., 2005; Logie and Stewart, 1995, ΦC31: Sharma et al., 2008). Cre has also been placed under the control of the tetracycline dependent expression system (Klehr-Wirth et al., 1997), rendering its transcription dependent on the presence

Recognition site	Sequence: Inverted Repeat	Sequence: Spacer	Sequence: Inverted Repeat	Reference
loxP **(also L2)**	ATAACTTCGTATA	ATGTATGC	TATACGAAGTTAT	**Hoess *et al.*, 1982**
lox **511 (L1)**	ATAACTTCGTATA	ATGTATaC	TATACGAAGTTAT	**Hoess *et al.*, 1986 Bethke and Sauer, 1997**
lox **5171**	ATAACTTCGTATA	ATGTgTaC	TATACGAAGTTAT	**Lee and Saito, 1998**
lox **2272 (*lox*2722)**	ATAACTTCGTATA	AaGTATcC	TATACGAAGTTAT	**Lee and Saito, 1998**
m2	ATAACTTCGTATA	AgaaAcca	TATACGAAGTTAT	**Langer *et al.*, 2002**
Lox **71 (LE mutant)**	taccgTTCGTATA	ATGTATGC	TATACGAAGTTAT	**Albert *et al.*, 1995**
Lox66 **(RE mutant)**	ATAACTTCGTATA	ATGTATGC	TATACGAAcggta	**Albert *et al.*, 1995**
loxP **257 (L3)**	ATAACTTCGTATA	AaGTcTcC	TATACGAAGTTAT	**Wong *et al.*, 2005**
FRT	GAAGTTCCTATTCC* GAAGTTCCTATTC	TCTAGAAA	GTATAGGAACTTC	**McLeod *et al.*, 1986**
F3	GAAGTTCCTATTCC* GAAGTTCCTATTC	TtcAaAtA	GTATAGGAACTTC	**Schlake and Bode, 1994**
F5	GAAGTTCCTATTCC* GAAGTTCCTATTC	TtcAaAAg	GTATAGGAACTTC	**Schlake and Bode, 1994**
FRT mutant -10 **(LE mutant)**	GAAGTTCaTATTC	TCTAGAAA	GTATAGGAACTTC	**Senecoff *et al.*, 1988**
FRT mutant +10 **(RE mutant)**	GAAGTTCCTATTC	TCTAGAAA	GTATAtGAACTTC	**Senecoff *et al.*, 1988**
attB	TCGAGTGAGGTGGAGTACGCGCCCGGGGAGCCCAAGGGCACGCCCTGGCACCCGCA			**Belteki *et al.*, 2003**
attP	CTAGACCCTACGCCCCCAACTGAGAGAACTCAAAGGTTACCCCAGTTGGGGCACG			**Belteki *et al.*, 2003**

Fig. 5 Recognition target sites of site-specific recombinases. *LoxP* and *FRT* sites (targets for Cre and Flp, respectively) contain two 13 bp inverted repeats flanking an 8 bp asymmetric spacer sequence. Nucleotides that deviate from the wildtype sequence are lowercase. The *FRT* wildtype, F3 and F5 sites contain an additional upstream repeat marked by *asterisks*. For simple excision reactions this repeat is dispensible. Unlike *LoxP* sites, the inverted repeats of Flp recognition targets differ in one nucleotide (*underlined*). *AttB* and *attP* are the heterotypic recombination targets of *Φ*C31. The core nucleotides where recombination occurs are underlined. By recombination so called *attL* and *attR* sites are generated which are refractory to any further reaction (adapted and modified from Branda and Dymecki, 2004).

or absence of tetracycline or its analogues (Klehr-Wirth et al., 1997; Saam and Gordon, 1999; Schonig et al., 2002). Another alternative for temporal regulation of SSRs is proposed by Jullien et al. (2003, 2007). There, the authors established a bipartite version of Cre termed DiCre which will only form a complete, active enzyme after induction by rapamycin. Furthermore, screenings for Cre-like enzymes were performed and yielded the so called Dre recombinase, which might complement the already successfully applied SSR systems (Sauer and McDermott, 2004).

3.2 Flip-In and RMCE as Versatile Applications of Site-Specific Recombinases

One of the advantages of using these site specific recombinases is that they can function in different living systems without the requirement of any accessory cofactors (van der Weyden et al., 2002), so that they can also be efficiently used in differentiated cell lines as opposed to homologous recombination.

The initial site-specific chromosomal integration experiments employing Cre and Flp were based on insertion of a single *LoxP* or *FRT* site in the mammalian genome followed by trapping of the rare integration events (this was referred to as "Flip-In") (O'Gorman et al., 1991; Schubeler et al., 1998; Koch et al., 2000) (Fig. 6a). This technique has three main limitations: (1) a low efficiency owing to the fact that once the transgene is integrated it is immediately excised due to the presence of identical target sites which act as substrates for another round of recombination. Also, since this excision is an intra-molecular reaction it is favoured over integration which is an inter-molecular reaction. (2) The entire plasmid with its prokaryotic vector sequence is integrated and (3) a positive selection marker gene is left behind in the chromosome after recombination.

To overcome the above mentioned limitations, site specific recombinases were manipulated to develop a clever and unique genome engineering strategy, i.e., RMCE. This technique was initially described by Waterhouse et al., to create large combinatorial phage antibody libraries using the Cre mediated cassette exchange system (Waterhouse et al., 1993). Since then, this strategy has found wide application in different cell lines utilising the Cre/*LoxP*, Flp/*FRT* and C31/*att* recombinase systems (Wirth et al., 2007).

The basic principle underlying RMCE is based on the fact that recombinase recognition target sites (RTs) possessing identical 8 bp spacer sequence will recombine efficiently whereas RTs differing in spacer sequence will not recombine or will recombine only at very low levels (Hoess et al., 1986; Senecoff and Cox, 1986; Senecoff et al., 1988; Lee and Saito, 1998). In other words, the spacer homology is a critical requirement for recombination between recognition target sites. Hence target sequences containing identical mutations in their spacer sequence will recombine but there will be no or less efficient recombination with other mutants or wild type sites. Also, a set of inverted identical recognition targets may be used for RMCE (Long et al., 2004; Liu et al., 2006). Several recognition target variants

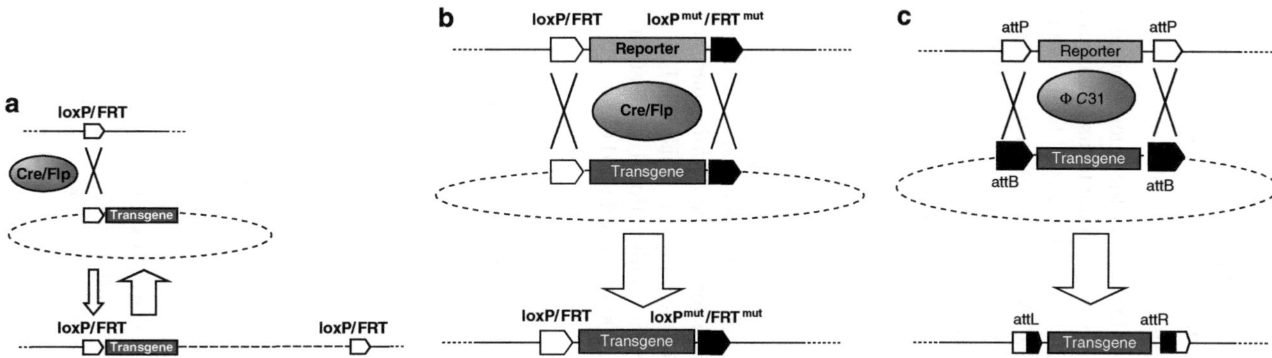

Fig. 6 Site-specific recombinase mediated genome engineering. (**a**) The initial generation of site-specific recombinase mediated chromosomal targeting (Flp-In™) was based on a cell harbouring a single recognition target. The introduction of both, the recombinase and an appropriate targeting vector results in the integration of the complete vector sequence, including all prokaryotic elements. Due to the thermodynamically preferred excision reaction, targeted subclones are rare as suggested by the size of the arrows. (**b**) Flp and Cre recombinases recognise their specific target sequences (*LoxP* and *FRT*, respectively) in the previously tagged genomic DNA and the incoming targeting vector. Suitable selection strategies can permit the recovery of the desired exchange event. Complete excision is prevented by using heterospecific (mutant) and non-interacting *LoxP* or *FRT* sites. (**c**) The *Φ*C31 recombinase mediates recombination between the heterotypic *attB* and *attP* sites. A cassette flanked by two *attB* sequences is integrated into the genomic site and hybrid *attL* and *attR* sites are generated which are not compatible for any further recombination events. Thus, integration of the desired cassette is strictly unidirectional.

were designed with mutations in their spacer sequences (Fig. 5). To date, the most commonly used *FRT* mutants are the F3 and F5 spacer variants which show no recombination with the wild-type *FRT* site (Schlake and Bode, 1994) and in case of the *LoxP* mutants, *LoxP* 2272 (Lee and Saito, 1998), m2 (Langer et al., 2002) and *LoxP* 257 (Wong et al., 2005) seem to show no cross-recombination.

The technique of RMCE involves two steps (Fig. 6b):

1. To tag the genomic locus of interest: In this step, the heterotypic and incompatible recognition targets are introduced into a genomic locus (please refer Section 2). This creates a cassette acceptor allele, i.e., the tagged locus can now be used to integrate and exchange different DNA cassettes of choice.
2. To target the genomic locus of interest: A targeting vector containing the desired transgene flanked by the same set of heterotypic recognition target sites can now replace the DNA region flanked by the recognition target sites in the tagged locus. This reaction is catalysed by the recombinase via a double reciprocal crossover recombination event. The overall strategy can hence be termed as the "Tag and Target" strategy (Baer and Bode, 2001).

Apart from Cre and Flp, *Φ*C31 integrase provides an alternative tool to perform RMCE. *Φ*C31 integrase from *Streptomyces* phage is a large serine recombinase that recombines DNA between two heterotypic attachment sites *attB* and *attP* (Thorpe and Smith, 1998). The resulting hybrid sites (*attL* and *attR*) after the recombination event can no longer act as substrates for the integrase. This *Φ*C31 integrase system is hence particularly useful for site-specific unidirectional integration in mammalian cells (Groth et al., 2000; Belteki et al., 2003; Keravala and Calos, 2008) (Fig. 6c).

3.3 Parameters Influencing the Performance of RMCE

In RMCE, to integrate any genetic information, a set of non-interacting recombinase recognition sites are employed as described above. These sites flank both an initially integrated "tagging" construct and a donor plasmid coding for the new gene – guaranteeing cassette exchange. However, the overall frequency of an exchange event is rather low and has to be selected for.

Several parameters influence efficiency of RMCE and by manipulating them the number of correct targeting events can be significantly enhanced. For example, different site-specific selection strategies have been devised to counterselect random integration of the targeting vector. One of these strategies allows the isolation of correctly targeted recombinants in one step. This relies on the complementation of a deleted version of a positive selection marker with its missing sequence from the incoming targeting plasmid (Verhoeyen et al., 2001; Wirth and Hauser, 2004; Schucht et al., 2006; Wallace et al., 2007). In a similar approach, an incoming promoterless positive selection marker is only expressed by a matching promoter provided by the parental locus after a correct integration (Cobellis et al., 2005). The positive selection strategies mentioned have been shown to work very efficiently

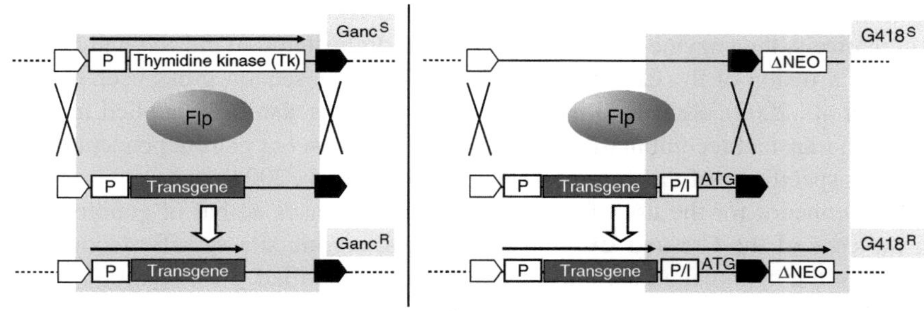

Fig. 7 Selection strategies to identify correctly targeted RMCE clones. The selection for successfully targeted clones can be performed via various strategies. The scheme summarizes two potential selection strategies (negative or positive, respectively), that could be applied. One option depicts the use of a tagging vector that harbours the thymidine kinase gene along with the heterologous recombinase recognition sites. Tagged cells will therefore be sensitive to gancyclovir (Ganc^s). Targeted integration is performed by co-transfecting a recombinase coding vector and a targeting vector that carries the same matching pair of recombinase recognition sites, flanking a transcription unit for the gene of interest. Correctly targeted RMCE cell clones that have specifically integrated the targeting sequence will now exhibit gancyclovir resistance (Ganc^r). The other option depicts the use of a tagging vector that additionally harbours an ATG-defective neomycin resistance gene. Parental cells will therefore be sensitive to G418 (G418^s). However, an incoming targeting vector carries an ATG codon positioned in frame with the defective neomycin resistance gene of the tagging construct is complemented. A promoter or internal ribosomal entry site (P/I) is also introduced to initiate downstream transcription/translation. Recombinant cell clones undergoing a correct RMCE event will now exhibit G418 resistance (G418^r). Combining both, the positive and negative selection methods leads to a more stringent strategy to obtain correctly targeted RMCE clones.

(Cobellis et al., 2005; Schucht et al., 2006) (Fig. 7). However, it should be noted that certain selection markers have been shown to act as transcriptional silencers in mammalian cells thereby decreasing expression levels (Artelt et al., 1991). One possible solution to overcome this could be its subsequent excision at a later stage. A negative selection approach such as utilising the thymidine kinase gene in the initial tagging construct has also been applied, thereby allowing to specifically eliminate all cells with the non-exchanged cassette (Toledo et al., 2006; Wong et al., 2005). Furthermore, random integration events can also be excluded by placing a diphteria toxin A gene in the non-exchanged region of the targeting plasmid (Araki et al., 2006). Interestingly, a reasonable efficiency of a correct RMCE event despite any site-specific selection has also been reported ranging from approximately 7% (Cobellis et al., 2005) to as high as 50% (Masui et al., 2005).

No systematic studies concerning the size limitation for RMCE have been performed. However, a recent publication demonstrated that even very large constructs (>100 kb) may be introduced by RMCE (Wallace et al., 2007). The authors replaced the murine α-globin regulatory domain with the human syntenic region and termed this technology Recombinase Mediated Genomic Replacement (RMGR).

Another issue to be considered is the co-delivery of the recombinase itself along with the targeting vector to be exchanged. Traditionally, expression of the enzyme is achieved by the transient transfection of the recombinase expression vector. Alternatively, the enzyme itself can be directly administered to the cell and for this purpose a fusion of the Cre protein with the HIV-TAT sequence has been created (Peitz et al., 2002). Retroviral pseudo-transduction has also been applied in order to obtain transient recombinase activity in a certain subset of cells dependent on the receptor specificity of this transport vehicle (Galla et al., 2004).

One concern for the use of site-specific recombinases is that of genotoxicity. In the case of the Cre and ΦC31 recombinases, cryptic site specific recognition targets in the mammalian chromosome have been identified. As a result of which, higher enzyme levels over an extended period of time have been shown to lead to chromosomal rearrangements and are thus harmful for the cells (Cre: Loonstra et al., 2001; Schmidt-Supprian and Rajewsky, 2007; ΦC31: Ehrhardt et al., 2006; Liu et al., 2006). Since cassette exchange only requires a transient pulse of recombinase activity, this is of no great relevance in the case of RMCE. Of note, no overt adverse effects of Flp recombinase have been reported so far.

4 RMCE for Biotechnological Applications

4.1 Protein Production – Developing RMCE for Protein Production

So far, this technology is still at the beginning to be exploited for biotechnological purposes. The following section will exemplify the potential of using RMCE as a tool for the establishment of production cell lines. Further, RMCE is discussed with respect to commonly used techniques like high-throughput screenings and gene amplification.

Since recombinant DNA technologies evolved 30 years ago, developments in protein expression systems and cell culture methods are of central interest for the biopharmaceutical industry. Important concerns for the development of a protein production cell line include the ability of a cell type to serve as a good producer cell line in terms of growth conditions and the yield and quality of the resulting recombinant molecule. The majority of clinical biopharmaceuticals are currently being produced in CHO cells and their Dihydrofolate reductase (DHFR)-deficient derivates, as well as in NS0 and HEK293 cells and the human retina-derived PER.C6® (Crucell, N.V., Netherlands). Much effort has been made during the last decades in the development of cell lines in terms of excellent safety profiles, scalability and productivity under serum-free culture conditions. The bottleneck of stable protein expression from such cell lines seems to be within the optimization of antibody expression technologies. There is an urgent need for systematic genetic approaches that allow simple screenings and result in desired expression of any therapeutically relevant protein.

One classical procedure for the establishment of a stable production cell line is to transfect a host cell (e.g., DHFR-deficient CHO cell lines) with plasmids containing the recombinant gene and its necessary regulatory elements along with a selectable marker gene. Commonly used selection genes are the DHFR and the glutamine synthetase (Kingston et al., 2002). After application of increasing amounts of methotrexate or methionine sulfoximine the production levels can increase due to amplification of the inserted genes. Although these selection methods are applicable to increase expression levels of antibody expression cassettes, their limitations have been described (Kim et al., 1998; Kim et al., 2001; Jun et al., 2006). High clone to clone variations as well as instability in expression levels after amplification have been shown to be due to severe genetic rearrangements during gene amplification as well as to the emergence of drug-resistance. Hence, the procedure of gene amplification needs an intense screening to identify a clone with the potential to express adequate amounts of protein after gene amplification, which remains stable over time.

In contrast to the drawbacks of gene-amplification such as genomic instability, followed by ambiguous expression levels, a single copy integration can yield to productivities that are competitive to state of the art industrial productivity levels (Yarranton, 1990). Still, the productivity as well as clonal stability can be influenced by the site of integration. Productivity will vary among clones and the identification of high producers will require intensive screenings for clones with high and stable expression patterns. As a consequence, for every new production clone the process development has to be re-established, which is time consuming and expensive. It is therefore desirable to reduce the production cost and the time needed for cell line development.

To this end, gene targeting strategies can be applied for expression of biopharmaceutical relevant proteins. Most efforts reported until today rely on the first-generation targeting systems with the above-mentioned limitations. The potential of this technology was exploited for the production of a human polyclonal anti-RhD antibody (Wiberg et al., 2006) by integrating 25 individual antibody expression cassettes into a defined FRT tagged integration site in CHO cells (Flp-In™ cell line; Invitrogen). An oligoclonal cell pool derived thereof provided a highly reproducible relative distribution of each antibody. Furthermore, comparable antibody expression levels could be achieved upon targeting at defined integration sites in CHO cells, thereby realising the concept for antibody production using Cre- (Kito et al., 2002) or Flp-mediated integration (Huang et al., 2007). Nevertheless, only marginal numbers of applicable clones could be obtained by both groups after sustained screenings for appropriate integration loci. These fulfilled the desired properties such as expression stability and the ability for gene amplification, which could be applied for transgene targeting and would express satisfactory levels of a desired antibody. So far, a potent integration site for efficient targeting of variable recombinant genes without the use of gene-amplification methods and their mentioned limitations is missing.

Based on the presented targeting strategies, the authors designed an improved strategy to discover a desired locus with specific characteristics suitable for the generation of a stable high-level antibody producing cell line. The first step is to genetically mark the cell line of choice with a screening cassette allowing site specific recombination to introduce the gene of interest by RMCE. Such screenings are mostly done by employing easily detectable marker genes (e.g., fluorescent proteins/selection markers), that allow a straight-forward screening procedure for high-expressing cells. However, this might not lead to the identification of a chromosomal locus that provides specific characteristics needed for the expression of a heterologous gene such as an antibody. Alternatively, a model protein for screening (e.g., a model antibody) could be used, which harbors the same or similar expression characteristics as the protein of interest that will be introduced into the pre-defined genomic locus during the second step. This should result in a cell line with optimal properties for the expression of any antibody of choice. After establishment of such a (master) cell line, that is suitable for production processes, the targeting of a given expression cassette can be accomplished rapidly by the above mentioned RMCE-targeting techniques. The use of a so called "master" cell line for subsequent targeting of variable customized expression cassettes is a rapid approach for the establishment of production cell lines for biopharmaceutical needs. After insertion of any targeting cassette, the terms for process development are matching those of the master cell line. Thus, the RMCE-technique provides an attractive and cost-effective solution for the development of a wide variety of production cell lines.

4.2 Virus Production – Applicability of RMCE Towards Safer and Efficient Production of Viral Vectors

The production of viruses, or viral vectors, can be seen as a special case of protein expression. In contrast to single protein expression, viruses consist of a number of proteins that have to be assembled to a functional unit – primarily to incorporate and protect genetic information and deliver it into target cells. The use of such production system requires a stoichiometric expression of all viral components (Yap et al., 2000).

The following section gives an overview on the use of RMCE as a tool towards safe and efficient generation of viral producing cell lines. It will focus on retroviral and adenoviral viruses, which are important vectors used in gene therapy trials and basic research.

4.2.1 Retroviral Vectors

Retroviral vectors constitute a powerful tool for stable gene transfer into mammalian cells. They can be used to efficiently infect cells of diverse origin of various species. Vectors derived from murine leukaemia virus (MLV) are used as gene delivery systems in clinical gene therapy trials. They have been the vectors of choice in *ex vivo* hematopoietic stem cell gene therapy and proven to be useful to correct several inherited diseases by integrating an expression unit for the therapeutic transgene(s) into the cellular genome (Cavazzana-Calvo et al., 2000a; Grez, 2006; Ott et al., 2006).

A number of packaging cell lines has been developed. These are based on mouse or human cell lines, which stably express the retroviral helper genes gag, pol and env and support high titer virus production (Miller et al., 1986, 1991, 1996; Danos and Mulligan, 1988; Markowitz et al., 1988; Cosset et al., 1995a, b; Davis et al., 1997; Ikeda et al., 2003). Conventionally, the three components are consecutively integrated into the producer cells via cotransduction of a selectable marker, leading to random integration of unpredictable copy numbers and yielding variable expression levels. Packaging cell lines are thus developed to achieve high, stoichiometric and stable expression of these three genes. Hence, the development of a high titer production cell line is time consuming and connected to tremendous screening efforts – a process that can take about several months, when conventional retrovirus packaging cell lines are used (Miller and Miller, 1993). As an alternative, transient transfection systems allow harvesting of infective particles two days after transfection (Landau and Littman, 1992; Finer et al., 1994; Soneoka et al., 1995; Naviaux et al., 1996). However, transient transfection production strategies harbour several pitfalls, as they cause batch to batch variability and are difficult to scale up. Therefore, stable well-characterised producer clones are desirable in clinical gene therapy.

The titre to be achieved from a given helper cell line strongly depends on the strength of expression of the viral vector. Since in classical settings this is done by random integration, the selection of a producer cell clone would ideally be based on a functional principle delivered by the transgene itself, e.g., if it confers a drug resistance. Alternatively, surface expression of markers supports the direct or indirect detection of a transgene product by antibodies and could serve as a method for isolation of highly expressing cells. If the transgene is undetectable or located intracellularly, the amount of work during cloning procedures increases dramatically, even more if there is no protein product, e.g., for vector-encoded therapeutic RNA.

Coexpression of a therapeutic and a selectable marker gene can be mediated by well-established methods, such as differential splicing, fusion of the two open reading frames using a self-digesting protease recognition site, e.g., FMDV-2A or construction of bicistronic expression units in which the expression of the two genes is translationally separated by an IRES element. However, coexpression of a marker gene and the transgene in the target cells may not be desirable in gene therapy approaches, because of possible side effects of the marker gene, e.g., possible immunogenicity potential.

Loew et al. (2004) have developed a strategy that relies on the reversible introduction of a marker gene flanked by recombinase recognition sites (*LoxP*) into the viral genome. After selection and titration of the "best" producer clone the marker gene can easily be removed from the provirus by Cre recombinase – mediated excision. Although this strategy clearly facilitates the isolation of virus producing cells coding for transgenes which are not or hardly detectable, it is connected to some limitations, such as the size of therapeutic RNA that can be packaged.

With the application of RMCE to a selected locus for retroviral vector insertion in packaging cells, flexible retrovirus producer cell line were established, that simplify the isolation of highly productive producer clones (Schucht et al., 2006; Coroadinha et al., 2006). This approach is less laborious and increases safety.

It is based on an advanced site-specific cassette replacement strategy, that combines retroviral tagging and a positive selection trap with the Flp/FRT recombination (Hauser et al., 2000; Verhoeyen et al., 2001).

To meet the above mentioned balanced expression of gag/pol, env and the retroviral vector for high titre production, the following strategy was applied: Firstly, an extensive screening for a single highly expressed and stable chromosomal integration site that particularly supports retroviral vector transcription was identified and tagged. Secondly, cells expressing balanced levels of gag/pol and env were created. The vector genome is integrated into the tagged (*FRT*-flanked) locus by RMCE. In order to select exclusively cells which underwent the site-specific recombination and to avoid contaminations of viruses released by the master cell line, the tagging cassette contains a transcriptional inactive selection marker that is only activated by correct site-specific integration. Thus the resulting producer cell clones are genetically identical and yield reproducibly high levels of viruses. Virus titres up to 2×10^7 IU/$10^6\times24$ h cells were achieved with high reproducibility (Schucht et al., 2006). Also, high-titre producer cells for a therapeutic vector that encodes the 8.9 kb collagen VII cDNA in a marker-free cassette were obtained within three weeks without screening. Since the master cell line is fully characterised with respect to retroviral vector production conditions, the establishment of a producer cell line is reduced to the replacement step of integrating the vector of interest.

RMCE has several benefits for virus production:

- Rapid establishment of high producer clones.
- Cultivation conditions can be established by the master cell lines. The short-term selection (3 weeks) will not alter the growth properties (medium requirements etc.).
- No transduction of harmful neighbouring sequences due to transcriptional read-through, thereby increasing safety.
- No need for any (screening) marker gene within the retroviral vector, thereby both, avoiding adverse effects and increasing the capacity for transduction of therapeutic gene(s).

4.2.2 Adenoviral Vectors

DNA viral vectors derived from adenoviruses have contributed to advances in both basic research and gene therapy. In contrast to retroviral vectors, they are non-integrative, thereby avoiding the risk of insertional mutagenesis. Helper-dependent (HD) adenoviral vectors, also called last-generation vectors are very attractive for gene therapy trials because of the highly reduced in vivo immune response compared to first- and second-generation adenoviral vectors. Adenoviruses can be administered to different organs, such as the liver, muscle or the central nervous system providing high-level and long-term transgene expression in different species including humans. However, the construction of adenoviruses, more precisely the insertion of a gene of interest into a specific site of the viral genome, has been hampered by the large size of the viral genome.

Nakano et al. (2005) have developed a method for introducing a certain gene into the replicating viral DNA without manipulating large sections of the viral genome. Using Cre mediated RMCE a viral cis-element of a recipient virus genome is excised, thereby inhibiting the packaging of this vector. A small donor plasmid, co-transfected with the first one, carries the gene of interest together with the cis-element – both flanked by two non-interacting *LoxP*-sites. Cre, stably expressed by the cell line, exchanges the *LoxP*-flanked region of the recipient DNA for the gene of interest and introduces the missing packaging signal. After three to four cycles of infection, the "purpose" vector was enriched up to 99.8%. The authors could show that RMCE is useful for generating large numbers of adenoviral vectors simultaneously thereby concluding that it is a promising method also for other DNA virus vectors.

4.3 *Optimization of Targeting Vector Design with Respect to the Chromosomal Integration Site*

It is important to consider the specific requirements of a particular integration site relating to the maximum level of recombinant protein production that can be achieved. In other words, molecular composition of the targeting vectors and the chromosomal integration site go hand-in-hand. As a paradigm, this concept will be discussed in terms of achieving optimum retroviral vector production. However, in theory this can be translated to any protein production system in general.

The vector titre produced by a producer cell line is strongly dependent on positive/negative cis-effects that are mediated by the chromosomal sequences presented in the vicinity of the integration site(s) of the retroviral construct. These cis-mediated effects have different outcomes depending on the molecular composition of the integrated vector(s). The establishment of a well-described cellular system in which the chromosomal locus can be exploited in order to maximize the performance of retroviral vectors with the most favourable design (for the defined chromosomal locus) emerges as mandatory. Besides the outstanding advantage of modular cell lines for production of clinically relevant retroviral vectors, they constitute an exceptional platform to systematically evaluate and directly compare different vector compositions in a specific chromosomal locus. This contributes to the development of a rational strategy for vector design (Gama-Norton et al., submitted, Loew et al., submitted). In the work by Gama-Norton et al., several gamma-retroviral vectors differing in the 5′ LTR promoter composition were targeted in two different well-defined chromosomal loci via RMCE. This approach allows to compare the performance in vector production of various retroviral vectors in the same chromosomal locus and to identify vectors that meet the requirements of the respective integration site. The panel of retroviral vectors tested presents the same relative hierarchy concerning to the ability to give rise to infectious retroviral particles. However, besides this consistent behaviour of vector performance in the two chromosomal loci, an integration site specific modulation on the titre is obvious, resulting in a locus specific "fine-tuning" with certain promoters.

This gives evidence that indeed, optimal combinations of specific integration sites and promoter content of a targeting vector have to be defined in order to maximize the level of recombinant proteins produced by those systems.

5 Perspectives

5.1 The Potential of RMCE for Re-Engineering a Targeted Genomic Locus

Nowadays expression vectors use strong viral or cellular promoters and enhancers to confer stable and high expression of a transgene. However, the performance of these regulatory elements strongly depends on the sequential surroundings and the epigenetic status of the integration site. The incorporation of protective cis-regulatory elements has been used to avoid this position effect. Such DNA elements should exhibit specific characteristics like being small in size, conferring stable and high expression levels and universal applicability so that they can be inserted into various commercially used cell lines and combined with many different promoters.

Several approaches have been described during the past few years where different epigenetic control elements were implicated to enhance gene expression (overview in Kwaks and Otte, 2006). One of the first applications of such a cis-acting element is the use of the human β-globin locus control region (LCR) that led to high and stable gene expression in mice when located 5′ to a transgene (Grosveld et al., 1987), but has limitations for standard approaches due to its size of about 16 kb. Another class of these control elements is chromosomal insulators. The best studied insulator is the chicken β-globin 5′ hypersensitive site 4 (cHS4) that blocks the positive influence of an enhancer element when placed between this enhancer and a promoter (Chung et al., 1993). The cHS4 has been shown to protect retroviral vectors from position effects when flanking a given expression cassette (Emery et al., 2000). It induces only moderate increase in gene expression in CHO-production cells due to its characteristic function to shield a gene from chromosomal surroundings, which also includes positive cis-effects (Izumi and Gilbert, 1999), but it conveys a more stable transgene expression. In comparison, scaffold/matrix attachment regions (S/MARs) not only separate a given transcriptional unit from its neighbors by creating independently regulated DNA loops, but also provide platforms for the assembly of factors enabling transcriptional events within a given domain (Bode et al., 1995). Different S/MAR-elements have been shown to increase the number of high-expressing clones either when transgenes were flanked or cotransfected with these elements. Hence, S/MARs are useful for the development of stably producing cell lines (Zahn-Zabal et al., 2001; Girod et al., 2007). Regulatory elements derived from (ubiquitously expressed) housekeeping genes called ubiquitous chromatin opening elements (UCOEs) have also been shown to protect transgenes from silencing and can increase expression levels up to 20-fold (Williams et al., 2005;

Zhang et al., 2007). A relatively new class of elements has been identified on the basis of their ability to block heterochromatin-associated repression. These STAR- (stabilizing and antirepressor-) elements also enhanced transgene expression in different cell lines in combination with different promoters (Kwaks et al., 2003).

A stringent comparison of the effects of these kinds of regulatory elements on diverse transgene expression cassettes harboring different specific promoters requires the integration of test systems at predefined genomic sites. The targeting techniques using site-specific recombination now enable these kinds of approaches. For the first time the performance of S/MAR elements and insulators at predefined genomic loci has been studied using RMCE-techniques, permitting a detailed comparison between the expression profiles of a reporter gene flanked either by insulator- or S/MAR-elements (Goetze et al., 2005). The investigation of five different loci within NIH3T3 cells indicated that both, the cHS4 insulator and the S/MAR-elements exert shielding function of the chromosomal surroundings leading to enhanced expression levels as compared to neutral flanking sequences at most but not all genomic integration sites.

Although the potential of some of these elements in different applications has been described, nearly nothing is known about the molecular impact of such epigenetic elements. There is a need to specify the functionality of a given element in a particular context of a transgene and its chromosomal surroundings. Targeted integration could comprise this more detailed analysis of an extensive amount of known as well as newly to identify regulatory elements, whose performances can be directly compared within a given genomic context. Therefore an experimental setup that allows the subsequent exchange of distinct of such elements using site-specific recombination techniques could be established. The emergence of a set of different potent recombinase recognition sites harbouring specific mutants allows subsequent cassette exchange within a given locus (Schlake and Bode, 1994; Missirlis et al., 2006). This ability to combine different sets of regulatory elements that might e.g., flank a transgene or can be strategically positioned within an expression cassette permits the specific engineering of a given genomic locus (Fig. 8). The overall result could be either that different genomic integration sites within a cell can provide a locus independent expression of a given protein or virus, or that each transgene to be expressed might need a specific regulatory element in a precisely defined position where it provides its stable expression regardless of the chromosomal surroundings.

Concerning the exploitation of a defined locus by RMCE for production of molecules of biotechnological interest it should be noted that the tagging cassette should be as similar as possible to the cassette to be targeted, in order to guarantee the maximal positive cis-effects that this locus could exert on expression. A defined locus is not necessarily universal, which means that the production of different biological products is not guaranteed even if the vector composition (i.e., regulatory sequences) is constant among the different vectors. An illustrative example is the exploitation of the loci screened for high retroviral expression as the ones described by Schucht et al., 2006 and Coroadinha et al., 2006, in which the production of recombinant protein and production of lentiviral vectors is highly compromised

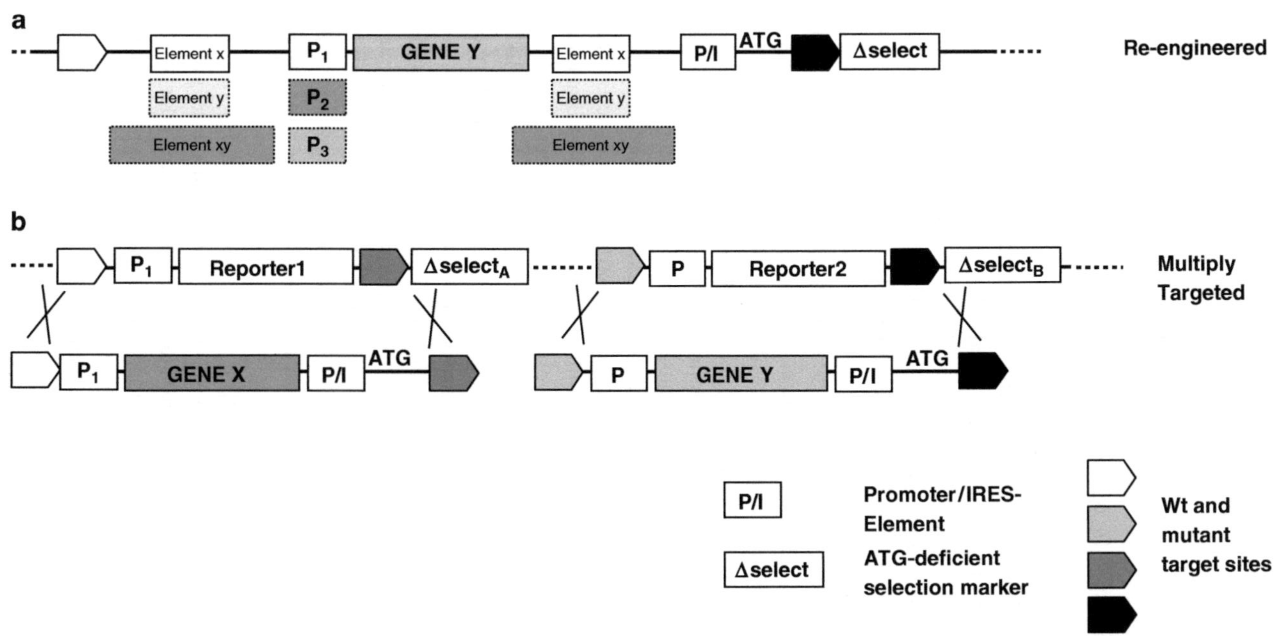

Fig. 8 Re-engineering of loci via RMCE. (**a**) RMCE allows re-engineering of a genomic locus towards optimal expression. Different genetic elements can be introduced / excised or exchanged to modulate the expression in a given locus. Additionally, transcription regulating elements like promoters can be tested without changing genomic influences. (**b**) The combination of different non-interacting recombinase sites or the combination of recombinase systems (e.g. Cre/*LoxP* and FLP/*FRT*) allows a flexible introduction of two and more expression cassettes within independent chromosomal integration sites.

(unpublished data from our group). Hence, and taking the example of the modular cell lines for retroviral production, the establishment of a lentiviral producer cell line should be achieved after the tagging of the genome with a cognate tagging vector and the optimal cell line for antibody production should be established upon integration of a tagging cassette expressing a recombinant protein. If these requirements are fullfilled, the exceptional advantages of the RMCE technique could be vastly applicable to the generation of cell lines suitable for production of a wide variety of molecules with biological significance.

5.2 The Next Generation – Circumventing the Need of Primary Genomic Modifications Employing Zinc Finger Recombinases

The principle of RMCE relies on the presence of recombination target sequences in the genome, which have been previously integrated by homologous recombination or upon random integration. Hence, current applications are limited to genetically modified cells in which these sites have been introduced into the genome. An attractive approach has been recently provided by the creation of chimeric recombinases that opens the possibility to target any desired endogenous sequence. For this purpose the resolvase/invertase family of serine recombinases (e.g. γ δ resolvases and Tn3 resolvase) is of particular interest since these recombinases are modular in both form and function: The N-terminal domain of these proteins contains all the residues known to be involved in the catalysis of recombination and the C-terminal domain is the primary determinant of sequence specific DNA binding (Yang and Steitz, 1995; Akopian et al., 2003, Akopian and Marshall Stark, 2005). Fusing the zinc-finger (ZF) DNA-binding domain of a transcription factor with a Tn3 resolvase mutant, Akopian et al. (2003) created a chimeric recombinase whose site specificity is determined by the DNA binding domain. This gave proof of principle that recombinases can be redirected to other sequences of choice. Other ZF domains have been fused to the Tn3 recombinase to create novel recombinases that function both in bacteria and in human cells (Gordley et al., 2007). The exploitation of the emerging rational design of ZF binding domains (see Section 2) for the generation of chimeric recombinases opens avenues for targeting recombinase activity to any sequences within natural genomes. Thereby, it is envisioned that engineered ZF recombinases will extend the current restrictions of site specific recombination methodologies such as RMCE.

Improving the specificity of recognition and recombination of these chimeric proteins, the optimized molecules may eventually mediate gene therapies, facilitate the genetic manipulation of model organisms and cells, and mature into powerful new tools for molecular biology and medicine. Despite the considerable efforts that are being taken in order to optimise tools that lead to site specific recombination in any endogenous context, their development is still under way and their applicability remain confined to experimental models for the time being.

6 Conclusions

Production of proteins and virus particles such as vaccines and (gene) therapeutics represents one major field of interest in biotechnology. For this purpose, a transgene usually has to be integrated into the genome of (mammalian) cells. When stably incorporating the respective transgene into chromosomal DNA, application of RMCE on pre-defined chromosomal loci overcomes the prominent disadvantages of random integration, i.e., unpredictable expression levels and potential lack of long term stability. These sites are selected either on the basis of known sequences with defined properties and thus become tagged by homologous recombination or Zinc Finger nucleases or recombinases; at the same time, when little is known about such sites, random tagging and screening for required properties is the more appropriate approach. It would be beneficial to accumulate information about genomic loci in producer cell lines like CHO that support high and sustained expression (although such knowledge will most likely be kept secret or become protected). Homologous recombination or ZF recombinases might be applied to target these sites. However, these methods are not applicable in every cell type or do not work very efficiently, respectively. Hence, it would be prudent to first integrate tags via homologous recombination or ZF recombinases to be able to target the supportive site by the more efficient RMCE.

The operations described in this review are based on single copy DNA integrates. With regard to safety, product consistency and operational transparency this is preferable over multi-gene integrates. However, given the single copy locus allows gene amplification, RMCE can be used to integrate relevant cassettes and amplification can follow, converting a defined single copy locus to a multi-copy amplicon.

The most frequently applied method to generate recombinant producer cell lines is to randomly integrate an expression cassette and to perform screenings for supportive chromosomal loci. This method may be accompanied by a tagging step for subsequent reuse by simply implementing the relevant elements. Thus, further screenings when producing new proteins or virus particles can be circumvented. Efficient selection strategies for RMCE allow for isolation of a high percentage of correctly targeted clones (Schucht et al., 2006 demonstrated that 100% can be reached). RMCE shortens the time required for producer cell lines from several months to approximately four weeks while providing predictable expression characteristics under well established conditions. On the whole, although new promising technologies for targeted integration are under way, RMCE is the most advantageous method to date and, surprisingly, is not yet being exploited to its full extent.

References

Akopian A, He J, Boocock MR, Stark WM (2003) Chimeric recombinases with designed DNA sequence recognition. *Proc Natl Acad Sci U S A* 100(15):8688–8691

Akopian, A., Marshall Stark, W. (2005) Site-specific DNA recombinases as instruments for genomic surgery. *Adv Genet* 55:1–23

Albert H, Dale EC, Lee E, Ow DW (1995) Site-specific integration of DNA into wild-type and mutant lox sites placed in the plant genome. *Plant J* 7(4):649–659
Alwin S, Gere MB, Guhl E, Effertz K, Barbas CF 3rd, Segal DJ, Weitzman MD, Cathomen T (2005) Custom zinc-finger nucleases for use in human cells. *Mol Ther* 12(4):610–617
Andrews BJ, Proteau GA, Beatty LG, Sadowski PD (1985) The FLP recombinase of the 2 micron circle DNA of yeast: interaction with its target sequences. *Cell* 40(4):795–803
Araki K, Araki M, Yamamura K (2006) Negative selection with the Diphtheria toxin A fragment gene improves frequency of Cre-mediated cassette exchange in ES cells. J Biochem (Tokyo) 140(6):793–798
Artelt P, Grannemann R, Stocking C, Friel J, Bartsch J, Hauser H (1991) The prokaryotic neomycin-resistance-encoding gene acts as a transcriptional silencer in eukaryotic cells. Gene 99(2):249–254
Baer A, Bode J (2001) Coping with kinetic and thermodynamic barriers: RMCE, an efficient strategy for the targeted integration of transgenes. Curr Opin Biotechnol 12(5):473–480
Bell AC, Felsenfeld G (1999) Stopped at the border: boundaries and insulators. Curr Opin Genet Dev 9(2):191–198
Belteki G, Gertsenstein M, Ow DW, Nagy A (2003) Site-specific cassette exchange and germline transmission with mouse ES cells expressing phiC31 integrase. Nat Biotechnol 21(3):321–324
Bethke B, Sauer B (1997) Segmental genomic replacement by Cre-mediated recombination: genotoxic stress activation of the p53 promoter in single-copy transformants. Nucleic Acids Res 25(14):2828–2834
Beumer K, Bhattacharyya G, Bibikova M, Trautman JK, Carroll D (2006) Efficient gene targeting in Drosophila with zinc-finger nucleases. Genetics 172(4):2391–2403
Bibikova M, Golic M, Golic KG, Carroll D (2002) Targeted chromosomal cleavage and mutagenesis in Drosophila using zinc-finger nucleases. Genetics 161(3):1169–1175
Bode J, Schlake T, Rios-Ramirez M, Mielke C, Stengert M, Kay V, Klehr-Wirth D (1995) Scaffold/matrix-attached regions: structural properties creating transcriptionally active loci. Int Rev Cytol 162A:389–454
Branda CS, Dymecki SM (2004) Talking about a revolution: the impact of site-specific recombinases on genetic analyses in mice. Dev Cell 6(1):7–28
Buchholz F, Angrand PO, Stewart AF (1998) Improved properties of FLP recombinase evolved by cycling mutagenesis. Nat Biotechnol 16(7):657–662
Cavazzana-Calvo M, Hacein-Bey S, de Saint Basile G, Gross F, Yvon E, Nusbaum P, Selz F, Hue C, Certain S, Casanova JL et al (2000a) Gene therapy of human severe combined immunodeficiency (SCID)-X1 disease. Science 288(5466):669–672
Cavazzana-Calvo M, Hacein-Bey S, de Saint-Basile G, Le Deist F, Fischer A (2000b) Gene therapy of severe combined immunodeficiencies. Transfus Clin Biol 7(3):259–260
Chung JH, Whiteley M, Felsenfeld G (1993) A 5′ element of the chicken beta-globin domain serves as an insulator in human erythroid cells and protects against position effect in Drosophila. Cell 74(3):505–514
Cobellis G, Nicolaus G, Iovino M, Romito A, Marra E, Barbarisi M, Sardiello M, Di Giorgio FP, Iovino N, Zollo M et al (2005) Tagging genes with cassette-exchange sites. Nucleic Acids Res 33(4):e44
Coroadinha AS, Schucht R, Gama-Norton L, Wirth D, Hauser H, Carrondo MJ (2006) The use of recombinase mediated cassette exchange in retroviral vector producer cell lines: predictability and efficiency by transgene exchange. J Biotechnol 124(2):457–468
Cosset FL, Morling FJ, Takeuchi Y, Weiss RA, Collins MK, Russell SJ (1995a) Retroviral retargeting by envelopes expressing an N-terminal binding domain. J Virol **69**(10):6314–6322.
Cosset FL, Takeuchi Y, Battini JL, Weiss RA, Collins MK (1995b) High-titer packaging cells producing recombinant retroviruses resistant to human serum. J Virol **69**(12):7430–7436.
Danielian PS, Muccino D, Rowitch DH, Michael SK, McMahon AP (1998) Modification of gene activity in mouse embryos in utero by a tamoxifen-inducible form of Cre recombinase. Curr Biol 8(24):1323–1326

Danos O and Mulligan RC (1988) Safe and efficient generation of recombinant retroviruses with amphotropic and ecotropic host ranges. Proc Natl Acad Sci USA **85**(17):6460–6464.

Davis JL, Witt RM, Gross PR, Hokanson CA, Jungles S, Cohen LK, Danos O, Spratt SK (1997) Retroviral particles produced from a stable human-derived packaging cell line transduce target cells with very high efficiencies. Hum Gene Ther **8**(12):1459–1467.

Doyon Y, McCammon JM, Miller JC, Faraji F, Ngo C, Katibah GE, Amora R, Hocking TD, Zhang L, Rebar EJ et al (2008) Heritable targeted gene disruption in zebrafish using designed zinc-finger nucleases. Nat Biotechnol 26(6):702–708

Dymecki SM (1996) A modular set of Flp, FRT and lacZ fusion vectors for manipulating genes by site-specific recombination. Gene 171(2):197–201

Ehrhardt A, Engler JA, Xu H, Cherry AM, Kay MA (2006) Molecular analysis of chromosomal rearrangements in mammalian cells after phiC31-mediated integration. Hum Gene Ther 17(11):1077–1094

Emery DW, Yannaki E, Tubb J, Stamatoyannopoulos G (2000) A chromatin insulator protects retrovirus vectors from chromosomal position effects. Proc Natl Acad Sci U S A 97(16):9150–9155

Feil R, Brocard J, Mascrez B, LeMeur M, Metzger D, Chambon P (1996) Ligand-activated site-specific recombination in mice. Proc Natl Acad Sci U S A 93(20):10887–10890

Feil R, Wagner J, Metzger D, Chambon P (1997) Regulation of Cre recombinase activity by mutated estrogen receptor ligand-binding domains. Biochem Biophys Res Commun 237(3):752–757

Festenstein R, Tolaini M, Corbella P, Mamalaki C, Parrington J, Fox M, Miliou A, Jones M, Kioussis D (1996) Locus control region function and heterochromatin-induced position effect variegation. Science 271(5252):1123–1125

Finer MH, Dull TJ, Qin L, Farson D, Roberts MR (1994) kat: a high-efficiency retroviral transduction system for primary human T lymphocytes. Blood **83**(1):43–50.

Galla M, Will E, Kraunus J, Chen L, Baum C (2004) Retroviral pseudotransduction for targeted cell manipulation. Mol Cell 16(2):309–315

Girod PA, Nguyen DQ, Calabrese D, Puttini S, Grandjean M, Martinet D, Regamey A, Saugy D, Beckmann JS, Bucher P et al (2007) Genome-wide prediction of matrix attachment regions that increase gene expression in mammalian cells. Nat Methods 4(9):747–753

Glaser S, Anastassiadis K, Stewart AF (2005) Current issues in mouse genome engineering. Nat Genet 37(11):1187–1193

Goetze S, Baer A, Winkelmann S, Nehlsen K, Seibler J, Maass K, Bode J (2005) Performance of genomic bordering elements at predefined genomic loci. Mol Cell Biol 25(6):2260–2272

Gordley RM, Smith JD, Graslund T, Barbas CF 3rd (2007) Evolution of programmable zinc finger-recombinases with activity in human cells. J Mol Biol 367(3):802–813

Grez M (2006) Breakthrough in gene therapy? Interview by Friederike Klein. MMW Fortschr Med 148(17):16

Grosveld F, van Assendelft GB, Greaves DR, Kollias G (1987) Position-independent, high-level expression of the human beta-globin gene in transgenic mice. Cell 51(6):975–985

Groth AC, Olivares EC, Thyagarajan B, Calos MP (2000) A phage integrase directs efficient site-specific integration in human cells. Proc Natl Acad Sci U S A 97(11):5995–6000

Hauser H, Spitzer D, Verhoeyen E, Unsinger J, Wirth D (2000) New approaches towards ex vivo and in vivo gene therapy. Cells Tissues Organs 167(2–3):75–80

Hoess RH, Ziese M, Sternberg N (1982) P1 site-specific recombination: nucleotide sequence of the recombining sites. Proc Natl Acad Sci U S A 79(11):3398–3402

Hoess RH, Wierzbicki A, Abremski K (1986) The role of the loxP spacer region in P1 site-specific recombination. Nucleic Acids Res 14(5):2287–2300

Huang Y, Li Y, Wang YG, Gu X, Wang Y, Shen BF (2007) An efficient and targeted gene integration system for high-level antibody expression. J Immunol Methods 322(1–2):28–39

Hunter NL, Awatramani RB, Farley FW, Dymecki SM (2005) Ligand-activated Flpe for temporally regulated gene modifications. Genesis 41(3):99–109

Ikeda Y, Takeuchi Y, Martin F, Cosset FL, Mitrophanous K, Collins M (2003) Continuous high-titer HIV-1 vector production. Nat Biotechnol **21**(5):569–572.

Izumi M, Gilbert DM (1999) Homogeneous tetracycline-regulatable gene expression in mammalian fibroblasts. J Cell Biochem 76(2):280–289

Jullien N, Sampieri F, Enjalbert A, Herman JP (2003) Regulation of Cre recombinase by ligand-induced complementation of inactive fragments. Nucleic Acids Res 31(21):e131

Jullien N, Goddard I, Selmi-Ruby S, Fina JL, Cremer H, Herman JP (2007) Conditional transgenesis using Dimerizable Cre (DiCre). PLoS ONE 2(12):e1355

Jun SC, Kim MS, Hong HJ, Lee GM (2006) Limitations to the development of humanized antibody producing Chinese hamster ovary cells using glutamine synthetase-mediated gene amplification. Biotechnol Prog 22(3):770–780

Keravala A, Calos MP (2008) Site-specific chromosomal integration mediated by phiC31 integrase. Methods Mol Biol 435:165–173

Kim YG, Cha J, Chandrasegaran S (1996) Hybrid restriction enzymes: zinc finger fusions to Fok I cleavage domain. Proc Natl Acad Sci U S A 93(3):1156–1160

Kim SJ, Kim NS, Ryu CJ, Hong HJ, Lee GM (1998) Characterization of chimeric antibody producing CHO cells in the course of dihydrofolate reductase-mediated gene amplification and their stability in the absence of selective pressure. Biotechnol Bioeng 58(1):73–84

Kim NS, Byun TH, Lee GM (2001) Key determinants in the occurrence of clonal variation in humanized antibody expression of cho cells during dihydrofolate reductase mediated gene amplification. Biotechnol Prog 17(1):69–75

Kingston RE, Kaufman RJ, Bebbington CR, Rolfe MR (2002) Amplification using CHO cell expression vectors. Current Protocols Mol Biol/edited by Frederick M Ausubel et al **Chapter 16**: Unit 16 23.

Kito M, Itami S, Fukano Y, Yamana K, Shibui T (2002) Construction of engineered CHO strains for high-level production of recombinant proteins. Appl Microbiol Biotechnol 60(4):442–448

Klehr-Wirth D, Kuhnert F, Unsinger J, Hauser H (1997) Generation of mammalian cells with conditional expression of cre recombinase. Trends Genetics TTO T40067 (http://tto.trendscom).

Koch KS, Aoki T, Wang Y, Atkinson AE, Gleiberman AS, Glebov OK, Leffert HL (2000) Site-specific integration of targeted DNA into animal cell genomes. Gene 249(1–2):135–144

Kwaks TH, Otte AP (2006) Employing epigenetics to augment the expression of therapeutic proteins in mammalian cells. Trends Biotechnol 24(3):137–142

Kwaks TH, Barnett P, Hemrika W, Siersma T, Sewalt RG, Satijn DP, Brons JF, van Blokland R, Kwakman P, Kruckeberg AL et al (2003) Identification of anti-repressor elements that confer high and stable protein production in mammalian cells. Nat Biotechnol 21(5):553–558

Lakso M, Sauer B, Mosinger B Jr, Lee EJ, Manning RW, Yu SH, Mulder KL, Westphal H (1992) Targeted oncogene activation by site-specific recombination in transgenic mice. Proc Natl Acad Sci U S A 89(14):6232–6236

Landau NR, Littman DR (1992) Packaging system for rapid production of murine leukemia virus vectors with variable tropism. J Virol 66(8):5110–5113

Langer SJ, Ghafoori AP, Byrd M, Leinwand L (2002) A genetic screen identifies novel non-compatible loxP sites. Nucleic Acids Res 30(14):3067–3077

Lee G, Saito I (1998) Role of nucleotide sequences of loxP spacer region in Cre-mediated recombination. Gene 216(1):55–65

Liu Q, Segal DJ, Ghiara JB, Barbas CF 3rd (1997) Design of polydactyl zinc-finger proteins for unique addressing within complex genomes. Proc Natl Acad Sci U S A 94(11):5525–5530

Liu K, Hipkens S, Yang T, Abraham R, Zhang W, Chopra N, Knollmann B, Magnuson MA, Roden DM (2006) Recombinase-mediated cassette exchange to rapidly and efficiently generate mice with human cardiac sodium channels. Genesis 44(11):556–564

Lloyd A, Plaisier CL, Carroll D, Drews GN (2005) Targeted mutagenesis using zinc-finger nucleases in Arabidopsis. Proc Natl Acad Sci U S A 102(6):2232–2237

Loew R, Selevsek N, Fehse B, von Laer D, Baum C, Fauser A, Kuehlcke K (2004) Simplified generation of high-titer retrovirus producer cells for clinically relevant retroviral vectors by reversible inclusion of a lox-P-flanked marker gene. Mol Ther 9(5):738–746

Logie C, Stewart AF (1995) Ligand-regulated site-specific recombination. Proc Natl Acad Sci U S A 92(13):5940–5944

Lombardo A, Genovese P, Beausejour CM, Colleoni S, Lee YL, Kim KA, Ando D, Urnov FD, Galli C, Gregory PD et al (2007) Gene editing in human stem cells using zinc finger nucleases and integrase-defective lentiviral vector delivery. Nat Biotechnol 25(11):1298–1306

Long Q, Shelton KD, Lindner J, Jones JR, Magnuson MA (2004) Efficient DNA cassette exchange in mouse embryonic stem cells by staggered positive-negative selection. Genesis 39(4):256–262

Loonstra A, Vooijs M, Beverloo HB, Allak BA, van Drunen E, Kanaar R, Berns A, Jonkers J (2001) Growth inhibition and DNA damage induced by Cre recombinase in mammalian cells. Proc Natl Acad Sci U S A 98(16):9209–9214

Mani M, Smith J, Kandavelou K, Berg JM, Chandrasegaran S (2005) Binding of two zinc finger nuclease monomers to two specific sites is required for effective double-strand DNA cleavage. Biochem Biophys Res Commun 334(4):1191–1197

Markowitz D, Goff S, Bank A (1988) A safe packaging line for gene transfer: separting viral genes on two different plasmids. J Virol 62:1120–1124

Masui S, Shimosato D, Toyooka Y, Yagi R, Takahashi K, Niwa H (2005) An efficient system to establish multiple embryonic stem cell lines carrying an inducible expression unit. Nucleic Acids Res 33(4):e43

McLeod M, Craft S, Broach JR (1986) Identification of the crossover site during FLP-mediated recombination in the Saccharomyces cerevisiae plasmid 2 microns circle. Mol Cell Biol 6(10):3357–3367

Meng X, Noyes MB, Zhu LJ, Lawson ND, Wolfe SA (2008) Targeted gene inactivation in zebrafish using engineered zinc-finger nucleases. Nat Biotechnol 26(6):695–701

Miller DG, Miller AD (1993) Inhibitors of retrovirus infection are secreted by several hamster cell lines and are also present in hamster sera. J Virol 67(9):5346–5352

Miller AD, Garcia JV, von Suhr N, Lynch CM, Wilson C, Eiden MV (1991) Construction and properties of retrovirus packaging cells based on gibbon ape leukemia virus. J Virol 65(5):2220–2224

Missirlis PI, Smailus DE, Holt RA (2006) A high-throughput screen identifying sequence and promiscuity characteristics of the loxP spacer region in Cre-mediated recombination. BMC Genomics 7:73

Mitchell RS, Beitzel BF, Schroder AR, Shinn P, Chen H, Berry CC, Ecker JR, Bushman FD (2004) Retroviral DNA integration: ASLV, HIV, and MLV show distinct target site preferences. PLoS Biol 2(8):E234

Moehle EA, Rock JM, Lee YL, Jouvenot Y, DeKelver RC, Gregory PD, Urnov FD, Holmes MC (2007) Targeted gene addition into a specified location in the human genome using designed zinc finger nucleases. Proc Natl Acad Sci U S A 104(9):3055–3060

Nakano M, Odaka K, Takahashi Y, Ishimura M, Saito I, Kanegae Y (2005) Production of viral vectors using recombinase-mediated cassette exchange. Nucleic Acids Res **33**(8): e76.

Naviaux RK, Costanzi E, Haas M, Verma IM (1996) The pCL vector system: rapid production of helper-free, high-titer, recombinant retroviruses. J Virol **70**(8):5701–5705.

O'Gorman S, Fox DT, Wahl GM (1991) Recombinase-mediated gene activation and site-specific integration in mammalian cells. Science 251(4999):1351–1355

Ott MG, Schmidt M, Schwarzwaelder K, Stein S, Siler U, Koehl U, Glimm H, Kuhlcke K, Schilz A, Kunkel H et al (2006) Correction of X-linked chronic granulomatous disease by gene therapy, augmented by insertional activation of MDS1-EVI1, PRDM16 or SETBP1. Nat Med 12(4):401–409

Pavletich NP and Pabo CO (1991) Zinc finger-DNA recognition: crystal structure of a Zif268-DNA complex at 2.1 A. *Science* **252**(5007):809–817.

Peitz M, Pfannkuche K, Rajewsky K, Edenhofer F (2002) Ability of the hydrophobic FGF and basic TAT peptides to promote cellular uptake of recombinant Cre recombinase: a tool for efficient genetic engineering of mammalian genomes. Proc Natl Acad Sci U S A 99(7):4489–4494

Porteus MH, Baltimore D (2003) Chimeric nucleases stimulate gene targeting in human cells. Science 300(5620):763

Porteus MH, Carroll D (2005) Gene targeting using zinc finger nucleases. Nat Biotechnol 23(8):967–973

Pruett-Miller SM, Connelly JP, Maeder ML, Joung JK, Porteus MH (2008) Comparison of zinc finger nucleases for use in gene targeting in mammalian cells. Mol Ther 16(4):707–717
Raymond CS, Soriano P (2007) High-efficiency FLP and PhiC31 site-specific recombination in mammalian cells. PLoS ONE 2:e162
Saam JR, Gordon JI (1999) Inducible gene knockouts in the small intestinal and colonic epithelium. J Biol Chem 274(53):38071–38082
Sauer B, Henderson N (1988) Site-specific DNA recombination in mammalian cells by the Cre recombinase of bacteriophage P1. Proc Natl Acad Sci U S A 85(14):5166–5170
Sauer B, McDermott J (2004) DNA recombination with a heterospecific Cre homolog identified from comparison of the pac-c1 regions of P1-related phages. Nucleic Acids Res 32(20):6086–6095
Schlake T, Bode J (1994) Use of mutated FLP recognition target (FRT) sites for the exchange of expression cassettes at defined chromosomal loci. Biochemistry 33(43):12746–12751
Schmidt-Supprian M, Rajewsky K (2007) Vagaries of conditional gene targeting. Nat Immunol 8(7):665–668
Schnütgen F, Stewart AF, von Melchner H, Anastassiadis K (2006) Engineering embryonic stem cells with recombinase systems. Methods Enzymol 420:100–136
Schonig K, Schwenk F, Rajewsky K, Bujard H (2002) Stringent doxycycline dependent control of CRE recombinase in vivo. Nucleic Acids Res 30(23):e134
Schubeler D, Maass K, Bode J (1998) Retargeting of retroviral integration sites for the predictable expression of transgenes and the analysis of cis-acting sequences. Biochemistry 37(34):11907–11914
Schucht R, Coroadinha AS, Zanta-Boussif MA, Verhoeyen E, Carrondo MJ, Hauser H, Wirth D (2006) A new generation of retroviral producer cells: predictable and stable virus production by Flp-mediated site-specific integration of retroviral vectors. Mol Ther 14(2):285–292
Schwenk F, Kuhn R, Angrand PO, Rajewsky K, Stewart AF (1998) Temporally and spatially regulated somatic mutagenesis in mice. Nucleic Acids Res 26(6):1427–1432
Senecoff JF, Cox MM (1986) Directionality in FLP protein-promoted site-specific recombination is mediated by DNA–DNA pairing. J Biol Chem 261(16):7380–7386
Senecoff JF, Rossmeissl PJ, Cox MM (1988) DNA recognition by the FLP recombinase of the yeast 2 mu plasmid. A mutational analysis of the FLP binding site. J Mol Biol 201(2):405–421
Sharma N, Moldt B, Dalsgaard T, Jensen TG, Mikkelsen JG (2008) Regulated gene insertion by steroid-induced PhiC31 integrase. Nucleic Acids Res 36(11):e67
Shimshek DR, Kim J, Hubner MR, Spergel DJ, Buchholz F, Casanova E, Stewart AF, Seeburg PH, Sprengel R (2002) Codon-improved Cre recombinase (iCre) expression in the mouse. Genesis 32(1):19–26
Soneoka Y, Cannon PM, Ramsdale EE, Griffiths JC, Romano G, Kingsman SM, Kingsman AJ (1995) A transient three-plasmid expression system for the production of high titer retroviral vectors. Nucleic Acids Res 23(4):628–633
Sternberg N, Sauer B, Hoess R, Abremski K (1986) Bacteriophage P1 cre gene and its regulatory region. Evidence for multiple promoters and for regulation by DNA methylation. J Mol Biol 187(2):197–212
Thorpe HM, Smith MC (1998) In vitro site-specific integration of bacteriophage DNA catalyzed by a recombinase of the resolvase/invertase family. Proc Natl Acad Sci U S A 95(10):5505–5510
Toledo F, Liu CW, Lee CJ, Wahl GM (2006) RMCE-ASAP: a gene targeting method for ES and somatic cells to accelerate phenotype analyses. Nucleic Acids Res 34(13):e92
Urnov FD, Miller JC, Lee YL, Beausejour CM, Rock JM, Augustus S, Jamieson AC, Porteus MH, Gregory PD, Holmes MC (2005) Highly efficient endogenous human gene correction using designed zinc-finger nucleases. Nature 435(7042):646–651
van der Weyden L, Adams DJ, Bradley A (2002) Tools for targeted manipulation of the mouse genome. Physiol Genomics 11(3):133–164
Verhoeyen E, Hauser H, Wirth D (2001) Evaluation of retroviral vector design in defined chromosomal loci by Flp-mediated cassette replacement. Hum Gene Ther 12(8):933–944
Vetter D, Andrews BJ, Roberts-Beatty L, Sadowski PD (1983) Site-specific recombination of yeast 2-micron DNA in vitro. Proc Natl Acad Sci U S A 80(23):7284–7288

Wallace HA, Marques-Kranc F, Richardson M, Luna-Crespo F, Sharpe JA, Hughes J, Wood WG, Higgs DR, Smith AJ (2007) Manipulating the mouse genome to engineer precise functional syntenic replacements with human sequence. Cell 128(1):197–209

Waterhouse P, Griffiths AD, Johnson KS, Winter G (1993) Combinatorial infection and in vivo recombination: a strategy for making large phage antibody repertoires. Nucleic Acids Res 21(9):2265–2266

Wiberg FC, Rasmussen SK, Frandsen TP, Rasmussen LK, Tengbjerg K, Coljee VW, Sharon J, Yang CY, Bregenholt S, Nielsen LS et al (2006) Production of target-specific recombinant human polyclonal antibodies in mammalian cells. Biotechnol Bioeng 94(2):396–405

Williams S, Mustoe T, Mulcahy T, Griffiths M, Simpson D, Antoniou M, Irvine A, Mountain A, Crombie R (2005) CpG-island fragments from the HNRPA2B1/CBX3 genomic locus reduce silencing and enhance transgene expression from the hCMV promoter/enhancer in mammalian cells. BMC Biotechnol 5:17

Wirth D, Hauser H (2004) Flp-mediated integration of expression cassettes into FRT-tagged chromosomal loci in mammalian cells. Methods Mol Biol 267:467–476

Wirth D, Gama-Norton L, Riemer P, Sandhu U, Schucht R, Hauser H (2007) Road to precision: recombinase-based targeting technologies for genome engineering. Curr Opin Biotechnol 18(5):411–419

Wong ET, Kolman JL, Li YC, Mesner LD, Hillen W, Berens C, Wahl GM (2005) Reproducible doxycycline-inducible transgene expression at specific loci generated by Cre-recombinase mediated cassette exchange. Nucleic Acids Res 33(17):e147

Wright DA, Townsend JA, Winfrey RJ Jr, Irwin PA, Rajagopal J, Lonosky PM, Hall BD, Jondle MD, Voytas DF (2005) High-frequency homologous recombination in plants mediated by zinc-finger nucleases. Plant J 44(4):693–705

Wu X, Li Y, Crise B, Burgess SM (2003) Transcription start regions in the human genome are favored targets for MLV integration. Science 300(5626):1749–1751

Yang W, Steitz TA (1995) Crystal structure of the site-specific recombinase gamma delta resolvase complexed with a 34 bp cleavage site. Cell 82(2):193–207

Yap MW, Kingsman SM, Kingsman AJ (2000) Effects of stoichiometry of retroviral components on virus production. J Gen Virol 81(Pt 9):2195–2202

Yarranton GT (1990) Mammalian recombinant proteins: vectors and expression systems. Curr Opin Biotechnol 1(2):133–140

Zahn-Zabal M, Kobr M, Girod PA, Imhof M, Chatellard P, de Jesus M, Wurm F, Mermod N (2001) Development of stable cell lines for production or regulated expression using matrix attachment regions. J Biotechnol 87(1):29–42

Zhang F, Thornhill SI, Howe SJ, Ulaganathan M, Schambach A, Sinclair J, Kinnon C, Gaspar HB, Antoniou M, Thrasher AJ (2007) Lentiviral vectors containing an enhancer-less ubiquitously acting chromatin opening element (UCOE) provide highly reproducible and stable transgene expression in hematopoietic cells. Blood 110(5):1448–1457

Importance of Genetic Environment for Recombinant Gene Expression

Alan J. Dickson

Abstract Production of biological (protein)-based therapeutics offers fundamental challenges. The ability to generate unique life-saving therapies has been engaged by many commercial concerns, mainly using eukaryotic (mammalian) cell culture platforms. Genes encoding the valuable biopharmaceuticals are introduced into the host cell where they integrate into the cellular genome. What goes on within the nucleus is no longer a black box but it has become clear over the last decade that we have not yet begun to fully appreciate the complexity of this sub-cellular compartment and there remains significant scope to further optimise this stage of commercial bioprocessing. This review highlights the current vision of the eukaryotic nucleus in relation to its role as the controller of expression of genes that are introduced for production of a desired product. The layers of interwoven complexity – eu- and hetero-chromatin, epigenetic marking of genes and genomes, nucleosomes, expression factories and chromosome territories – will be described. As this knowledge base has accumulated it has led to the use of approaches that seek to maximise the expression of introduced genes, using a rationalised understanding of nuclear architecture and higher level genome regulation.

The past decade has seen step-changes in our perception of the eukaryotic nucleus in terms of structural environments and, consequently, the potential for previously unconsidered modes of regulation of gene expression. Driven by technological developments, that have permitted increased understanding of nuclear structure, we perceive that there are layers of complexity in eukaryotic transcription control that may have the potential to either thwart or enhance cell gene engineering. The existence of regulation at the level of nuclear structure and genomic environment has relevance to approaches that utilise eukaryotic cells as hosts for expression of exogenous genes (as in the use of mammalian cells as "factories" for production of biopharmaceuticals). This level of regulation has consequences for the extent

A.J. Dickson
Faculty of Life Sciences, The Michael Smith Building, University of Manchester, Oxford Road, Manchester M13 9PT, UK
e-mail: alan.dickson@manchester.ac.uk

M. Al-Rubeai (ed.), *Cell Line Development*, Cell Engineering 6,
DOI 10.1007/978-90-481-2245-5_4, © Springer Science+Business Media B.V. 2009

and stability of expression of genes introduced into cells (either genes that encode for the desired biopharmaceutical or genes encoding for proteins predicted to enhance the "factory" activity of the cell) and for endogenous cellular genes (for which expression may be modified in response to incorporation of foreign genes into specific areas of the genomic environment, with potential consequences for cellular function).

This review will highlight current understanding of the structural relationships between chromosomes, genes and the physical entity that comprises the eukaryotic nucleus. Current perceptions have been developed from information obtained from a number of experimental eukaryotic systems. After describing the generic model for relationships between nuclear structure and gene regulation, I will discuss the implications for production of biopharmaceuticals in relation to commercially-relevant eukaryotic cells (predominantly Chinese Hamster ovary, CHO, and NS0 myeloma). Within this context there are very clear linkages between our increasing understanding of genomic environment and current developments related to incorporation of specific DNA sequences within expression vectors for use with mammalian cell lines (Chapter 1, this volume). This will be discussed, as appropriate, along with the forward vision of how information on genomic environment may be used for further rationalised optimisation of future expression platforms.

1 Genes and the Eukaryotic Nucleus: From the Simple Onwards

Our early perception of the molecular organisation of the nucleus was one in which DNA was integrated with proteins into a complex (chromatin) with the chromatin presenting well-defined structures, the chromosomes, for a short part of the cell division cycle (mitosis). Through the rest of the cell division cycle (interphase) chromatin failed to exist in well-defined chromosome structures and formed a diffuse structure envisaged to spread throughout the entire nucleus. Throughout interphase nuclei can be seen to exhibit a degree of structural organisation, with a differential diffuseness being observed for nucleoli, which provide a functional sub-compartment in the nucleus associated with production of ribosomal RNA transcripts. This provided an early model for physical partitioning of functions linked to specific gene expression within nuclear sub-structures and compartments. Further early views on nuclear compartmentation exist in relation to concepts of euchromatin and heterochromatin, areas of chromatin associated with active or inactive transcription, respectively, with a general acceptance that the periphery of the nucleus was associated with gene-poor chromatin and that gene-rich chromatin was localised to the internal areas of the nucleus (Boyle et al., 2001). The concept of eu- and hetero-chromatin as physical entities still presents a working statement on gene activity within the eukaryotic cell nucleus and links to the developing hypotheses on more detailed nuclear structures and epigenetic control mechanisms (Fig. 1; van Driel et al., 2003; Razin et al., 2007). Further evidence for structural organisation within the nucleus has been generated from defined protein complexes

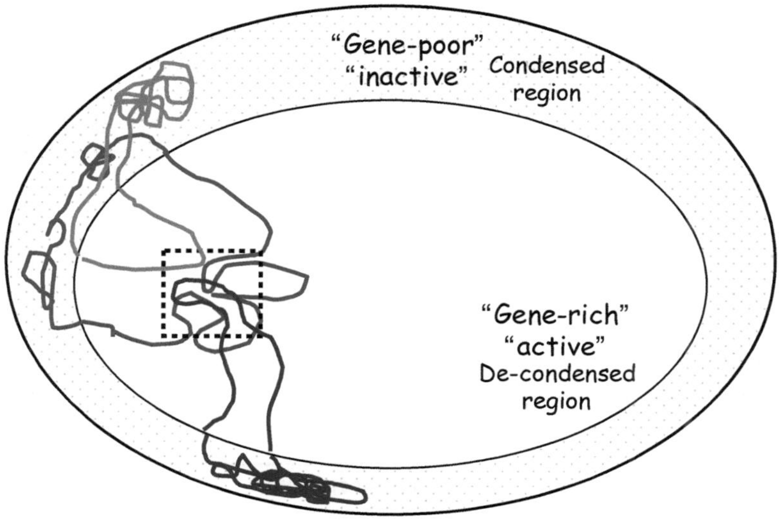

Fig. 1 Spatial sub-compartmentation of the eukaryotic nucleus. This diagram illustrates structural and functional compartmentation with the distinctive packaging of genomic material in the peripheral region (defined as "gene-poor" or "inactive") associated with heterochromatin and the central region (defined as "gene-rich" or "active") associated with euchromatin. The three *coloured lines* represent DNA strands (loops) from different chromosomes clustering into a transcription factory complex in which RNA generation will take place. As described in the text, such looping permits regulated selection and co-ordination of expression of specific genes (from chromosomes that reside in spatially distinct areas within the nucleus) in response to environmental, developmental and other regulatory inputs

(e.g. Cajal Bodies, PML [promyelocytic leukemia] bodies; Cioce and Lamond, 2005) that exist in limited numbers in eukaryotic nuclei and may act as functional/spatial markers of nuclear compartments.

1.1 Chromatin Organisation: Nucleosomes and Epigenetic Events

Nucleosomes form the basic sub-unit of DNA/protein chromatin and subsequently nucleosomes are arranged together into higher levels of structure that underlie euchromatin and heterochromatin organisation (Fig. 2; Ferreira et al., 1997; Dillon, 2008; Misteli, 2007). In addition to the core histones within nucleosomes, other proteins involved in transcription and its regulation (potentially including those involved in physical partitioning) will interact with nucleosomes (Razin et al., 2007). The interactions (for example to enhance or inhibit transcription) that occur are subject to control and potential docking interactions are directed by covalent modification to histones or DNA (epigenetic marking) of specific nucleosomes/genes (Fig. 2; Hake and Allis, 2006). A significant number of epigenetic marking

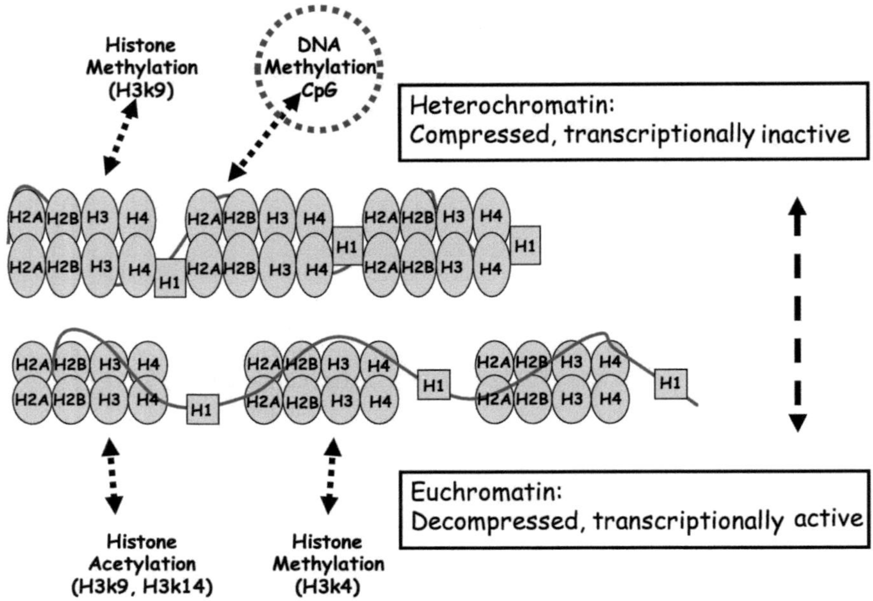

Fig. 2 Epigenetic regulation of chromatin accessibility. The transition of specific chromatin regions between euchromatin (more open/accessible) and heterochromatin (more closed/inaccessible) states is subject to regulation by specific covalent modifications to histones and DNA. Illustrative examples are shown for covalent modifications (via changes to acetylation to the side chains of specific lysine residues in histones or methylation of arginine or lysine residues of histone or bases in DNA) that occur in response to the regulatable activities of histone acetylases/deacetylases or protein and DNA methylases/demethylases. Multiple and specific epigenetic modifications may be found within any defined chromatin region and the extent and patterning of modification will define the accessibility and, hence, potential transcriptional activity of that region

events have been defined but it is appropriate to note that the precise consequence of specific covalent changes may not be predictable and will depend on the overall surrounding context (Dillon, 2008; Table 1; Fig. 3).

Epigenetic mechanisms (at the level of covalent modifications to histones and DNA) have been correlated to control of transcription from specific genes (or groups of adjacent genes) in relation to developmental and environmental signals (Mellor et al., 2008). Heterochomatic silencing is of particular relevance to cell engineering, with this concept suggesting that exogenous genes may be inserted to areas of relatively active transcription but are subsequently switched off as the epigenetic status of the surrounding genes define covalent modifications of histones or DNA associated with the incorporated gene (Ahmed and Brickner, 2007). This addresses a potential fluidity to activity of genes incorporated into the genome of a eukaryotic cell and links to the potential for use of insulator (e.g. UCOE; Antoniou et al., 2003)/localisation (e.g. S/MAR; Zahn-Zabal et al., 2001) elements with optimised expression vectors.

Table 1 Examples of major epigenetic modifications and consequences for gene expression

Chromatin Target	Modification	Effect on Gene Expression
DNA	Methylation of bases (C in CG or CNG)	Repression
Histone H3	Acetylation of –	
	Lys 9 (k9)	Activation
	Lys 14 (k14)	Activation
Histone H3	Methylation of –	
	Lys 4 (k4)	Activation
	Lys 9 (k9)	Repression
	Lys 27 (k27)	Repression

This table gives an outline summary of major types of epigenetic covalent modifications that occur within the eukaryotic genome and indicates the expected consequence for genes marked in the manner presented for the examples. More detailed reviews of this area (including modification by phosphorylation, ubiquitinylation and sumoylation can be found in articles by Hake and Allis (2006), Berger (2007), Razin et al. (2007) and Mellor et al. (2008).

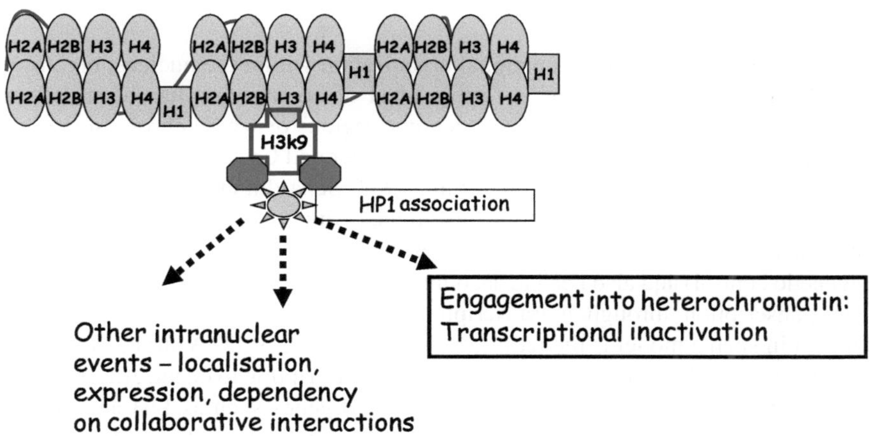

Fig. 3 Epigenetic marking provides "docking points" for proteins that define the accessibility and expression status of genes. The methylation of lysine 9 (k9) of histone H3 provides a platform for formation of multi-protein complexes that engage molecular and structural events within the nucleus. In the case of methylated, H3k9, heterochromatin-like protein 1 (HP1) promotes the association/incorporation of the methylated histone area within heterochromatin areas of the nucleus. A growing series of proteins that influence the stabilisation of chromatin areas or expression are able to engage with the epigenetic marks on chromatin (Berger, 2007)

1.2 Beyond the Nucleosome: Into Expression Factories

As mentioned previously, the nucleolus presents a defined area of the nucleus associated with expression of genes encoding ribosomal RNA transcripts. As a compartment, it is also associated with specific proteins including RNA polymerase I that generate multi-protein complexes that co-operate towards a specific function and this has been extended to implicate a series of functional nuclear organisational sub-compartments such as replication "factories" (Cook, 2002) and transcription factories – with the latter associated with transcription of specific genes (or sets of genes; Jackson et al., 1993; Cook, 1999; Faro-Trindade and Cook, 2006; Carter et al., 2008; Mitchell and Fraser, 2008). Hypotheses and models based around this scenario suggest that such factories give greater efficiency to transcription and allow selective control of expression of genes of linked functional significance (Xu and Cook, 2008), potentially from different chromosomes. Such a model requires significant flexibility within the overall chromatin structure, with loops of DNA being recruited to (or potentially racked through) a multi-enzyme complex sub-organelle that acts as an organisational centre for expression of transcripts of protein-encoding genes (van Driel et al., 2003; Fraser, 2006).

1.3 Beyond the Nucleosome: Into Chromosome Territories

In addition to the concept of functional organisation within the nucleus, organised around transcriptional factories, a further level of organisation has been defined related to the manner in which genes, genome regions or whole chromosomes are spatially fixed within defined nuclear areas (Cremer and Cremer, 2001). Such areas are referred to as territories (Cremer et al., 1982; Schardin et al., 1985) and these have been mapped by hybridisation techniques (Fig. 1). Territories will engage with other structural and regulatory features (such as nucleosome organisation and epigenetic control) and also need to account for the spatial requirements of regulated gene transcription through a transcription factory model (Brink et al., 2006). Inherent in such a model is the requirement that looping of DNA strands for active transcription would be a likely pre-requisite for transcription to occur. Alternatively, it has been shown (for developmentally-regulated genes such as IgH and Hox1B loci) that gene positioning within the territory map may alter with conditions such as cell differentiation status (Kosak et al., 2002; Chambeyron and Bickmore, 2004; Stadler et al., 2004; Pennisi, 2006; Ragoczy et al., 2006; Osborne et al., 2007; Meaburn and Misteli, 2008) and this may relate to mechanisms of gene silencing and activation that extend the simple epigenetic mechanism of histone or DNA modification. However, within this complex network of events, it is equally feasible that epigenetic marking and nuclear spatial positioning of genes are related events (with either being primary or secondary events). A further consideration of territories, shown by 3C and/or 4C technology (chromosome conformation capture/-on-chip), is that spatial localisation of genes is not random and that partner sequences can be brought into close proximity from distinctly different chromosomal areas (either from the same chromosome or from other chromosomes; Simonis et al., 2006; Branco and Pombo, 2006; Osborne et al., 2007; Rafalska-Metcalf and Janicki, 2007; Wallace and Felsenfeld, 2007). Within the complex structure of a

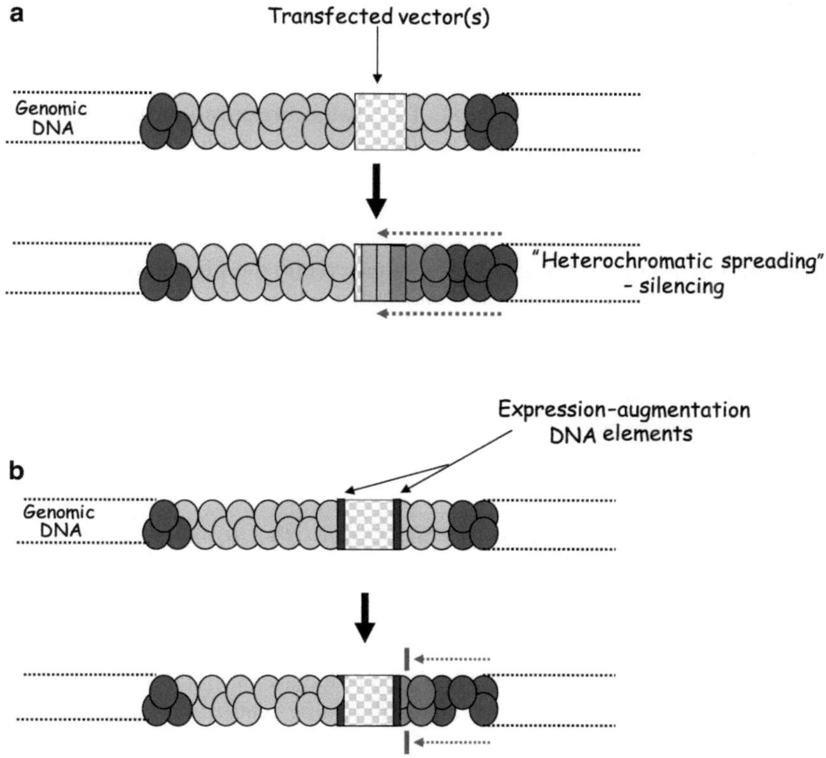

Fig. 4 Heterochromatic spreading: silencing of integrated transgenes and the actions of expression-augmentation DNA elements. Diagrammatically, the insertion of a transgene (e.g. a recombinant gene for a commercially valuable biopharmaceutical) into the host cell genome is seen to disrupt the existing chromatin organisation (**a**). Although initially expressed due to increased accessibility of the chromatin area to the transcriptional machinery (e.g. favouring looping into transcription factories via liberation from less accessible status), the original epigenetic status (directed via the "red" inactive chromatin region marking) will be able to reassert itself across the genomic area. This, normally referred to as heterochromatic spreading, will promote (via epigenetic marking of the inserted gene sequences) a reversion to a less active status, in which the newly-marked genomic region will return to the default accessibility status of the original chromatin. The inclusion of expression-augmentation DNA elements within the transgene vector (**b**) operates to prevent heterochromatic spreading, through a variety of potential mechanisms. Mechanisms could result in enhanced (and more permanent) localisation within "active" genome regions, positioning of gene sequences in environments removed from the chromatin modifying enzymes that silence chromatin activity or direct insulation (buffering) from the actions of the modifying enzymes

nucleus, underpinned by a potential nuclear scaffold, that emerges we can envisage that there may be ownership of specific chromosomes or loops of chromosomes by distinct nuclear areas (and/or transcription factories) (Spilianakis et al., 2005).

2 What Does this Mean for Recombinant Gene Expression?

From the discussions above it will be clear that the eukaryotic cell nucleus is not a simple repository for genetic material. The spatial organisation has been clearly defined across a range of levels and features but understanding what it means for gene expression – and, in particular, for genes that are introduced into the host cell genome – remains to be determined in detail. Questions remain whether the spatial organisation of the genome environment is a determinant in the regulation of gene expression or if expression (regulated by other events) modifies a dynamic organisation to give the spatial patterns observed. Whatever may be the answer(s) to such questions, there is no doubt that this presents a challenge in the optimisation of production of recombinant cell lines of commercial value. To illustrate this, I will focus on the implications for the CHO and NS0 expression platforms although the comments will have equal relevance to other eukaryotic expression platforms.

2.1 CHO and NS0 Myeloma Cell Lines: Production of Biopharmaceuticals

A very significant percentage of biopharmaceuticals have been or are being produced in either CHO (using dihydrofolate reductase or glutamine synthetase as selection markers) or NS0 myeloma (using glutamine synthetase) cell lines (Barnes et al., 2000; Wurm, 2004; Butler, 2005; Birch and Racher, 2006). Stable cell lines (in which the recombinant gene vector has been integrated into the host cell genome with a selection marker to generate successful transformants) are used for commercial-scale production. Transfectants are routinely screened in large numbers, by various approaches, to identify cells that produce sufficient quantities of the desired product (Birch and Racher, 2006). The screening process will have identified cells that have incorporated the recombinant gene into an area of, at least, reasonable transcriptional activity. Subsequently cell lines undergo expansion, scale-up and product harvest.

2.2 Consequences of Genomic Environment Towards Successful or Enhanced Production of Biopharmaceuticals

The nature of the organisational structures and regulatory properties of the nuclear genomic environment imposes a number of implications for commercial success for biopharmaceutical production (Kwaks and Otte, 2006). Chief amongst these is a lack of knowledge about how the expression vectors in general use integrate into the host cell genome. The concept of "hot spots" within the genome environment are likely to equate to integration within genes, genome regions or chromosome territories that have positional advantages for engaging with transcription factories

(Misteli, 2007). Screening of (very) large numbers of transformants presents an approach to find cells in which integration has been to a (relative) "hot spot" but this presents a time-consuming process that is unique to each cell line development programme. As an alternative, the genomic nature of expression "hot spots" has been sought in a number of approaches and host cell lines have been generated that have targeted integration to genomic areas with good transcriptional activity (Fukushige and Sauer, 1992; Koduri et al., 2001; Huang et al., 2007).

Inherent within the concept of "hot spots", epigenetic marking (to histones or DNA) that occurs to recombinant genes incorporated into the host cell genome may limit expression (due to effects on accessibility to or interaction with the appropriate nuclear environment). Marking can be modified by altering the activities of enzymes that add or remove specific epigenetic marks. This can be achieved by use of small molecules (Backliwal et al., 2008) or by genetic engineering of the cellular phenotype (Kwaks et al., 2005). Sodium butyrate (an inhibitor of histone deacetylase, HDAC) provides a good example of the first approach and this, and other HDAC inhibitors, has been reported to enhance the yield of recombinant protein from mammalian cells (Chun et al., 2003; Backliwal et al., 2008; Jiang and Sharfstein, 2008). Key questions remain in relation to the mapping of specific epigenetic marks on integrated genes and their surrounding regions and how specific histone modifications/DNA methylation patterns relate to expression. Treatment with HDAC inhibitors has effects beyond that on desired gene expression, including "non-specific" effects on cell proliferation at effective concentrations but precise dissection of the relationships between specific epigenetic modifications and genomic environment offers powerful opportunities towards rational design of expression platforms (Yee et al., 2007).

Stability of expression remains a problem in cell line development programmes and knowledge of the overall genomic environment places that into context (Berger, 2007; Fraser and Bickmore, 2007). The initial site of insertion of recombinant genes (whether random or targeted to an area of high transcriptional activity) may facilitate high initial expression but epigenetic mechanisms (via heterochromatic spread) may switch off (silence) transcription to revert to the "expected" status of DNA within that specific chromosome territory (Fig. 4; Barnes et al., 2004). A variety of DNA elements have been incorporated into vectors with a view to minimising epigenetic silencing/targeting integration to nuclear areas of "effective" transcription and enhancing expression (Needham et al., 1992; Zahn-Zabal et al., 2001; Kim et al., 2004; Girod et al., 2005; Williams et al., 2005; Girod et al., 2007; Otte et al., 2007). Given the complex nature of the genomic environment a variety of mechanisms could be suggested for success of such elements, ranging from actions as barrier elements/insulators (to prevent a creeping epigenetic spread) to favouring a repositioning within a favourable chromosome territory (removing the recombinant genes from an environment in which enzymes involved in epigenetic modification operate). To date there have been a very limited number of experiments that have examined the positioning or status of recombinant genes in either CHO or NS0 myeloma cells lines to make clear statements of mechanisms that prevent silencing. Several studies have defined the location of recombinant gene sequences to specific chromosomal regions (by fluorescent in situ hybridisation techniques) for engineered CHO cell lines, in basal conditions and in response to methotrexate amplification/selection (Dixkens et al., 1998; Kim and Lee, 1999; Yoshikawa et al., 2000; Derouazi et al., 2006). Relationships between recombinant gene positioning and expression (and/or stability of expression) remain correlative rather than causative and any linkages between sites of insertion and expression (e.g. telomere-type vs non-telomere-type; Yoshikawa et al., 2000) have still to be directly proven. Whatever mechanism is stated to contribute to the effect of any specific vector-regulatory element, the situation may well be a mixture of several events.

Instability of recombinant gene expression may also arise from loss of gene copy (speculated to be partly determined by the chromosomal localisation of amplified genes; Kaufman, 1990) rather that via silencing of maintained genes and, in some cases, this loss of expression has been linked to rearrangements of the recombinant vectors (Barnes et al., 2003). Using DHFR vectors along with CHO cells, transfectants selected on first round screening normally undergo subsequent amplification with methotrexate (Kaufman, 1990). The amplified cell lines express increased gene copy numbers due to a break-fusion-bridge (or related) mechanism (Lo et al., 2002) and this is likely to involve gene loops brought together within chromosome territories. The positioning of the recombinant sequences within specific chromosome regions may favour chromosomal translocations and rearrangements to a greater extent and this is a further aspect that genomic environment plays in determination of cell line quality (Meaburn et al., 2006).

3 Can We Use Knowledge from this Emerging Area to Develop Better Expression Processes?

In short the answer to the question posed here must be "yes" but there are several caveats. A knowledge of the DNA sequence within the genome of a cell line under study presents a starting point to understand the potential for regulatory phenomena but this is finessed by the exquisite structural features that package and control expression of genes in the nucleus. The eukaryotic nucleus and its spatial organisation is being worked on very actively in a number of laboratories but for the major host cell used for production of biopharmaceuticals (CHO) we have limited, but growing, public genomic data (Wlaschin and Hu, 2007). We are at a stage where our increasing knowledge is indicating just how much we do not know or understand. There are several areas that have the potential to offer fundamental advances. For example, can we design a CHO cell (or other host cell)-specific expression vector that targets a chromosome territory that offers the highest possible transcription, retains stability and does not interfere with any other gene function? Are specific gene types expressed best when targeted via specific chromosome territories or transcription factories? Can we subvert the need to make stable cell lines by more effective use of transient expression vectors (but how do transient expression vectors or indeed non-integrated expression vectors engage with the organisation with the eukaryotic nucleus)? CHO cells are likely to remain a platform of choice for the foreseeable future.

Over the years CHO cells have undergone many changes to their genome and present an unusual karyotype (Dixkens et al., 1998; Derouazi et al., 2006). What does this mean for the structural (and functional) organisation of the chromosome territories? Would there be potential to select for clones in host cell populations that were better in relation to nuclear handling of recombinant genes – due to increased numbers, or activity, of transcription factories or differentially-defined chromosome territories? This is a fascinating, but complex, topic that provides a physical explanation for many of the observations surrounding the quality of cell lines used for production of recombinant gene products. As a starting point for successful cell line generation – rapid and robust reassurance of high level, stable transcription – the knowledge from this research theme underpins the likelihood of successful cell line development programmes.

References

Ahmed S, Brickner JH (2007) Regulation and epigenetic control of transcription at the nuclear periphery. Trends Genet 23:396–402

Antoniou M, Harland L, Mustoe T et al (2003) Transgenes encompassing dual-promoter CpG islands from the human TBP and HNRPA2B1 loci are resistant to heterochromatin-mediated silencing. Genomics 82:269–279

Backliwal G, Hildinger M, Kuettel I et al (2008) Valproic acid: A viable alternative to sodium butyrate for enhancing protein expression in mammalian cell cultures. Biotechnol Bioeng 101:183–189

Barnes LM, Bentley CM, Dickson AJ (2000) Advances in animal cell recombinant protein production: GS-NS0 expression system. Cytotechnology 32:109–123

Barnes LM, Bentley CM, Dickson AJ (2003) Stability of protein production from recombinant mammalian cells. Biotechnol Bioeng 81:631–639

Barnes LM, Bentley CM, Dickson AJ (2004) Molecular definition of predictive indicators of stable protein expression in recombinant NS0 myeloma cells. Biotechnol Bioeng 85:115–121

Berger SL (2007) The complex language of chromatin regulation during transcription. Nature 447:407–412

Birch JR, Racher AJ (2006) Antibody production. Adv Drug Deliv Rev 58:671–685

Boyle S, Gilchrist S, Bridger JM et al (2001) The spatial organization of human chromosomes within the nuclei of normal and emerin-mutant cells. Hum Mol Genet 10:211–219

Branco MR, Pombo A (2006) Intermingling of chromosome territories in interphase suggests a role in translocations and transcription-dependent associations. PLoS Biol 4:780–788

Brink MC, van der Velden Y, de Leeuw W et al (2006) Truncated HP1 lacking a functional chromodomain induces heterochromatization upon in vivo targeting. Histochem Cell Biol 125:55–61

Butler M (2005) Animal cell cultures: Recent achievements and perspectives in the production of biopharmaceuticals. Appl Microbiol Biotechnol 68:283–291

Carter DRF, Eskiw C, Cook PR (2008) Transcription factories. Biochem Soc Trans 36:585–589

Chambeyron S, Bickmore WA (2004) Chromatic decondensation and nuclear reorganisation of the HoxB locus upon induction of transcription. Genes Dev 18:1119–1130

Chun B-H, Park S-Y, Chung N et al (2003) Enhanced production of recombinant B-domain deleted factor VIII from Chinese hamster ovary cells in propionic and butyric acids. Biotechnol Lett 25:315–319

Cioce M, Lamond AI (2005) Cajal bodies: A long history of discovery. Annu Rev Cell Dev Biol 21:105–131

Cook PR (1999) Organisation of replication and transcription. Science 284:1790–1795
Cook PR (2002) Predicting three-dimensional genome structure from transcriptional activity. Nat Genet 32:347–352
Cremer T, Cremer C (2001) Chromosome territories, nuclear architecture and gene regulation in mammalian cells. Nat Rev Genet 2:292–301
Cremer T, Cremer C, Schneider T et al (1982) Analysis of chromosome positions in the interphase nucleus of Chinese hamster cells by laser UV-microirradiation experiments. Hum Genet 62:201–209
Derouazi M, Martinet D, Besuchet Schmutz N et al (2006) Genetic characterisation of CHO production hosts DG44 and derivative recombinant cell lines. Biochem Biophys Res Commun 340:1069–1077
Dillon N (2008) The impact of gene location in the nucleus on transcriptional regulation. Dev Cell 15:182–186
Dixkens C, Posseckert G, Keller T et al (1998) Structural analysis of the amplified IFN-β and DHFR genes in a Chinese hamster ovary cell line using multicolour FISH analysis. Chromosome Res 6:329–332
Faro-Trindade I, Cook PR (2006) Transcription factories: Structures conserved during differentiation and evolution. Biochem Soc Trans 341:1133–1137
Ferreira J, Paolella G, Ramos C et al (1997) Spatial organisation of large-scale chromatin domains in the nucleus: A magnified view of single chromosome territories. J Cell Biol 139:1597–1610
Fraser P (2006) Transcriptional control thrown in a loop. Curr Opin Genet Dev 16:490–495
Fraser P, Bickmore W (2007) Nuclear organisation of the genome and the potential for gene regulation. Nature 447:413–417
Fukushige S, Sauer B (1992) Genomic targeting with a positive selection LOX integration vector allows highly reproducible gene expression in mammalian cells. Proc Natl Acad Sci USA 89:7905–7909
Girod PA, Zhan-Zabal M, Mermod N (2005) Use of the chicken lysozyme 5′ matrix attachment region to generate high producer CHO cell lines. Biotechnol Bioeng 91:1–11
Girod PA, Nguyen DQ, Calabrese D et al (2007) Genome-wide prediction of matrix attachment regions that increase gene expression in mammalian cells. Nat Methods 4:747–753
Hake SB, Allis CD (2006) Histone H3 variants and their potential role in indexing mammalian genomes: The "H3 barcode hypothesis". Proc Natl Acad Sci USA 103:6428–6435
Huang Y, Li Y, Wang YG et al (2007) An efficient and targeted gene integration system for high level antibody expression. J Immunol Methods 322:28–39
Jackson DA, Hassan AB, Errington RJ et al (1993) Visualisation of focal sites of transcription within human nuclei. EMBO J 12:1059–1065
Jiang Z, Sharfstein ST (2008) Sodium butyrate stimulates monoclonal antibody over-expression in CHO cells by improving gene accessibility. Biotechnol Bioeng 100:189–194
Kaufman RJ (1990) Selection and co-amplification of heterologous genes in mammalian cells. Methods Enzymol 185:537–566
Kim SJ, Lee GM (1999) Cytogenetic analysis of chimeric antibody-producing CHO cells in the course of dihydrofolate reductase-mediated gene amplification and their stability in the absence of selective pressure. Biotechnol Bioeng 64:741–749
Kim JM, Kim JS, Park DH et al (2004) Improved recombinant gene expression in CHO cells using matrix attachment regions. J Biotechnol 107:95–105
Koduri RK, Miller JT, Thammana P (2001) An efficient homologous recombination vector pTV(I) contains a hot spot for increased recombinant protein expression in Chinese hamster ovary cells. Gene 280:87–95
Kosak ST, Skok JA, Medina KL et al (2002) Subnuclear compartentalisation of immunoglobulin loci during lymphocyte development. Science 296:158–162
Kwaks THJ, Otte AP (2006) Employing epigenetics to augment the expression of therapeutic proteins in mammalian cells. Trends Biotechnol 24:137–142

Kwaks T, Sewalt RGAB, van Blokland R et al (2005) Targeting of a histone acetyltransferase domain to a promoter enhances protein expression levels in mammalian cells. J Biotechnol 115:35–46

Lo AW, Sabatier L, Fouladi B et al (2002) DNA amplification by breakage/fusion/bridge cycles initiated by spontaneous telomere loss in a human cancer cell line. Neoplasia 4:531–538

Meaburn KJ, Misteli T (2008) Locus-specific and activity-independent gene repositioning during early tumourogenesis. J Cell Biol 180:39–50

Meaburn KJ, Misteli, T, Soutoglou E (2006) Spatial genome organization in the formation of chromosomal translocations. Seminar on Cancer Biology published online October 26, 2006. 10.1016/j.sem-cancer.2006.10.008

Mellor J, Dudek P, Clynes D (2008) A glimpse into the epigenetic landscape of gene regulation. Curr Opin Genet Dev 18:116–123

Misteli T (2007) Beyond the sequence: Cellular organization of genome function. Cell 128:787–800

Mitchell JA, Fraser P (2008) Transcription factories are nuclear subcompartments that remain in the absence of transcription. Genes Dev 22:20–25

Needham M, Gooding C, Hudson K et al (1992) LCR/MEL – A versatile system for high-level expression of heterologous proteins in erythroid cells. Nucleic Acids Res 20:997–1003

Osborne CS, Chkalova L, Mitchell JA et al (2007) Myc dynamically and preferentially relocates to a transcription factory occupied by Igh. PLoS Biol 5:e192

Otte AP, Kwaks THJ, van Blokland RJM et al (2007) Various expression-augmenting DNA elements benefit from STAR-select, a novel stringency selection system for protein expression. Biotechnol Prog 23:801–807

Pennisi E (2006) Molecular biology – Genes commute to factories before they start work. Science 312:1304

Rafalska-Metcalf JU, Janicki SM (2007) Show and tell: Gene expression in living cells. J Cell Sci 120:2301–2307

Ragoczy T, Bender MA, Telling A et al (2006) The locus control region is required for association of the murine β-globin locus with engaged transcription factories during erythroid maturation. Genes Dev 20:1447–1457

Razin SV, Iarvaia OV, Sjakste N et al (2007) Chromatin domains and regulation of transcription. J Mol Biol 369:597–607

Schardin M, Cremer T, Hager HD et al (1985) Specific staining of human chromosomes in Chinese hamster-X-man hybrid cell lines demonstrates interphase chromosome territories. Hum Genet 71:281–287

Simonis M, Klous P, Splinter E et al (2006) Nuclear organisation of active and inactive chromatin domains uncovered by chromosome conformation capture-on-chip (4C). Nat Genet 38:1348–1354

Spilianakis CG, Lalloti MD, Town T et al (2005) Interchromosomal associations between alternatively expressed loci. Nature 435:637–645

Stadler S, Schnapp V, Mayer R et al. (2004) The architecture of chicken chromosome territories changes during differentiation. BMC Cell Biol 5: Art No. 44

Van Driel R, Fransz PF, Verschure PJ (2003) The eukaryotic genome: A system regulated at different hierarchical levels. J Cell Sci 116:4067–4075

Wallace JA, Felsenfeld G (2007) We gather together: Insulators and genome organization. Curr Opin Genet Dev 17:400–407

Williams S, Mustoe T, Mulcahy T et al. (2005) CpG island fragments from the HNRPA2B1/CBX3 genomic locus reduce silencing and enhance transgene expression from the hCMV promoter/enhancer in mammalian cells. BMC Biotechnol 5: Art. No. 17

Wlaschin KF, Hu W-S (2007) A scaffold for the Chinese hamster genome. Biotechnol Bioeng 98:429–439

Wurm FM (2004) Production of recombinant protein therapeutics in cultivated mammalian cells. Nat Biotechnol 22:1393–1398

Xu M, Cook PR (2008) Similar active genes cluster in specialised transcription factories. J Cell Biol 181:615–623

Yee JC, de Leon Gatti M, Philp RJ et al (2007) Genomic and proteomic exploration of CHO and hybridoma cells under sodium butyrate treatment. Biotechnol Bioeng 99:1186–1204

Yoshikawa T, Nakanishi F, Ogura Y et al (2000) Amplified gene location in chromosomal DNA affected recombinant protein production and stability of amplified genes. Biotechnol Prog 16:710–715

Zahn-Zabal M, Kobr M, Girod PA et al (2001) Development of stable cell lines for production or regulated expression using matrix attachment regions. J Biotechnol 87:29–42

Expression Vector Engineering for Recombinant Protein Production

Helen Kim, John Laudemann, Jennitte Stevens, and Michelle Wu

Abstract The first step in the process of generating a high expressing mammalian cell line for production of therapeutic recombinant protein is developing a robust expression vector that is compatible with the host cell line of choice. Transcription of the recombinant gene will largely depend on the strength of the expression vector and the site of genomic integration of the vector. The promoter, enhancer, poly-A sequence, and the stringency of the selection marker will all contribute to the overall strength of the expression vector, although the degree of contribution from each component may vary. In addition, use of various genomic DNA elements such as Expression Augmenting Sequence Element (EASE), scaffold- or matrix-attachment regions (S/MAR elements), Insulators, or Universal Chromatin Opening Element (UCOE), can further improve recombinant protein expression by protecting transgenes from inactivation by flanking chromatin sequences. The use of chromatin remodeling genomic DNA elements can also reduce variations in the expression levels of the same construct in individual clones, thus facilitating often labor-intensive clone screening process. Similar to promoters and enhancers, chromatin remodeling elements contain binding sites for transcription factors, however, these transcription factors tend to have relaxed sequence requirements or low-affinity binding characteristics individually, which can result in stable DNA interactions through cooperative binding with other transcriptional factor partners (Zuckerkandl and Villet, 1988).

In addition to the choice of promoter, enhancer, or genomic DNA elements used in the vector, the order, orientation, and distances between the individual components can have a profound effect on the final activity of the vector, and subsequently expression of the recombinant protein. This is not surprising since cooperative effects of relatively low-affinity DNA binding factors will depend on proper position or distribution of each binding site, compatible with precisely positioned high order structure, required for the transcription initiation complex.

H. Kim (✉), J. Laudemann J. Stevens, and M. Wu
Department of Protein Science, Amgen Inc, Thousand Oaks, California, USA
e-mail: ykim@amgen.com

M. Al-Rubeai (ed.), *Cell Line Development*, Cell Engineering 6,
DOI 10.1007/978-90-481-2245-5_5, © Springer Science+Business Media B.V. 2009

1 Position Effects of the Genomic Sequence Elements

The S/MAR elements are sequences that can behave as chromatin borders and protect transfected genes from the surrounding chromatin influences. In vivo, these sequences are shown to be essential for gene activation as well as maintenance of correct patterns of gene expression. Deletion of S/MAR sequences can increase transcription of a neighboring gene, suggesting that the S/MAR elements can act as a negative regulator in their natural state, similar to Insulator sequences (Conlon and Meyer, 2006). In this publication, the authors examined transcription of a downstream gene to assess the effect of deleting a S/MAR element, however, analysis of upstream gene transcription would have provided an additional insight into the overall and directional activities of this particular S/MAR element. In vitro, use of such elements in expression vectors can generate higher proportion of transfected cells with improved expression levels, with reduced clonal variability and improved stability (reviewed in Kim, 2007). These effects are thought to stem from the ability of S/MAR elements to protect transgene from the transcriptionally repressive influences from the surrounding DNA upon integration into the genome. These negative influences include activities of external silencers as well as heterochromatinization (Bode et al., 2000).

We reported previously that vectors with S/MAR at the 5′ border of the expression cassette consistently produce CHO cell lines expressing significantly higher levels of recombinant protein, compared with vectors with S/MAR at the 3′ border of the expression cassette (Kim, 2007). Our findings are consistent with results from other published studies using retroviral vectors and S/MARs derived from the human IFNβ gene locus (Shübeler et al., 1996), demonstrating that the ability of S/MARs to increase transcription depended on the distance of S/MAR and its neighboring promoter and enhancer, within a retroviral vector. These results clearly demonstrate the importance of distance and the 3-dimensional relationship between the S/MARs and other regulatory elements in the vector. In addition to position-dependent S/MAR activity, we also observed different activities of S/MAR elements depending on the orientation of these elements within a given position (Kim, 2007). These differences were not as dramatic as differences arising from having S/MAR elements in different positions within the vector, however, the trend of activity differences appeared to be consistent for each element we tested in our laboratory. The mechanism underlying the orientation-dependent activity of S/MAR elements may be similar or related to the mechanism underlying the directional flow of transcription in vivo, which appears to be based on strand bias or the chromatin structure of its neighboring DNA sequences (Evans, 2008). Strand bias can be thought as any strand-specific activity, based on the unequal DNA base composition of a particular strand, where %G is not equal to %C or %A is not equal to %T, respectively. For eukaryotes, strand bias has been observed in regions of origins of replication or transcription start sites and transcribed genes (Aerts et al., 2004).

In the mouse genome, transcription start sites cluster towards the A+/T+(more As than Ts) boundary with a bias to the downstream side of the boundary and avoid the T+/A+(more Ts than As) boundaries. In addition, the probability of gene expression often depends on the direction of the transcription unit in reference to the A+/T+and the T+/A+boundaries (Evans, 2008). Interestingly the A+/T+boundaries are SATB1 binding regions, whereas the T+/A+boundaries are not (Dickinson et al., 1992; de Belle et al., 1998). SATB1 (special AT-rich binding protein 1) was originally identified as a protein that recognizes double-stranded DNA with a high degree of base-unpairing, referred to as base-unpairing regions, or BURs, often present in S/MAR elements (Dickinson et al., 1992). SATB1 is the most well-characterized MAR-binding protein (MBP) to date, and recent studies have shown that SATB1 can regulate gene expression in vivo by altering chromatin structure, nuclear architecture, and the epigenetic profiles of the target sequences (reviewed in Galande et al., 2007; Han et al., 2008; Shutao et al., 2003; Wen et al., 2005; Yasui et al., 2002). In addition to playing a role in establishing a defined chromatin structure, SATB1 is shown to be important in maintaining a permissive transcriptional environment for defined regions in the genome, leading to coordinated expression of target genes and contributing to tissue-specific gene expression (Cai et al., 2006). Recently, SATB2, a close homolog of SATB1, has been identified as a gene mutated in human patients with the cleft palate syndrome (FitzPatrick et al., 2003), and later shown to also regulate coordinated gene expression by interacting with S/MAR elements, contributing to, in this case, tissue-specific CNS gene expression (Britanova et al., 2005). It is then, plausible to hypothesize that the sequences within S/MAR elements, such as the SATB1 binding sites may influence structure and the spatial organization of neighboring vector sequences and ultimately the resulting protein expression, when integrated into the chromosome of the host cell line. Therefore, the physical position of the gene within the vector and in reference to the S/MAR element would determine the efficiency of its transcription, and the expression level of one's recombinant protein of choice.

2 Position Effects of the Selection Cassette

A critical vector component that can affect expression of a recombinant gene is the selection cassette. Stringency of selection can dramatically affect protein expression of the resulting transfected pools and individual clones, since applying more stringent selection can lead to the identification of transfected cells with higher expression levels of the selection gene and the recombinant gene of interest. Increasing selection stringency is a commonly used method to achieve high levels of transgene expression and a strong selection process can often supersede expression differences resulting from other components within the vector. Expression vector evaluation studies done my laboratory have shown that differences in expression levels by using two different promoters can be masked upon application of a significantly more stringent selection pressure (data not shown).

When the expression of the selection marker is attenuated, the cell must have increased copy numbers of the selection cassette integrated and expressed, in order to compensate and survive. Cells with low copy numbers of integrants, and therefore lower copy of drug resistant marker genes, are eliminated, resulting in overall higher gene expression (van Blokland et al., 2007; reviewed in Warm, 2004). It is important to note that not all integrated plasmids will be expressed and the status of expression will depend on the site of genomic integration site. In this case, addition of various genomic sequence elements described earlier in this review can help overcome some of the negative effects of the genomic environments, thereby leading to better correlation between integrated gene copy number and the expression level. The stringency of a selection cassette can be increased by treating cells with methotrexate (MTX) when using the DHFR system, with methionine sulphoximine (MSX) when using the GS system, or with very high concentrations of most of the antibiotic agents routinely used for stable cell line generation. Additional methods of increasing stringency of selection include attenuating the strength of the selection cassette itself, by utilizing a weak, minimal promoter and the polyA sequence within the selection cassette or utilizing a weakened start site of the selection gene itself. A recent work by van Blokland et al. described another high stringent selection system in which a selection marker with an attenuated start codon was placed upstream of the gene of interest, containing an optimal start codon, in a bicistronic configuration. In our laboratory, since we observed differential activities of various vector components based on their positions within the vector context, we investigated effects of relative positions of the selection cassette in an expression vector, and whether this change can lead to different stringency of selection and ultimately different protein expression. To this end, we constructed two vectors that both contain a CMV5 promoter/enhancer, driving a GFP expression, and a puromycin selection cassette driven by a SV40 early enhancer/promoter, located either at the 5′ of the CMV5-GFP cassette (pMW14), or directly at the 3′ of the CMV5-GFP expression cassette (pMW15). These vectors were stably transfected into HEK293 cells. Upon transfection, we consistently observed reduced numbers of puromycin resistant colonies using the expression vector pMW15, compared to when cells were transfected with the expression vector pMW14 (Fig. 1). Moreover, the reduction in the number of puromycin resistant colonies was accompanied by a significant increase in the mean intensity of the GFP positive profile of the stably transfected pools (Fig. 2). In this study, having the selection unit at the 3′ border of the expression cassette appeared to increase the stringency of selection, compared to having the selection unit at the 5′ border of the expression cassette, possibly due to reduced transcription of the puromycin resistance gene, resulting from transcription interference. Similarly, Eszterhas et al. (2002) reported differential gene expression when they placed two nearly identical transcription units in different orientations, tandem, divergent, or convergent, into a host cell genome. In this study, the suppression was most severe with the convergent orientation, and least severe with the divergent orientation. Although these experiments examined effects of two nearby transcription units within the genomic context rather than a vector context, it’s reasonable to speculate that similar mechanisms of transcriptional interference contribute to the regulation of the genes being tested in both systems.

Transcription Interference is described as a general phenomenon whereby an active transcription unit disrupts the activity of another unit in a nearby locus.

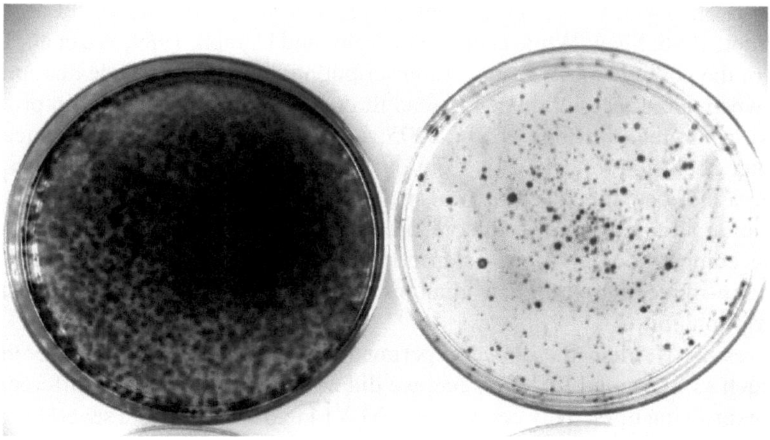

Fig. 1 Examination of stable HEK293F transfections demonstrates vectors pMW14 and pMW15 lead to different numbers of puromycin resistant colonies, suggesting different levels of puromycin-mediated selection stringency. Briefly, the HEK293F cells were transfected with either pMW14 or pMW15, and equal numbers of transfected cells were plated in a 10 cm followed by selection with 1ug/mL puromycin. 2 week post selection, the cells were stained with methylene blue and plates were photographed

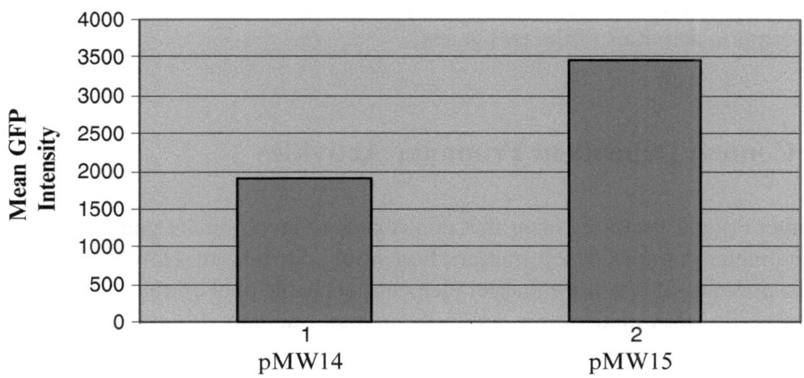

Fig. 2 Expression analysis of stably transfected HEK293F cells show higher mean GFP intensity of puromycin selected pools using the vector pMW15 compared to the vector pMW14. Mean GFP intensity was measured by a Caliber flow cytometry method

This is a phenomenon which appears to be conserved in evolution and was first observed in maize (Fincham and Sastry, 1974; McClintock, 1968), and later in other systems such as yeast, viruses, bacteria, plants, flies, and mammals (Bateman and Paule, 1988; Cook et al., 1996; Corbin and Maniatis, 1989; Cullen et al., 1984; Eggermont and Proudfoot, 1993; Emerman and Temin, 1984; Esperet et al., 2000;

Fiering et al., 1993; Greger et al., 1998; Greger et al., 2000; Martens et al., 2005; Proudfoot, 1986; Shaw-White et al., 1993; Vales and Darnell, 1989; Wu et al., 1990). Some of the possible mechanisms of transcription interference include disruption of the downstream transcription due to read through transcription such as in promoter occlusion model (Martens et al., 2005; Greger et al., 1998). The concept of promoter occlusion model stems from our observations that the mammalian RNA polymerase II enzyme proceeds past the polyA site and does not have a specific termination site (reviewed in Rosonina et al., 2006). Other potential mechanisms include competition for limited transcription factors by the two promoters, and chromatin-induced topological changes in the DNA caused by the upstream transcription unit.

Interestingly, when the above experiment was repeated using a different test gene such as a secreted IgG1Fc gene, we did not observe significant differences in stable expression of Fc, between vectors pMW14 and 15. There are several possible explanations for the different results we obtained, using GFP and the IgG1Fc as test genes. One possibility is that different levels of transcription of Fc RNA may not correlate to different levels of protein as it did with GFP if the post transcriptional processing of Fc is sufficiently more robust so that it overcomes the differences in the mRNA levels. Alternatively, we may not detect differences in the protein levels if Fc protein is more stable than the GFP protein, since our method of measuring protein expression relies on accumulation of proteins over several days of production. Lastly, stabilities of the two different mRNA species may be different and the potential differences of the two mRNA stabilities may mask differences in the transcription activities of the two genes.

3 Context Dependent Promoter Activities

Another critical vector element that contributes to recombinant gene expression is the promoter, without which transcription would not initiate. However, depending on the presence of relevant enhancer elements, available pool of transcription factors, or a specific DNA methylation status, promoters can be highly context-dependent, as demonstrated by their in vivo tissue- and cell-type specific activities (McCown et al., 1996; Klein et al., 1998). The tissue- and cell-type specific promoter activities are due to different availability of relevant transcription factors in different tissue and cell types. Moreover, the organization and the location of transcriptionally active and inactive compartments, and the available pools of transcription factors can vary in the eukaryotic cell nucleus depending on different stages of cell cycle and differentiation, thus spatial context of the gene within the nucleus can be important in modulating expression (reviewed in Francastel et al., 2000).

For the purpose of recombinant protein expression in cultured cells, there are multiple promoters that are available for long term stable expression, including several cellular gene promoters such as promoters derived from genes that encode β-actin, elongation factor-1α (EF-1α), and ubiquitin, as well as viral element-derived promoters such as the widely used cytomegalovirus (CMV) immediate-early promoter/enhancer and the Simian Virus 40 (SV40) early promoter/enhancer series (reviewed in Yew, 2005). Many of these promoters have been modified by addition of various enhancer or intronic sequence elements, for improved strength. Other available promoters include composite promoters derived from fusions of two different naturally occurring promoter sequences or fusions of a promoter and an enhancer sequences. Finally there are the "more" synthetic promoters which are usually generated by combinatorial methods using libraries containing a large array of transcription factor binding site containing oligonucleotides (Li et al., 1999; Edelman et al., 2000). Generating a synthetic promoter can be used as a method to create novel promoters, tailored to one's specific needs, such as developing promoters suited for a specific host cell type or a specific culture and production condition. The success of this type of approach depends on several critical parameters including having a well designed library and efficient high-throughput selection processes. Other tailored approaches for identifying promoters that are designed for high expression in their particular expression system include generating and screening a host cell-derived genomic DNA library for promoters that are highly active in their host cell line of choice (Pontiller et al., 2008). Others have used regulatory DNA sequences of a gene that are highly expressed during a recombinant protein fed-batch bioreactor process to identify promoter sequences that are highly active during production period, thus leading to high expression (Prentice et al., 2007).

As described earlier in this review, the activities of these tailored promoters can be further improved by addition of enhancer and intronic sequences, or various genomic DNA elements such as EASE, S/MAR, Insulators, or UCOE. Interestingly, but not surprisingly, studies from our laboratory comparing various combinations of promoters and genomic chromatin elements revealed that these effects are highly context dependent. In these studies, we compared the activities of different proprietary genomic elements (designated here as A, B and C) in different host cell lines: HEK293, CHOs and CHOad. In the first experiment, we compared the activities of element A and B in the context of the optimized expression vector for each host, using GFP expression as a read out. In the CHOad cell line, we found that both elements resulted in higher expression than the respective control vector alone, with element A providing a significantly higher increase in gene expression of recombinant human IgG1Fc, over element B (Fig. 3a). When these vectors were tested in both the adherent CHOad or the suspension CHOs cells, we found that the two vectors ranked opposite from one another in these two CHO lines (Fig. 3b). The difference between the two vectors was approximately twofold from one another in each cell line. As observed previously, both A and B elements provided significant benefit in expression over the control vector which did not contain a genomic sequence element. We were surprised to find that the element A did not provide detectable improvement in reporter gene expression in either CHO cells when it was combined with an alternate CMV-derived promoter, although the identical construct resulted in twofold increase in expression of the transfected pool, over control, when it was introduced into a HEK293 cell line (data not shown). These results suggested that the apparent lack of the element A activity in the CHO

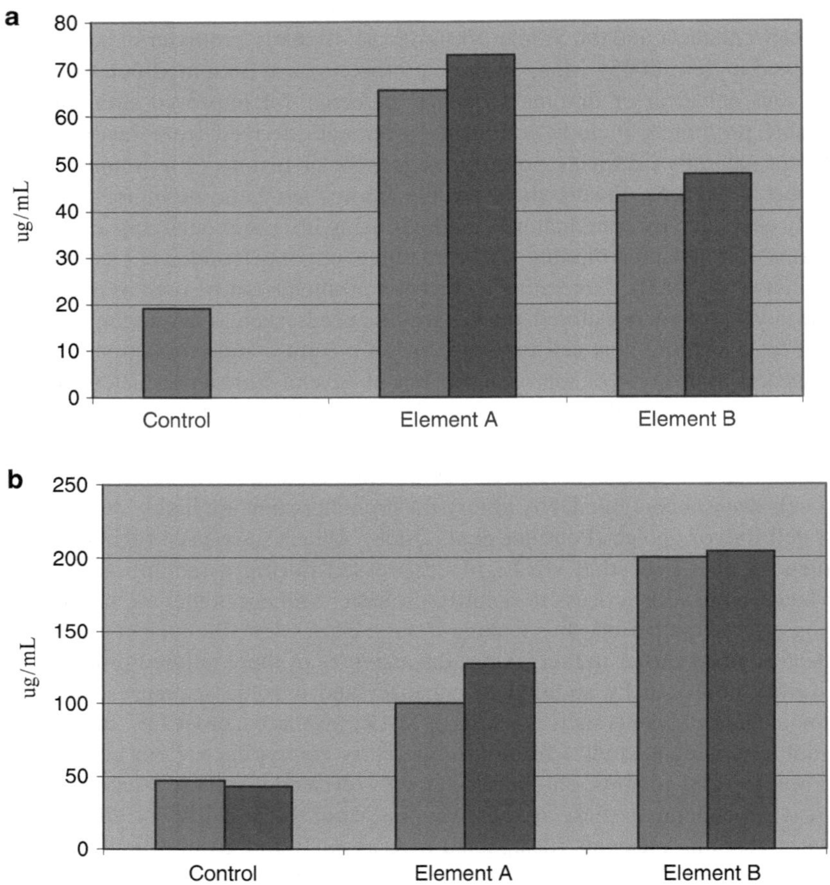

Fig. 3 (**a**) Analysis of stably transfected adherent CHOad cells show increased expression levels of recombinant human IgG1Fc when either the genomic Element A or B is incorporated into the vector, with the Element A providing a significantly higher increase. (**b**) When vectors containing either the Element A or the Element B were tested in the suspension CHOs cells, two vectors ranked opposite from one another from the adherent CHOad line, with the Element B providing a significantly higher increase in the expression of recombinant human IgG1Fc

cells was due to the interaction between the host cell line and the promoter rather than the inherent activity of the genomic element A, due to possible lack of relevant transcription factors in the CHO host that may be necessary for optimal interaction between the CMV-derived promoter and the genomic element A. Conversely, we observed relatively poor expression of a reporter gene, when the expression construct containing element B was introduced into HEK293 cells, although the identical construct led to several fold improvement in expression in both the CHOad and the CHOs host cells. Again, when we used an expression construct containing the element B with an alternate CMV-derived promoter, we were able to increase reporter gene expression over the previous vector. However, the increase was not significant over a control vector without element B in the HEK293 cells, suggesting the previous poor expression with the construct containing element B in the HEK293 cells was due to the interaction between the host cell line and the promoter. The chromatin element B appears to be neutral and does not seem to have either a positive or a negative activity in the HEK293 cell line we tested. Additional experiments in our laboratory suggested that the genomic element C can enhance CMV-derived promoters more robustly than SV40-derived promoters (data not shown). These results taken together, suggest that promoters as well as various genomic elements can confer dramatically different activities, depending on the context of other components in the vector and the transcriptional environment of the host cell. Moreover, genomic chromatin elements, as well as promoters, appear to possess tissue- and cell-specific activity, as indicated by the lack activity of element B in a HEK293 cell line, clearly suggesting not all genomic chromatin elements work in a similar manner. Interestingly, results from other studies indicate that even a same family of genomic chromatin elements, such as the family of S/MAR elements, and further S/MAR elements which are structurally and biochemically identical, can act differently, based on context (Heng et al., 2004). Studies by Heng et al. suggest that although S/MAR elements are necessary for matrix attachment and loop formation, simply having a S/MAR element within the DNA sequence may not be sufficient. Further, whether a S/MAR element is able to attach to the matrix and form a chromatin loop appear to depend on a currently unknown regulatory environment in the host cell line, including the genomic integration site. If we assume that a direct interaction of S/MAR element with the nuclear matrix is important for gene expression, and if we further assume that the ability of S/MAR elements to interact with the nuclear matrix is dependent on the genomic integration site, such results provide an insight into why we still observe clonal variation, albeit less, even when the expression construct contains a validated S/MAR element. The authors also suggest that certain S/MAR interaction with the nuclear matrix can be dynamic whereas other S/MAR interactions ca be more static or fixed, and the duration and the strength of S/MAR interaction with the nuclear matrix may correlate with the ability of a particular S/MAR to improve gene expression.

In addition to tissue- and cell-specificity, we can further speculate that the activity of genomic chromatin elements are dependent on the activity of their partnering promoters, although the converse may not be true, however, it will require more studies using additional genomic element/promoter combinations with a wider range or host cell systems to test this hypothesis.

Finally, in addition to the presence of specific transcription factor binding sites and DNA sequences, spacing between relevant transcription factor binding sites and modules appear to be critical in promoter activity, as demonstrated by studies described in Hartenbach and Fussenegger (2006), which reported a twofold increase in reporter gene expression in CHO-K1 cells by introducing a single nucleotide within a synthetic promoter (Hartenbach and Fussenegger, 2006).

4 Conclusion

Eukaryotic gene regulation is a complex, multi step process, involving highly ordered, specialized nuclear architecture and the dynamic interplay between these structures and nuclear proteins. Similarly, transcription complexes regulating artificial protein expression units within a vector are also highly ordered structures, and several lines of published studies have demonstrated that changing the order, orientation, or distances between the transcription factor binding sites can have a profound effect on gene expression. Furthermore, differences in the transcriptional machinery found in individual cell lines can lead to dramatically different outcomes from the same vector in different cell lines. It's not difficult to imagine the importance of cellular micro-environment, since binding of transcription factors is generally thought to be highly transient process, and stable association of transcription factors and the chromatin is achieved only when the transcription factor is incorporated into an active complex (Misteli, 2007). Taken together, development of an efficient vector requires proper arrangement of relevant vector components to maximize their activities, as well as productive interactions between different components within the vector. Additionally, the host cell environment should be taken into consideration when developing or choosing a proper vector system for recombinant protein expression.

Acknowledgements The authors would like to acknowledge and thank Amgen Protein Science Department for supporting our work.

References

Aerts S, Thijs G, Dabrowski M, Moreau Y, De Moor B (2004) Comprehensive analysis of the base composition around the transcription start site in Metazoa. *BMC Genomics* 5:34

Baskar JF, Smith PP, Nilaver G, Jupp RA, Hoffmann S, Peffer NJ, Tenney DJ, Colberg-Poley AM, Ghazal P, Nelson JA (1996) The enhancer domain of the human cytomegalovirus major immediate early promoter determines cell-type specific expression in transgenic mice. *J Virol* 70:3207–3214

Bateman E, Paule M (1988) Promoter occlusion during ribosomal RNA transcription. *Cell* 54:985–992

Bode J, Benham C, Knopp A, Mielke C (2000) Transcriptional augmentation: Modulation of gene expression by scaffold/matrix-attached regions (S/MAR elements). *Crit Rev Eukaryot Gene* Expr 10:73–90

Britanova O, Akopov S, Lukyanov S, Gruss P, Tarabykin V (2005) Novel transcription factor Satb2 interacts with matrix attachment region DNA elements in a tissue-specific manner and demonstrates cell-type-dependent expression in the developing mouse CNS. *Eur J Neurosci* 2:658–668

Cai S, Han HJ, Kohwi-Shigematsu T (2003) Tissue-specific nuclear architecture and gene expression regulated by SATB1. *Nat Genetics* 34:42–51

Cai S, Lee CC, Kohwi-Shigematsu T (2006) SATB1 packages densely looped, transcriptionally active chromatin for coordinated expression of cytokine genes. *Nat Genetics* 38:1278–88

Conlon TM, Meyer KB (2006) The chicken Ig light chain 3′-enhancer is essential for gene expression and regulates gene conversion via the transcription factor E2A. *Eur J Immunol* 36:139–48

Cook W, Wobbe K, Boni J, Coen D (1996) Regulation of neighboring gene expression by the herpes simplex virus type 1 thymidine kinase gene. Virology 218:193–203

Corbin V, Maniatis T (1989) Role of transcriptional interference in the *Drosophila melanogaster Adh* promoter switch. Nature 337:279–282

Cullen B, Lomedico P, Ju G (1984) Transcriptional interference in avian retroviruses—implications for the promoter insertion model of leukemogenesis. Nature 307:241–245

de Belle I, Cai S, Kohwi-Shigematsu T (1998) The genomic sequences bound to special AT-rich sequence-binding protein 1 (SATB1) in vivo in Jurkat T cells are tightly associated with the nuclear matrix at the bases of the chromatin loops. J Cell Biol 141:335–348

Dickinson LA, Joh T, Kohwi Y, Kohwi-Shigematsu T (1992) A tissue-specific MAR/SAR DNA-binding protein with unusual binding site recognition. Cell 70:631–645

Edelman GM, Meech R, Owens GC, Jones FS (2000) Synthetic promoter elements obtained by nucleotide sequence variation and selection for activity. Proc Natl Acad Sci USA 97:3038–3043

Eggermont J, Proudfoot N (1993) Poly(A) signals and transcriptional pause sites combine to prevent interference between RNA polymerase II promoters. EMBO J 12:2539–2548

Emerman M, Temin HM (1984) Genes with promoters in retrovirus vectors can be independently suppressed by an epigenetic mechanism. Cell 39:449–467

Esperet C, Sabatier S, Deville M, Ouazana R, Bouhassira E, Godet J, Morle F, Bernet A (2000) Non-erythroid genes inserted on either side of the human HS-40 impair activation of its natural alpha-globin gene targets without being themselves preferentially activated. J Biol Chem 275:25831–25839

Eszterhas SK, Bouhassira EE, Martin DI, Fiering S (2002) Transcriptional interference by independently regulated genes occurs in any relative arrangement of the genes and is influenced by chromosomal integration position. Mol Cell Biol 22:469–479

Evans KJ (2008) Strand bias structure in mouse DNA gives a glimpse of how chromatin structure affects gene expression. BMC Genomics 9:16

Fiering S, Kim CG, Epner EE, Groudine MG (1993) An "in-out" strategy using gene targeting and *flp* recombinase for the functional dissection of complex DNA regulatory elements: analysis of the β-globin locus control region. Proc Natl Acad Sci USA 90:8469–8473

Fincham J, Sastry G (1974) Controlling elements in maize. Ann Rev Genet 8:15–50

FitzPatrick DR, Carr IM, McLaren L, Leek JP, Wightman P, Williamson K, Gautier P, McGill N, Hayward C, Firth H, Markham AF, Fantes JA, Bonthron DT (2003) Identification of SATB2 as the cleft palate gene on 2q32–q33. Hum Mol Genet 12:2491–2501

Francastel C, Schübeler D, Martin D, Groudine M (2000) Nuclear compartmentalization and gene activity. Nat Rev Mol Cell Biol 1:137–143

Galande S, Purbey PK, Notani D, Kumar PP (2007) The third dimension of gene regulation: organization of dynamic chromatin loopscape by SATB1. Current Opin Genet Dev 17:408–14

Greger I, Demarchi F, Giacca M, Proudfoot N (1998) Transcriptional interference perturbs the binding of Sp1 to the HIV-1 promoter. Nucleic Acids Res 26:1294–1300

Greger I, Aranda A, Proudfoot N (2000) Balancing transcriptional interference and initiation on the GAL7 promoter of *Saccharomyces cerevisiae*. Proc Natl Acad Sci USA 97:8415–8420

Han HJ, Russo J, Kohwi Y, Kohwi-Shigematsu T (2008) SATB1 reprograms gene expression to promote breast tumour growth and metastasis. Nature 452:187–93

Hartenbach S, Fussenegger M (2006) A novel synthetic mammalian promoter derived from an internal ribosome entry site. Biotechnol Bioeng 95:547–59

Heng HH, Goetze S, Ye CJ, Liu G, Stevens JB, Bremer SW, Wykes SM, Bode J, Krawetz SA (2004) Chromatin loops are selectively anchored using scaffold/matrix-attachment regions. J Cell Sci 117:999–1008

Kim HY (2007) Use of DNA insulator elements and scaffold/matrix-attached regions for enhanced recombinant protein expression. In: Butler M (ed) Cell Culture and upstream processing. Taylor Francis, London, UK, pp 19–36

Klein RL, Meyer EM, Peel AL, Zolotukhin S, Meyers C, Muzyczka N, King MA (1998) Neuron-specific transduction in the rat septohippocampal or nigrostriatal pathway by recombinant adeno-associated virus vectors. Exp Neurol 150:183–194

Li X, Eastman EM, Schwartz RJ, Draghia-Akli R (1999) Synthetic muscle promoters: activities exceeding naturally occurring regulatory sequences. Nat Biotechnol 17:241–245

Loser P, Jennings GS, Strauss M, Sandig V (1998) Reactivation of the previously silenced cytomegalovirus major immediate early promoter in the mouse liver: involvement of NfkappaB. J Virol 72:180–190

Martens JA, Wu PY, Winston F (2005) Regulation of an intergenic transcript controls adjacent gene transcription in *Saccharomyces cerevisiae*. Genes Dev 19:2695–2704

McClintock B (1968) The states of a gene locus in maize. Carnegie Inst. Wash. Year Book 66:20–28

McCown TJ, Xiao X, Li J, Breese GR, Samulski RJ (1996) Differential and persistent expression patterns of CNS gene transfer by an adeno-associated virus (AAV) vector. Brain Res 713:99–107

Misteli T (2007) Beyond the Sequence: Cellular organization of genome function. Cell 128:787–800

Pontiller J, Gross S, Thaisuchat H, Hesse F, Ernst W (2008) Identification of CHO endogenous promoter elements based on a genomic library approach. Mol Biotechnol 39:135–139

Prentice HL, Tonkin CJ, Caamano L, Sisk WP (2007) High level expression of proteins using sequences from the ferritin heavy chain gene locus. J Biotechnol 128:50–60

Proudfoot N (1986) Transcriptional interference and termination between duplicated α-globin gene constructs suggests a novel mechanism for gene regulation. Nature 322:562–565

Rosonina E, Kaneko S, Manley J (2006) Terminating the transcript: breaking up is hard to do. Genes Dev 20:1050–1056

Shaw-White J, Denko N, Albers L, Doetschman T, Stringer J (1993) Expression of the *lacZ* gene targeted to the HPRT locus in embryonic stem cells and their derivatives. Transgenic Res 2:1–13

Shübeler D, Mielke C, Maass K, Bode J (1996) Scaffold/matrix-attached regions act upon transcription in a context-dependent manner. Biochemistry 35:11160–11169

Vales L, Darnell J (1989) Promoter occlusion prevents transcription of adenovirus polypeptide IX mRNA until after DNA replication. Genes Dev 3:49–59

van Blokland HJ, Kwaks TH, Sewalt RG, Verhees JA, Klaren VN, Siersma TK, Korse JW, Teunissen NC, Botschuijver S, van Mer C, Man SY, Otte AP (2007) A novel, high stringency selection system allows screening of few clones for high protein expression. J Biotechnol 128:237–245

Warm F (2004) Production of recombinant protein therapeutics in cultivated mammalian cells. Nat Biotechnol 22:1393–1398

Wen J, Huang S, Rogers H, Dickinson LA, Kohwi-Shigematsu T, Noguchi CT (2005) SATB1 family protein expressed during early erythroid differentiation modifies globin gene expression. Blood 105:3330–3339

Wu J, Grindlay G, Bushel P, Mendelsohn L, Allan M (1990) Negative regulation of the human ε-globin gene by transcriptional interference: role of an Alu repetitive element. Mol Cell Biol 10:1209–1216

Yasui D, Miyano M, Cai S, Varga-Weisz P, Kowhi-Shigematsu T (2002) SATB1 targets chromatin remodeling to regulate genes over long distances. Nature 419:641–645

Yew NS (2005) Controlling the kinetics of transgene expression by plasmid design. Adv Drug Deliv Rev 57:769–80

Zuckerkandl E, Villet R (1988) Generation of high specificity of effect through low-specificity binding of proteins to DNA. FEBS Lett 231:291–298

Cell Xpress^{TM} Applications in Development and Characterization of Biopharmaceutical Recombinant Protein Producing Cell Lines

Jennifer R. Cresswell, Nan Lin, Genova A. Richardson, and Kevin J. Kayser

Abstract The biopharmaceutical industry is focused on the development of quality processes for producing therapeutic proteins and monoclonal antibodies in mammalian cells. The first step in the process development workflow is cell line development. This chapter focuses on applications developed on the LEAP^{TM} (Laser-Enabled Analysis and Processing) instrument to expedite the cell line development process. Applications reviewed include expression optimization, high throughput single cell clone isolation and cell population characterization.

1 Introduction

The development of high producing mammalian cell lines that will be used to manufacture recombinant therapeutic proteins can be a time-consuming, expensive, and variable process. Multiple parameters significantly impact cell line quality and development timeline including expressed protein properties, parental cell line, expression system, transfection strategies, selection methods, and clone selection (Browne and Al-Rubeai, 2007; Dinnis and James, 2005; Wurm, 2004). Additionally, if a high performance cell line is not identified early in the process, the financial impact to the company can be considerable. Biopharmaceutical companies seek to continuously improve cell line development workflows, thereby decreasing variability in the system and increasing the frequency of high value clones. Within the workflow, clone selection typically requires the most time and labor to accomplish. The traditional method of single-cell cloning, limiting dilution (Schreiner et al., 1989; Stein et al., 1983; Underwood and Bean, 1988; Yang et al., 1992), in its most basic form requires minimal equipment cost and is technically simple to perform. However, limiting dilution cloning requires expansion to verify clonality and to evaluate secretion capabilities of the clones.

J.R. Cresswell (✉), N. Lin, G.A. Richardson, and K.J. Kayser
Cell Sciences and Development, SAFC Biosciences, 2909 Laclede Ave.
St. Louis, MO 63103., USA
e-mail: Jennifer.Crosswell@sial.com

M. Al-Rubeai (ed.), *Cell Line Development*, Cell Engineering 6, 109
DOI 10.1007/978-90-481-2245-5_6, © Springer Science+Business Media B.V. 2009

Industrial and academic researchers have successfully developed and adapted new methods in response to the challenges described. With an emphasis on early secretion assessment and high throughput methods, cell line development scientists seek to reduce labor and timelines, while ensuring high quality manufacturing cell lines. Among the emerging technologies, Cell Xpress™ was developed to improve the cell line selection process. Cell Xpress is a module of the LEAP™ instrument (Laser-Enabled Analysis and Processing, Cyntellect, San Diego, CA) that utilizes F-theta optics and high-speed galvanometer mirrors for high throughput, in situ cell imaging and laser-mediated cell elimination for multiple applications (Hanania et al., 2005; Koller et al., 2004). The instrument includes bright field and multi-channel fluorescence imaging as well as two laser options for cell processing (UV-355 nm, Green-532 nm). Automated Laser Targeting, Cell Counting, Cell Viability and Cell Xpress are among the software modules equipped by the system.

We have developed methods using the LEAP instrument and applied them at multiple points in the cell line development process (Lin et al., 2008). In the following sections of this review, the applications are described and compared to other clone selection approaches. These applications facilitate expression construct optimization, high throughput screening, single cell clone isolation, and characterization of clonal cell populations from early secretion performance.

2 Applications of Cell Xpress in Mammalian Cell Line Generation

The Cell Xpress module of the LEAP system allows the evaluation of secreted IgG associated with individual viable cells. In brief, cells are plated on 384- or 96-well C-lect™ plates (Cyntellect) with a proprietary capture reagent (Fig. 1). During an overnight incubation, the IgG secreted by the cells is bound to the capture reagent on the plate surface in the vicinity of the secreting cell. The next day, a fluorescent-conjugated anti-IgG detection reagent is applied which binds to the secreted IgG. Typically, the secreted antibody is bound to the plate immediately surrounding the cell, and after staining is visualized as a fluorescent halo. Live cells are labeled with a fluorescent viable cell dye, such as CellTracker™ Green (CTG, Invitrogen™, Molecular Probes®, Eugene, OR).

After washing to remove unbound fluorescent reagents, the plate is loaded into the LEAP instrument, and images from the IgG detection reagent and the live cell dye are acquired. The Cell Xpress algorithm identifies each viable cell in the well and calculates the intensity of the secretion halo surrounding that cell. The Cell Xpress secretion assay may be performed to obtain secretion data for analysis of a cell population. This secretion information can then be used in tandem with the instrument's laser processing capabilities to eliminate undesirable cells from the population. Cell line development applications of the Cell Xpress assay are described in the subsections below. Protocols for these applications are detailed elsewhere (Lin et al., 2008).

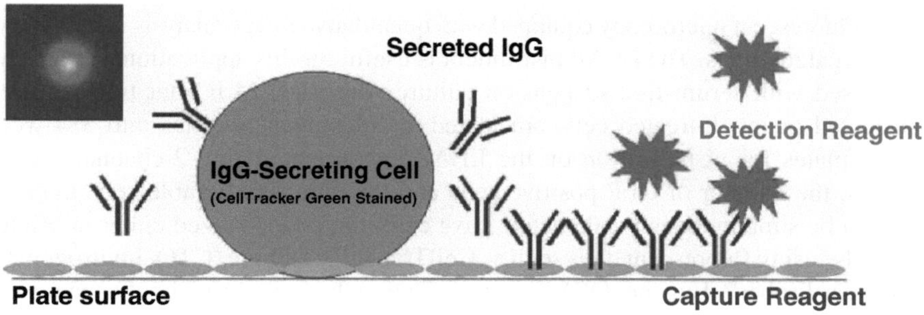

Fig. 1 Schematic illustration of Cell Xpress secretion assay. Cells are incubated in the presence of capture reagent overnight. The secreted recombinant IgG is captured in close proximity of the cell, creating a halo. The cells are then stained using a fluorescent live cell dye (CTG), and the secreted and membrane-bound recombinant IgG is detected using a fluorescent detection reagent. The *inset* (*upper left*) is a sample image of an IgG-secreting cell stained using this assay. The *brighter area* in the center is the overlap of the live cell dye and the detected membrane-associated antibody. The *darker shaded halo* surrounding the cell represents the captured and detected IgG produced by this cell.

2.1 Transfection Protocol Optimization Using Transfection Efficiency

A successful transfection is the fundamental step to any recombinant cell line generation process. Gene delivery methods such as lipofection, electroporation, and calcium phosphate precipitation, facilitate the delivery of recombinant protein expression vectors across the cell membrane and into the nucleus through DNA trafficking mechanisms that are not completely elucidated (Twyman, 2005). The number of cells that received plasmid DNA compared to the total number of viable cells transfected is known as transfection efficiency (Chenuet et al., 2008). Optimization of transfection protocols through transfection efficiency is critical for robust transfectant pool generation. A high level of recombinant protein production is a function of multiple systems within the cell. For example, integration of the plasmid DNA in transcriptional hot spots is a rare event. High transcription from open chromatin loci, in combination with highly functional translational, post-translational and secretory machineries, would lead to a high producing clone (Dinnis and James, 2005). Robust growth and enhanced metabolic traits are indispensable in manufacturing for high recombinant protein yield. Only high transfection efficiencies would lead to a large number of high recombinant protein producing clones, considering the low frequency of a clone possessing most of these favorable characteristics.

It is common practice to use a reporter protein such as GFP for estimating transfection efficiency. In this case, the ratio of GFP positive cells to viable cells is calculated and used as transfection efficiency (Derouazi et al., 2006; Liu et al.,

2008). Fluorescent microscopy equipped with quantitative image analysis is necessary for such calculations. The LEAP instrument is useful for this application, especially when used with serum-free suspension cultures. In brief, 24 h after transfection, transfected or mock-treated cells are plated at equivalent densities into 384-well C-Lect plates for visualization on the LEAP instrument. Using 2-channel image analysis, the number of GFP positive cells and the number of viable cells in each well can be simultaneously calculated. Live cells can be visualized either in bright field or by using fluorescent dyes such as CellTracker™ Orange (CTO, Invitrogen™, Molecular Probes®, Eugene, OR) (Fig. 6.2). Due to the high throughput capacity of LEAP, a large number of cells can be evaluated with ease. As a result, the statistical robustness is comparable to Fluorescence-Activated Cell Sorting (FACS) analysis and surpasses other microscopic methods. The fluorescence intensity value of each GFP positive cell is also recorded and can be exported for further analysis. In addition to CHO cells, this method is adaptable to other cell types commonly used in industrial mammalian cell culture such as NS0, HEK293 (unpublished data), and EB_x® avian embryonic stem cells (Vivalis, Nantes, France) (Davis, 2007).

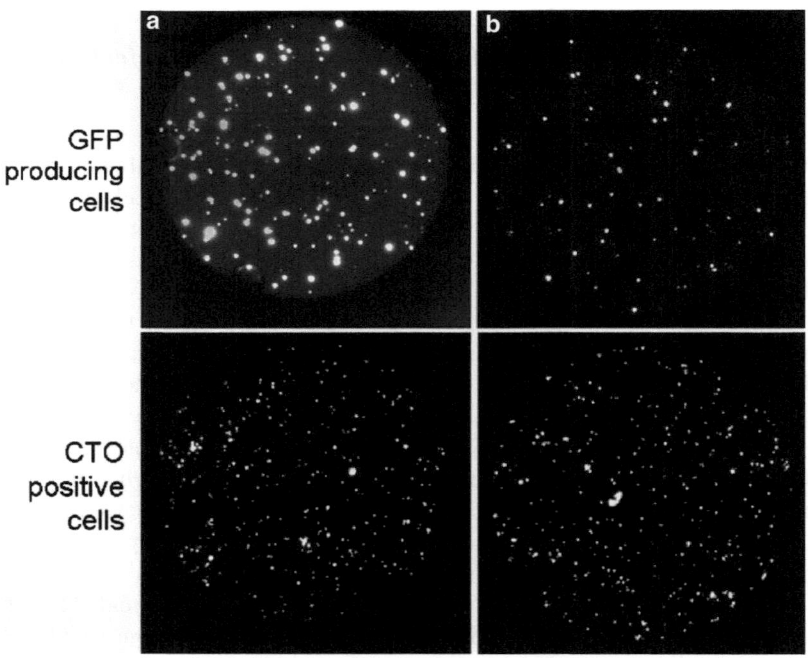

Fig. 2 Representative fluorescent images of transient GFP protein production. *Top panel*: GFP protein fluorescence. The same exposure and gain settings were used for all images. Non-transfected controls (not shown) revealed no background GFP fluorescence. *Bottom panel*: CellTracker Orange (CTO) staining for viable cells. Note the higher level of GFP protein fluorescence in cell line (**a**) versus cell line (**b**).

2.2 *Transient Transfection Levels and Manufacturability of Recombinant Therapeutic IgG*

Transient transfection is commonly used for monitoring gene transfer and recombinant protein expression in mammalian cells. Several hours after transfection, host cells begin to express recombinant proteins. Typically recombinant protein secreted in the supernatant becomes detectable by ELISA 24–72 hours after transfection. Confirmation of expression and validation of the vector constructs are hence achieved. Without replicating or integrating into the host cell genome, the plasmid DNA is diluted as cells divide. As a result, the level of the transiently expressed protein usually peaks around 3–5 days and decreases rapidly unless appropriate selection pressure for mammalian cells is applied to the host cells. Such selection preserves cells with integrated expression vector and kills the rest, resulting in a stably transfected population (Twyman, 2005; Wurm and Petropoulos, 1994).

Traditionally, transient transfection is evaluated by the following methods: (1) quantitation of recombinant protein levels in spent media and cell lysate via ELISA or western blotting, or (2) assays to determine bioactivity of the recombinant protein. Albeit well established, ELISA can be labor intensive due to the extensive dilution and washing steps, and usually takes at least two full days to perform. In addition, collecting supernatant from a heterogeneous population can only give the average expression level of the recombinant IgG. Low expression levels can be attributed to deficiencies in expression construct design, recombinant protein sequence, and low transfection efficiency. As a solution, many researchers include a transfection control, normally a reporter protein vector (e.g., GFP or β-galactosidase) in a parallel transfection reaction (Chen et al., 1999; Wurm and Bernard, 1999). This approach leads to additional experimental steps or extra recombinant proteins for the host cells to express.

The Cell Xpress secretion assay allows quantitative in situ detection of secreted recombinant IgG. Within 48 hours after transfection, transfected populations are plated and analyzed using Cell Xpress procedures. Fluorescence intensity of the secretion halos (Fig. 1) associated with live cells is quantified by image analysis. The procedure requires overnight incubation and approximately 1–3 labor hours to complete in the following day, which is less labor intensive than ELISA. More importantly, Cell Xpress enables assessment of IgG secretion of individual cells. This provides an accurate calculation of transfection efficiency (the ratio of cells associated with detectable halos and total number of viable cells) and hence makes it possible to distinguish effects associated with transfection efficiency from expression constructs when performing transient transfection evaluations (Fig. 3). In order to subtract background fluorescence and normalize between experiments and cell lines, appropriate parental cells should be included on the same plate as controls.

It has become common practice to transiently assess recombinant protein manufacturability during an early phase of target development. Cell line recombinant protein secretion can be affected by factors such as amino acid sequences or codon usage (Carton et al., 2007; Kalwy et al., 2006; Kim et al., 1997). Cell Xpress allows

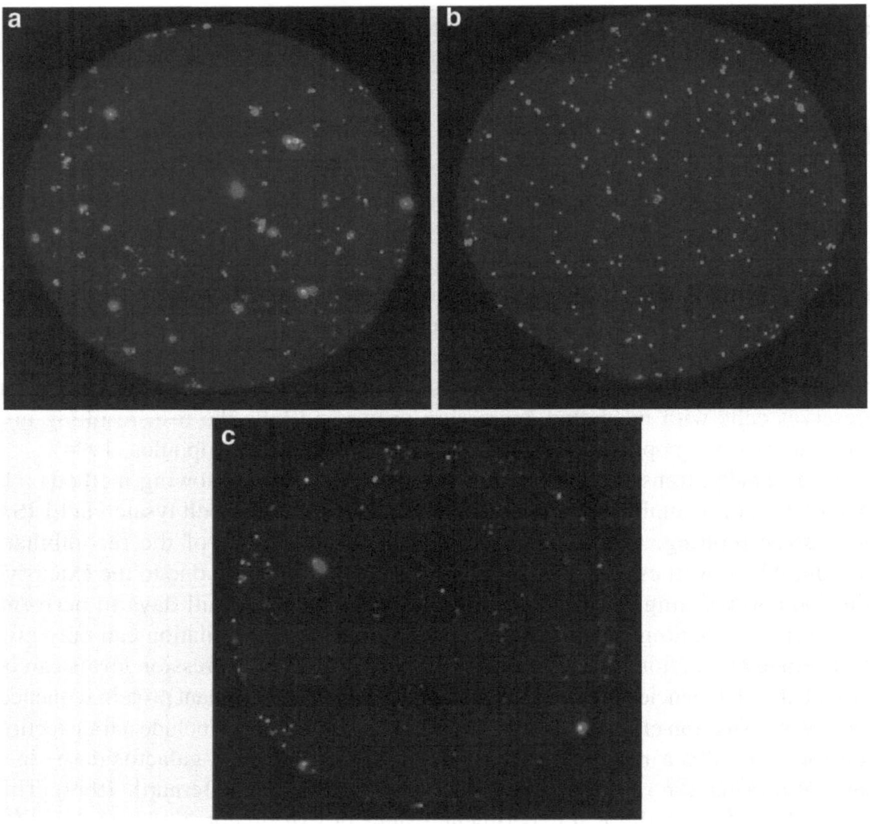

Fig. 3 Representative well images in transient transfection evaluations using Cell Xpress. All three panels are images of transient transfections using the same recombinant human IgG. Transfections (**a**) and (**b**) used the same expression construct but in two different parental CHO cell lines. Transfection (**a**) had significantly higher transfection efficiency than transfection (**b**) as demonstrated by the greater number of live cells associated with IgG secretion halos. Transfection (**c**) was performed using the same parental cell line as transfection (**a**) but a different expression vector. Note the difference in expression levels resulting from different expression vectors, indicated by the size and fluorescence intensity of the secretion halos.

parallel assessment of multiple candidate IgGs for the same target and identification of the most efficiently expressed and secreted. For example, a recombinant mouse IgG was expressed in the CHO K1 parental cell line. The mRNA of both heavy and light chain of this IgG was detectable using RT-PCR (data not shown), however, Cell Xpress secretion analysis indicated absence of detectable secretion halos 48 hours post-electroporation (Fig. 4). In further investigation, intracellular staining using a fluorescent-conjugated anti-mouse IgG detection reagent revealed that the recombinant IgG was present in the cytosol without being successfully secreted (Fig. 4). These results implied that future optimization and trouble-shooting may

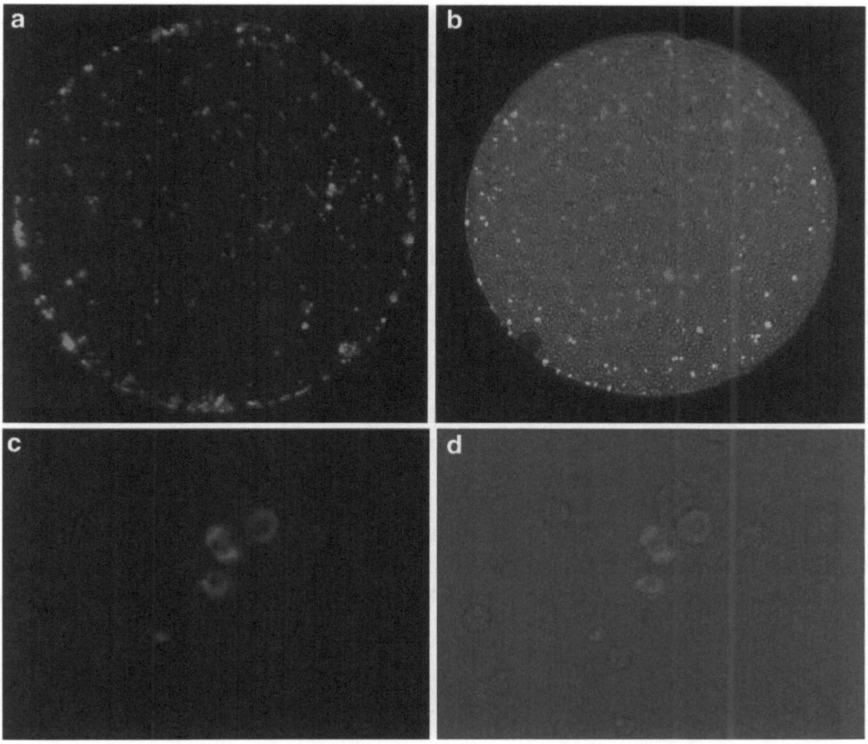

Fig. 4 Expression optimization of a recombinant mouse IgG using Cell Xpress. (**a**) Representative well image using the Cell Xpress secretion assay to detect secreted recombinant mouse IgG in a stably transfected population. No secretion halos were detected. 4X magnification. Fluorescence indicates CTG staining. (**b**) Cells from the same population as (**a**) were fixed, permeabilized and stained using the same detection reagent. Magnification 4X. Overlay of PE fluorescence and phase contrast images. Intracellular PE fluorescence indicates the presence of cytosolic recombinant mouse IgG. (**c**) Fixed, permeabilized and stained cells from (**b**). 20X magnification. PE fluorescence. (**d**) Overlay of (**c**) and phase contrast image. 20X magnification. PE fluorescence was localized in the cytosol.

include approaches such as optimization of the signal peptide to facilitate secretion. In contrast, *panel a* in Fig. 3 represents a candidate IgG that is secreted, indicating positive manufacturability.

Similarly, comparative assessment of expression vector components such as promoters and enhancer elements can be performed using Cell Xpress in transient expression. Instead of using non-secreted reporter proteins, recombinant IgGs can be directly used for such assessments, making the results more relevant and reproducible in early phase biopharmaceutical development. Comparisons among parental cell lines in terms of transfectability using the same expression vectors are also possible (Richardson, 2007).

2.3 Cell Population Enrichment by Cell Xpress Laser Processing

Cell population enrichment is routinely practiced on either a stably transfected or a clonal population by eliminating the non- or low-secreting cells of the population to improve the efficiency of subsequent single-cell isolation. One of our primary applications of Cell Xpress is to enrich stably transfected pools using laser processing. In brief, laser conditions (including energy level, pulse number, and process iterations) are optimized for maximum targeted cell elimination via photothermal or photochemical mechanisms (Koller et al., 2004). The optimized laser conditions should result in minimal micro-cavitation or cell movement, which could result in damage to non-targeted cells. The cell populations are then plated and processed with the optimized laser processing conditions. Multiple washes are performed post laser processing to remove all debris. Several days after processing, cells are pooled into a 6-well plate. When cells reach an appropriate density, the culture is expanded to shaker culture, and reevaluated by Cell Xpress secretion assay and HPLC to confirm the success of enrichment.

As depicted in Fig. 5, cells from a stably transfected population visualized on the LEAP instrument demonstrated a biphasic distribution of halo intensities (Secretion Area Average Intensities, or SAAI) prior to enrichment. Following laser

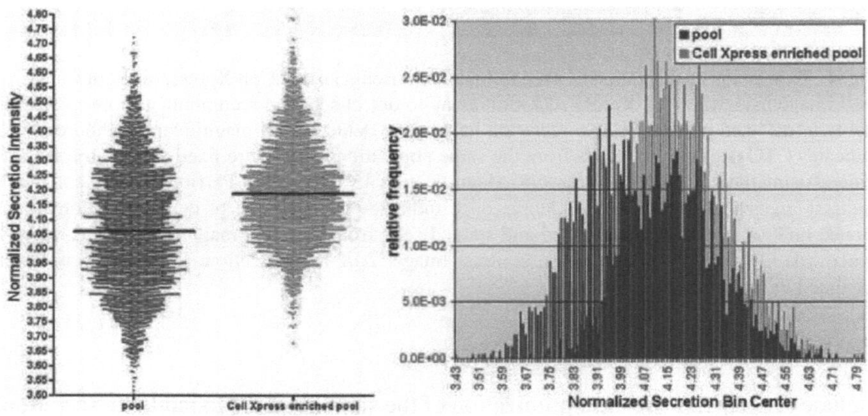

Fig. 5 Secretion heterogeneity of a transfected pool and a laser enriched transfected pool. On the *left* is a scattered dot plot of the secretion halo intensity (SAAI) in the original pool and the enriched pool. Each point on the graph represents the SAAI of a single cell within the population. The *black horizontal line* represents the mean SAAI for a given population. Note that the mean SAAI for the enriched pool is significantly higher in the enriched pool than the original pool, suggesting that laser enrichment isolated higher producing cells. On the *right* is the same data plotted in a relative frequency histogram. The original pool population exhibits a biphasic distribution. The *left peak* represents the low and non-producers within the population. The *right peak* represents the higher producers isolated by laser processing. The Cell Xpress enriched pool only has one peak primarily composed of higher secreting cells.

processing and expansion, the enriched population had more Gaussian distribution with significantly elevated mean and median SAAI.

We have observed that cell populations with greater heterogeneity (higher coefficient of variation and biphasic distribution) achieve more improvement after laser enrichment. In general, transfected pools give better enrichment results than clonal populations, especially highly homogeneous clones (data not shown).

FACS is another tool for cell population enrichment of selected phenotypes and employed by both industry and academia (Carroll and Al-Rubeai, 2004; Yoshikawa et al., 2001). However, the only current direct staining method to detect secreted IgG is gel encapsulation, which is rather cumbersome to perform. The other methods to detect secreted IgG, such as fluorescein-MTX (F-MTX) and surface matrix capture, rely on indirect staining (Bohm et al., 2004; Borth et al., 2000; Carroll and Al-Rubeai, 2004; DeMaria et al., 2007; Kromenaker and Srienc, 1994; Yoshikawa et al., 2001). Alternatively, Cell Xpress enables direct in situ IgG secretion assessment in a single staining step prior to population enrichment.

2.4 Single Cell Cloning Using Cell Xpress

In a clonal population that produces a recombinant protein, all cells arise from the same genetic background, which leads to more reproducible growth, productivity, and product quality. It is essential to manufacture each protein product from one clonal production cell line, unless equivalent quality is proven from another clonal cell line.

Single cell cloning is highly labor-intensive, time consuming and the most rate-limiting step of the cell line development process. Traditional limiting dilution protocols (Fuller et al., 2001; Galfre and Milstein, 1981) rely on manual microscopic verification and provide little image documentation of clone history during expansion. Another drawback of the limiting dilution methods is that clonality is based on statistical calculations (Coller and Coller, 1983; Coller and Coller, 1986) and requires visual verification of clonality. It is only feasible to manually screen large number of plates using phase-contrast microscopy when the cell expands to at least an 8-cell cluster. Morphology of the cluster and number of clusters are the sole determinant of clonality. Repeated sub-cloning by limiting dilution is often practiced in order to ensure clonality and to support regulatory filings, which increases the already lengthy timelines of cell line development.

We have applied Cell Xpress to automate the traditional limiting dilution protocols. Cell Xpress single cell cloning enables higher throughput by using 384-well plates instead of 96-well plates. Image documentation of a clone begins on Day 0, immediately after plating, with the aid of a fluorescent viable cell dye, and continues throughout the entire expansion process in 384-well plates. Early outgrowth rates can be determined based upon cell density in bright-field imaging. When coupled with Cell Xpress secretion analysis in the single-cell stage, this method provides direct measurement of IgG secretion at the earliest point of clone expansion.

All of the above contribute to improved discovery rate of high-secreting clones with reduced labor and time.

The workflow of Cell Xpress single cell cloning is as follows. First, the transfected pool is enriched to isolate and expand the high-producing cells within the population. After confirmation of successful enrichment, the enriched transfected pool is plated in a 384-well plate to achieve one cell per well. Plated cells are imaged on LEAP using either one-color in case of CTG stained cells or two-color in case of single-cell Cell Xpress evaluation. The higher producing clones can then be selected for expansion and further analysis.

An increasing number of researchers now utilize automated high throughput plating such as FACS to supplement or replace traditional limiting dilution, which provides improved clonality assessment and reduced labor (Borth et al., 2000; Carroll and Al-Rubeai, 2004; Yoshikawa et al., 2001; Zeyda et al., 1999). However, the pressure and velocity of sorting may apply some shear stress to cells, even if the intensity of such stress is to be studied (Dean and Hoffman, 2007; Shapiro, 2003). Sorting conditions therefore often require optimization to achieve high recovery. Conversely, Cell Xpress single-cell cloning typically exposes the cells to less shear stress. Additionally, Cell Xpress can be coupled at the single-cell stage with direct in situ IgG secretion detection.

3 Evaluation of Clonal Recombinant Cell Line Secretion

In addition to applications in cell line generation process, Cell Xpress can be used as an evaluation tool for IgG secretion after clone expansion to shake flask stage. The mean SAAI values have significant linear correlation to the productivity of the clones as measured by HPLC or other equivalent methods. The correlation coefficient (R^2) values of population mean or median SAAI vs. maximum productivities in non-fed shake flask cultures range from 0.7 to 0.95 with different cell types and recombinant IgGs (Fig. 6). With a 24-hour turnaround, Cell Xpress secretion analysis serves as preliminary screening that precedes and supplements the traditional growth and expression assay.

We have used the Cell Xpress secretion analysis to select candidate clones immediately after expansion to shake flask cultures. As depicted in Fig. 7, the highest productive primary clone among the clones studied was sub-cloned. The sub-clones were compared to the primary clone in mean SAAI, and only the sub-clones with mean SAAI>90% of the primary were banked. The top 28 clones (from 84 candidate sub-clones, approximately top 30%) proceeded to 12-day growth and productivity assays. The results indicated that 26 out of 28 clones (93%) that were selected based on Cell Xpress demonstrated improved peak volumetric productivity comparing to the primary clone. This approach reduced labor and overall time by only maturing the best candidate clones to growth and productivity evaluations.

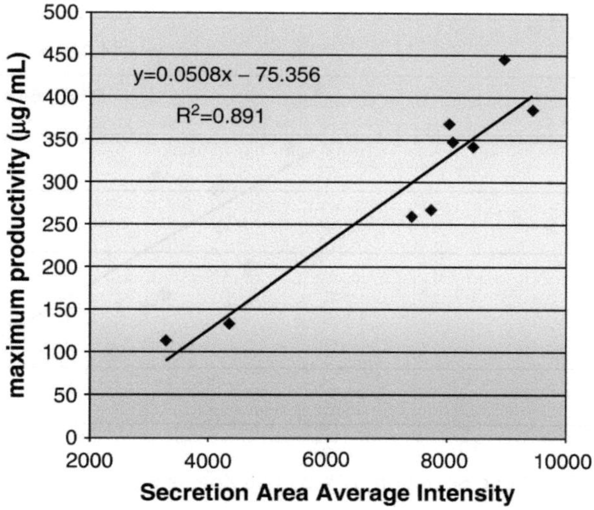

Fig. 6 Cell Xpress secretion correlates to productivity. The mean secretion level of a population (SAAI) directly correlates with maximum shake flask productivity. The Cell Xpress vs. HPLC productivity relationship of ten clonal cell lines is plotted.

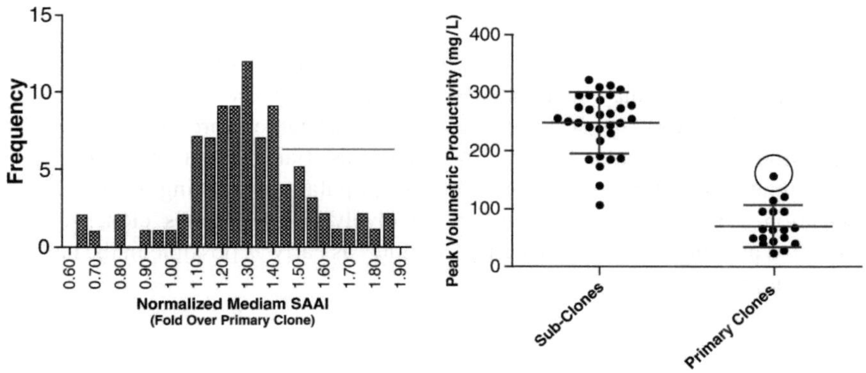

Fig. 7 Selection of top F_c-fusion protein producing clones using Cell Xpress secretion analysis. *Left panel*: a histogram distribution of normalized median SAAI of all candidate clones expanded to 125 mL shake flask cultures. *Horizontal solid line* indicates the clones selected for a 12-day growth and productivity assay. *Right panel*: peak volumetric shake flask productivity of the candidate clones selected via Cell Xpress. All sub-clones were derived from the highest producing primary clone (*circled*). The mean plus or minus the standard deviation of all clones screened are plotted as lines with error bars.

Because Cell Xpress collects single cell secretion data, it allows assessment of population distribution in addition to population secretion mean. Secretion distributions can be characterized based upon CV% and distribution profile, Gaussian or biphasic.

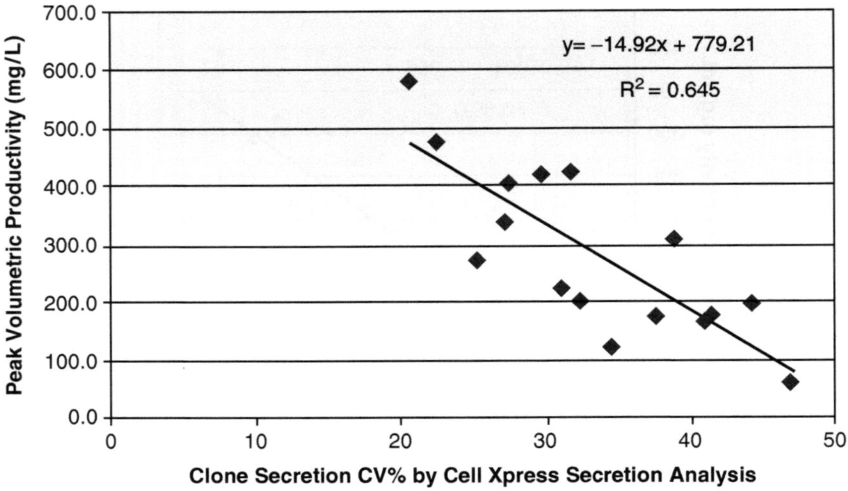

Fig. 8 Cell Xpress secretion and volumetric productivity. The inverse correlation between secretion CV% (x-axis) and peak volumetric productivity (y-axis) in shake flask cultures. Data derived from 16 CHO clones producing a recombinant human IgG derived from one transfection and evaluated using Cell Xpress.

High heterogeneity is often due to a high percentage of low or non-producing cells in the population and may be associated with low volumetric productivity (Fig. 8). Corisdeo et al. (2008) reported that cell lines with unstable expression had biphasic distribution of intracellular IgG by FACS analysis. Early secretion heterogeneity may be related to drift in expression levels of a population, resulting in lowered IgG production over time. Once identified, relatively heterogeneous clones can be eliminated as candidates for production cell lines as early as possible in the cell line evaluation process.

4 Discussion

4.1 Early Secretion Analysis

Predictive modeling for early clone selection has been an attractive topic in cell line development. In theory, genotypes and phenotypes early in clone expansion may relate to mechanisms of recombinant protein transcription, translation, post-translational modification, and secretion. The ability to accurately forecast the performance of a clone in bioreactor from its early characteristics would significantly improve the efficiency of cell line generation. Efforts have been made to determine the predictive power of productivity in various stages of expansion (Porter et al., 2007).

Other groups have identified cell line characteristics that may be used as specific early markers of cell line productivity and expression stability. For instance, there have been several reports on correlations between gene copy numbers, relative mRNA levels of heavy and light chains of IgG and long-term expression stability in NS0 cells (Barnes et al., 2001; Barnes et al., 2003; Barnes et al., 2004; Barnes et al., 2007; Barnes et al., 2006). Such correlative studies have also been conducted in CHO cell lines (Jiang et al., 2006; Jiang and Sharfstein, 2008; Jun et al., 2006; Kim et al., 2001).

Cell Xpress secretion analysis during expansion may be developed into an effective tool for predictive modeling. Since only a small number of cells are required, single cell secretion of a clone can be evaluated using Cell Xpress as early as from 6-well plate phase without sacrificing a large fraction of the expanding culture. Mean secretion data can be supplemented by heterogeneity analysis as an indicator of future clone performance. During clonal expansion, subpopulations of non- or low-secreting cells may overtake the population and result in poor final clone performance. Early secretion heterogeneity analysis may identify such subpopulations as they emerge, a capability not found with traditional supernatant analysis.

We have begun to evaluate Cell Xpress as a predictive tool during early expansion phase, but have observed a poor correlation of 6-well stage SAAI vs. shake flask productivity. Factors that affect this correlation may include the following: (1) expression instability leading to drastic drop in productivity from 6-well to shake flask cultures or (2) clonal drift in expression during adaptation to suspension culture. Further investigation and optimization may isolate such confounding factors and improve accuracy of predictive modeling with Cell Xpress.

4.2 Cell Cycle and Secretion Analysis

One of the important considerations for the endpoint secretion analysis methods, as opposed to a full-length growth and productivity analysis, is cell cycle specificity of antibody secretion. Possible artifacts associated with such specificity in non-synchronized populations may affect accuracy of clone ranking. Recombinant IgG expression is believed to have a cell cycle preference in hybridomas (Charlet et al., 1995; Park and Ryu, 1994), NS0 myeloma lines, and recombinant CHO cells (Swiderek and Al-Rubeai, 2007). Based on such preference, cell cycle arrest in G1 by over-expression of cyclin-dependent kinase (cdk) inhibitors in various cell types has been well described in literature to increase antibody production (Bi et al., 2004; Mazur et al., 1998; Meents et al., 2002; Seifert and Phillips, 1999).

We address cell cycle specificity by isolating the factors contributing to the variability of mean or median SAAI measured using Cell Xpress. We observed that the variability within the same clone analyzed is approximately 20% (unpublished data) among replicate wells, plates and experiments. Colchicine induced G2/M arrest (Tobey et al., 1990) led to increase in mean population SAAI. However, we observed no significant change in CV% between the G2/M arrested culture and the non-synchronized culture (Fig. 9), suggesting that cell cycle is not significantly

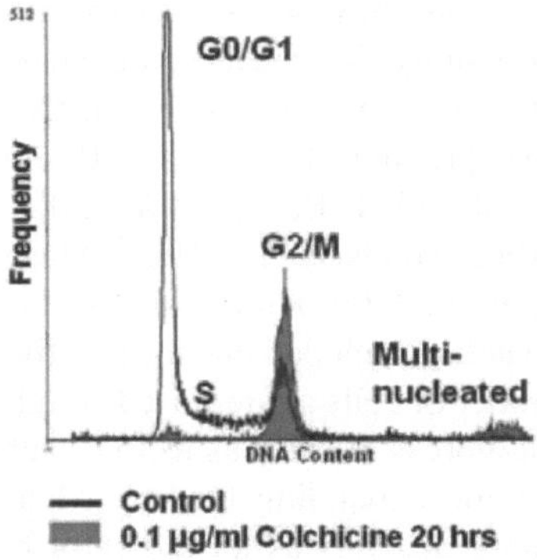

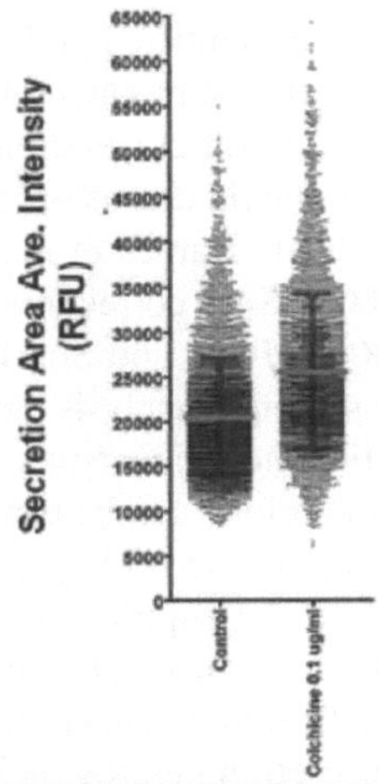

Cell Cycle Phase	Control	Colchicine 0.1 ug/ml
G0/G1	59.69%	3.61%
S	30.05%	0
G2/M	10.25%	96.39%

	Control	Colchicine 0.1 ug/ml
Number of values	4131	3145
Mean SAAI	20555	25578
CV%	31.95%	33.84%

Fig. 9 Cell cycle and Cell Xpress secretion analysis. Cell Xpress secretion analysis of a CHO cell line producing a recombinant human IgG under non-synchronized and G2/M arrested conditions. G2/M arrest led to increase in mean SAAI but no significant change in secretion CV% compared to the non-synchronized culture. *Left panel*: FACS cell cycle analysis of the G2/M arrested cells (solid histogram, treated by 0.1 µM colchicine for 20 h prior to Cell Xpress analysis) and non-treated control (line histogram). Percent of 10,000 cells analyzed in each sample in G0/G1, S, G2/M phases were determined by ModFit LT (Verity Software House, Topsham, ME) and summarized in the *left table*. *Right panel*: Scattered dot plot of Secretion Area Average Intensity (SAAI) of the control and colchicine treated samples. *Solid line* indicates the population average. *Error bars* indicate standard deviations. Mean and SD values of SAAI are summarized in the *right table.*

contributing to the secretion heterogeneity observed. Additional experiments are required to fully examine the effect of cell cycle on secretion heterogeneity.

4.3 Future Applications of Cell Xpress

Recombinant IgGs and F_c-fusion proteins constitute the main therapeutic protein market (Holliger and Hudson, 2005; Reichert et al., 2005). Other therapeutic proteins, such as cytokines, growth factors, and recombinant enzymes, take a smaller part of the market but often pose greater challenges in expression and manufacturing. In light of facing some of the challenges, we are currently working to adapt the Cell Xpress method to detect secretion of non-IgG recombinant proteins. Using one capture antibody and one detection antibody in designs similar to a standard ELISA, one can achieve in situ detection of the secreted recombinant protein of interest.

Similar to FACS, Cell Xpress can also acquire data in a single-cell and multi-parameter fashion. Quantitative analyses with cell surface or intracellular staining to detect proteins of interest is the logical next step. Population enrichment based on these proteins of interest will follow in the ongoing development. Additionally, quantitative analyses of organelles can be coupled with existing secretion analysis, which will address questions such as metabolic state of cells (Hinterkorner et al., 2007).

5 Conclusions and Summary

The pressure for faster timelines and higher productivities will continue to drive new solutions for biotherapeutic protein production. High throughput cloning methods and early identification of candidate clones will be key factors in these shortened timelines. Cell Xpress enables evaluating secretion properties of individual cells in a high throughput manner, which makes it a valuable tool for multiple applications during the mammalian cell line development process. Our initial work with the Cell Xpress has shed light on previously overlooked characteristics of production clones. Cell Xpress analysis, combined with other research approaches, affords us the opportunity to further study the mechanisms of recombinant protein production. Future effort will undoubtedly refine the current applications and demonstrate the potential utility of this emerging technology.

Acknowledgments We would like to thank Erika Holroyd, Kathleen Roeder and Angela Davis for laboratory support.

References

Barnes LM, Bentley CM, Dickson AJ (2001) Characterization of the stability of recombinant protein production in the GS-NS0 expression system. Biotechnol Bioeng 73:261–270

Barnes L.M, Bentley C.M, Dickson A.J. (2003) Stability of recombinant protein production in the GS-NS0 expression system is unaffected by cryopreservation. Biotechnol Prog 19:233–237

Barnes L.M, Bentley C.M, Dickson A.J. (2004) Molecular definition of predictive indicators of stable protein expression in recombinant NS0 myeloma cells. Biotechnol Bioeng 85:115–121

Barnes L.M, Moy N, Dickson A.J. (2006) Phenotypic variation during cloning procedures: analysis of the growth behavior of clonal cell lines. Biotechnol Bioeng 94:530–537

Barnes L.M, Bentley C.M, Moy N, Dickson A.J (2007) Molecular analysis of successful cell line selection in transfected GS-NS0 myeloma cells. Biotechnol Bioeng 96:337–348

Bi J.X, Shuttleworth J, Al-Rubeai M (2004) Uncoupling of cell growth and proliferation results in enhancement of productivity in p21CIP1-arrested CHO cells. Biotechnol Bioeng 85:741–749

Bohm E, Voglauer R, Steinfellner W, Kunert R, Borth N, Katinger H (2004) Screening for improved cell performance: selection of subclones with altered production kinetics or improved stability by cell sorting. Biotechnol Bioeng 88:699–706

Borth N, Zeyda M, Kunert R, Katinger H (2000) Efficient selection of high-producing subclones during gene amplification of recombinant Chinese hamster ovary cells by flow cytometry and cell sorting. Biotechnol Bioeng 71:266–273

Browne S.M, Al-Rubeai M (2007) Selection methods for high-producing mammalian cell lines. Trends Biotechnol 25:425–432

Carroll S, Al-Rubeai M (2004) The selection of high-producing cell lines using flow cytometry and cell sorting. Expert Opin Biol Ther 4:1821–1829

Carton J.M, Sauerwald T, Hawley-Nelson P, Morse B, Peffer N, Beck H, Lu J, Cotty A, Amegadzie B, Sweet R (2007) Codon engineering for improved antibody expression in mammalian cells. Protein Expr Purif 55:279–286

Charlet M, Kromenaker SJ, Srienc F (1995) Surface IgG content of murine hybridomas: direct evidence for variation of antibody secretion rates during the cell cycle. Biotechnol Bioeng 47:535–540

Chen R, Greene E.L, Collinsworth G, Grewal J.S, Houghton O, Zeng H, Garnovskaya M, Paul R.V, Raymond J.R. (1999) Enrichment of transiently transfected mesangial cells by cell sorting after cotransfection with GFP. Am J Physiol 276:F777–F785

Chenuet S, Martinet D, Besuchet-Schmutz N, Wicht M, Jaccard N, Bon AC, Derouazi M, Hacker D.L, Beckmann J.S, Wurm F.M. (2008) Calcium phosphate transfection generates mammalian recombinant cell lines with higher specific productivity than polyfection. Biotechnol Bioeng 101:937–945

Coller H.A, Coller BS (1983) Statistical analysis of repetitive subcloning by the limiting dilution technique with a view toward ensuring hybridoma monoclonality. Hybridoma 2:91–96

Coller H.A, Coller B.S (1986) Poisson statistical analysis of repetitive subcloning by the limiting dilution technique as a way of assessing hybridoma monoclonality. Methods Enzymol 121:412–417

Corisdeo S, Cassel M.J, Kinney C.S, Ganguly S, Kraichely K.M. Use of flow cytometry to screen and predict stability of candidate manufacturing cell lines. 2008; Conference Presentation. Bioprocess International Annual Meeting. Anaheim, CA.

Davis L. Optimization of electroporation and clone selection of EB14 chicken embryonic stem cells to express recombinant monoclonal antibodies. 2007; Conference Poster. IBC Cell Line Development and Engineering. San Diego, CA.

Dean P.N, Hoffman R.A (2007) Overview of flow cytometry instrumentation. Curr Protoc Cytom Chapter 1: Volume 1 Unit1 1.

DeMaria C.T, Cairns V, Schwarz C, Zhang J, Guerin M, Zuena E, Estes S, Karey K.P.(2007) Accelerated clone selection for recombinant CHO CELLS using a FACS-based high-throughput screen. Biotechnol Prog 23:465–472

Derouazi M, Flaction R, Girard P, de Jesus M, Jordan M, Wurm F.M. (2006) Generation of recombinant Chinese hamster ovary cell lines by microinjection. Biotechnol Lett 28:373–382

Dinnis D.M, James D.C. (2005) Engineering mammalian cell factories for improved recombinant monoclonal antibody production: lessons from nature? Biotechnol Bioeng 91:180–189

Fuller S.A, Takahashi M, Hurrell J.G. (2001) Cloning of hybridoma cell lines by limiting dilution. Curr Protoc Mol Biol Chapter 11: http://mrw.interscience.wiley.com/emrw/9780471142720/cp/cpmb/toc Unit11 8.

Galfre G, Milstein C (1981) Preparation of monoclonal antibodies: strategies and procedures. Methods Enzymol 73:3–46

Hanania EG, Fieck A, Stevens J, Bodzin L.J, Palsson B.O, Koller M.R. (2005) Automated in situ measurement of cell-specific antibody secretion and laser-mediated purification for rapid cloning of highly-secreting producers. Biotechnol Bioeng 91:872–876

Hinterkorner G, Brugger G, Muller D, Hesse F, Kunert R, Katinger H, Borth N (2007) Improvement of the energy metabolism of recombinant CHO cells by cell sorting for reduced mitochondrial membrane potential. J Biotechnol 129:651–657

Holliger P, Hudson P.J. (2005) Engineered antibody fragments and the rise of single domains. Nat Biotechnol 23:1126–1136

Jiang Z, Sharfstein S.T. (2008) Sodium butyrate stimulates monoclonal antibody over-expression in CHO cells by improving gene accessibility. Biotechnol Bioeng 100:189–194

Jiang Z, Huang Y, Sharfstein S.T. (2006) Regulation of recombinant monoclonal antibody production in chinese hamster ovary cells: a comparative study of gene copy number, mRNA level, and protein expression. Biotechnol Prog 22:313–318

Jun S.C, Kim M.S, Hong H.J, Lee G.M. (2006) Limitations to the development of humanized antibody producing Chinese hamster ovary cells using glutamine synthetase-mediated gene amplification. Biotechnol Prog 22:770–780

Kalwy S, Rance J, Young R (2006) Toward more efficient protein expression: keep the message simple. Mol Biotechnol 34:151–156

Kim C.H, Oh Y, Lee T.H. (1997) Codon optimization for high-level expression of human erythropoietin (EPO) in mammalian cells. Gene 199:293–301

Kim N.S, Byun T.H, Lee G.M. (2001) Key determinants in the occurrence of clonal variation in humanized antibody expression of cho cells during dihydrofolate reductase mediated gene amplification. Biotechnol Prog 17:69–75

Koller M.R, Hanania E.G, Stevens J, Eisfeld T.M, Sasaki G.C, Fieck A, Palsson B.O. (2004) High-throughput laser-mediated in situ cell purification with high purity and yield. Cytometry A 61:153–161

Kromenaker S.J, Srienc F (1994) Stability of producer hybridoma cell lines after cell sorting: a case study. Biotechnol Prog 10:299–307

Lin N, Cresswell J.R, Richardson G.A, Gerber M.A, Kayser K.J (2008) Methods and applications of laser-enabled analysis and processing (LEAP). Curr Protoc Cytom Chapter 2: Volume 1 Unit2 14.

Liu C, Dalby B, Chen W, Kilzer J.M, Chiou H.C. (2008) Transient transfection factors for high-level recombinant protein production in suspension cultured mammalian cells. Mol Biotechnol 39:141–153

Mazur X, Fussenegger M, Renner W.A, Bailey J.E. (1998) Higher productivity of growth-arrested Chinese hamster ovary cells expressing the cyclin-dependent kinase inhibitor p27. Biotechnol Prog 14:705–713

Meents H, Enenkel B, Werner R.G, Fussenegger M (2002) p27Kip1-mediated controlled proliferation technology increases constitutive sICAM production in CHO-DUKX adapted for growth in suspension and serum-free media. Biotechnol Bioeng 79:619–627

Park S.H, Ryu D.D (1994) Cell cycle kinetics and monoclonal antibody productivity of hybridoma cells during perfusion culture. Biotechnol Bioeng 44:361–367

Porter A, Barnes L, Dickson A, Rancher A. Behaviour of GS-CHO cell lines in a selection strategy. 2007; Conference Presentation. ESACT 2007. Dresden, Germany.

Reichert J.M, Rosensweig C.J, Faden L.B, Dewitz M.C (2005) Monoclonal antibody successes in the clinic. Nat Biotechnol 23:1073–1078

Richardson G.A. An evaluation of the intrinsic IgG production capabilities of different Chinese hamster ovary parental cell lines. 2007; Conference Presentation. ESACT 2007. Dresden, Germany.

Schreiner C.L, Bauer J.S, Danilov Y.N, Hussein S, Sczekan M.M, Juliano R.L. (1989) Isolation and characterization of Chinese hamster ovary cell variants deficient in the expression of fibronectin receptor. J Cell Biol 109:3157–3167

Seifert D.B, Phillips J.A. (1999) The production of monoclonal antibody in growth-arrested hybridomas cultivated in suspension and immobilized modes. Biotechnol Prog 15:655–666

Shapiro H.M. (2003) Practical flow cytometry. Chapter 1. Overture. Wiley, Hoboken, NJ, pp 170–171

Stein L.D, Ledgley C.J, Sigal N.H. (1983) Patterns of isotype commitment in human B cells: limiting dilution analysis of Epstein Barr virus-infected cells. J Immunol 130:1640–1645

Swiderek H, Al-Rubeai M (2007) Functional genome-wide analysis of antibody producing NS0 cell line cultivated at different temperatures. Biotechnol Bioeng 98:616–630

Tobey R.A, Oishi N, Crissman H.A .(1990) Cell cycle synchronization: reversible induction of G2 synchrony in cultured rodent and human diploid fibroblasts. Proc Natl Acad Sci USA 87:5104–5108

Twyman R.M. (2005) Gene transfer to animal cells. BIO Scientific. Oxford, UK. pp 1–18.

Underwood P.A, Bean P.A. (1988) Hazards of the limiting-dilution method of cloning hybridomas. J Immunol Methods 107:119–128

Wurm F.M. (2004) Production of recombinant protein therapeutics in cultivated mammalian cells. Nat Biotechnol 22:1393–1398

Wurm F.M, Bernard A (1999) Large-scale transient expression in mammalian cells for recombinant protein production. Curr Opin Biotechnol 10:156–159

Wurm F.M, Petropoulos C.J (1994) Plasmid integration, amplification and cytogenetics in CHO cells: questions and comments. Biologicals 22:95–102

Yang F, Cai S, Xu L, Feng L (1992) Establishment of hybridoma cell line secreting specific monoclonal antibodies against turnip mosaic virus and analysis of properties of the McAb. Chin J Biotechnol 8:247–254

Yoshikawa T, Nakanishi F, Ogura Y, Oi D, Omasa T, Katakura Y, Kishimoto M, Suga KI (2001) Flow cytometry: an improved method for the selection of highly productive gene-amplified CHO cells using flow cytometry. Biotechnol Bioeng 74:435–442

Zeyda M, Borth N, Kunert R, Katinger H (1999) Optimization of sorting conditions for the selection of stable, high-producing mammalian cell lines. Biotechnol Prog 15:953–957

Selection Methods for High-Producing Mammalian Cell Lines

S.M. Browne and M. Al-Rubeai

Abstract The continually expanding market for biotherapeutics such as recombinant proteins that are produced in mammalian cell cultures and the relatively high clinical doses required of these therapeutics is predicted to lead to a bioreactor capacity crunch. Current estimates suggest that by the turn of the decade worldwide bioreactor capacity, currently standing at approximately 500,000 L, will no longer be able to meet demand. Many advances have been made in process design and medium formulation yet there is still scope to improve specific productivities of these manufacturing cell lines. An important step in this process is the selection of a high producing clone from cell lines that are often highly heterogeneous with regard to productivity this can be a difficult task however given the sheer volume of cells that need to be screened. Here we summarise some of the various methods currently available for the isolation of highly productive clonal cell lines.

1 Introduction

Biopharmaceuticals represent over 20% of all NMEs (New Medical Entities) approved by the EU and US regulatory agencies since 2000 (Walsh, 2006). This class of drug consists of recombinant proteins and monoclonal antibodies and although it includes nucleic-acid based products the vast majority of the group are protein based, with approximately 70% of these being glycoproteins. Over half of the biopharmaceuticals approved in recent years have been produced in mammalian cell lines, predominantly in CHO (Chinese Hamster Ovary) cell lines and murine myeloma cell lines such as NS0 and SP2/0. Although mammalian cells lack the vigour and production capacity of other systems such as bacteria or yeast they are currently the system of choice for the production of larger, more complex proteins

S. Browne and M. Al-Rubeai (✉)
School of Chemical and Bioprocess Engineering, and Conway Institute of Biomolecular and Biomedical Research, University College Dublin, Belfield, Dublin 4, Ireland
e-mail: m.al-rubeai@ucd.ie

M. Al-Rubeai (ed.), *Cell Line Development*, Cell Engineering 6,
DOI 10.1007/978-90-481-2245-5_7, © Springer Science+Business Media B.V. 2009

due to their similarity to human cells. They are the only system that can naturally carry out the required post-translational modifications and specific glycosylations needed if these proteins are intended for use as a therapeutic product. Lack of fidelity in glycosylation patterns, in particular N-linked glycosylation, can affect efficacy, stability and immunogenicity of a protein and lead to rapid clearance *in vivo* (Sethuraman and Stadheim, 2006). Although attempts have been made to develop systems that replicate mammalian glycosylation patterns in yeast, insect and plant cells (Hamilton et al., 2006; Hollister and Jarvis, 2001; Cox et al., 2006) this is a difficult process as it involves the removal of innate glycosylation reactions and the introduction of human reactions. To date none have been approved for therapeutic production and mammalian cell lines remain very much to the fore.

However, a feature of mammalian cells is an innate variability/heterogeneity that they display within cell lines in a range of growth characteristics. This heterogeneity has long been evident in mammalian cell culture. As far back as the 1950s in early attempts at isolation of single cells it was noted that clones derived from tissue-culture stocks showed marked differences in nutritional requirements (Sato et al., 1957). Today, many industrially important cell lines such as CHOs, NS0s and hybridomas show a large amount of heterogeneity in features such as growth rate, maximum viable cell number, and cumulative cell time and most importantly in specific production of recombinant protein (Barnes et al., 2001, 2006; Kim et al., 1998, 2001; Marder et al., 1990).

As a consequence of this heterogeneity it is required that any producing cell line of a recombinant protein is clonal, that is, a cell line derived from a single cell (ICH Guideline, 1996). As the recombinant protein of interest is to be produced in living cells there is the possibility that its coding sequence can undergo mutations, thus various guidelines need to be followed to ensure sequence integrity is maintained until the end of culture. Also, as a consequence of this heterogeneity and the impact it can have on specific production values within cell lines there is a need to screen transfected populations for a suitable high producer as the manufacturing cell line.

Many commonly used methods for selecting high-producing cell lines are time-consuming and tedious, most are based on the traditional limiting dilution format. Added to this is the fact that high producers exist in cell populations at very low frequencies – productivity generally follows a log distribution with the majority of cells in a population hovering somewhere near the mean and a small number showing increased specific production. Although effective vector design and gene amplification can increase the incidence of high producers within populations of cells still tens of thousands of clones should ideally be screened to increase the chances of finding a line with favourable characteristics. Most commonly used methods of selection can, in practical terms, only screen a few hundred.

Biopharmaceutical production in mammalian cells is an expensive process with long timelines for process development. Generation of an appropriate host cell line for production needs to be carried out as efficiently as possible without subsequent compromise in quality or productivity. The goal is to create a stable cell line that consistently secretes product at a high level and also exhibits a suitable growth profile – reaching a high maximum viable cell number and maintaining a high level of viability within the bioreactor to extend culture duration and consequently product yield. The timelines in developing such a cell line are extensive and are often prolonged by the inclusion of gene amplification, single cell suspension adaptation or adaptation to protein-free/chemically defined medium. Although these timelines can be decreased by using platform technologies including pre-adapted host cell lines, transfection with the gene of interest and selection of a high-producer with adequate growth profiles is often the longest step in process development.

With approximately 500 new biopharmaceuticals in the pipeline and the relatively large clinical doses that these therapies require production capacity, currently standing at approximately 500,000 L (Butler, 2005), will soon encounter difficulties in meeting the demands of this growing market. Although optimisation of production processes over the years through the development of optimized media formulations, gene-expression systems and process design have resulted in product yields in the region of 5 g/L there is still an urgent need for increased specific productivities from cells i.e. higher yields from smaller/shorter production runs. For this reason, there is a need to develop methods for the selection of high producers in a relatively simple, efficient and cost effective manner. In this chapter we will look at selection methods for high producing cells currently available ranging from early methods of single cell isolation and limiting dilution formats through to high-throughput flow cytometric and fluorescence based methods and recently fully automated systems.

2 Traditional Cloning Methods

2.1 Single Cell Isolation

One of the first difficulties encountered when selecting for a high producing cell line is the isolation of a single cell/clone. Early attempts to isolate single cells were unsuccessful. Cells, even those derived from established continuous cell lines, grow poorly at low densities, and below a minimum density threshold either will not divide or enter senescence after a limited number of divisions (Sanford et al., 1961; Ham and McKeehan, 1979).

The main factors in the reluctance of isolated cells to undergo division are nutritional requirements and the absence of conditioning/growth factors. Medium that is suitable for optimum growth at high cell density is not sufficient for the growth of isolated cells and similarly specific nutrients not required by cells in regular culture are essential for growth and expansion of single cells (Sato et al., 1957). Under normal culture conditions cells are exposed to high levels of various hormones and growth factors, and to growth signals from adjacent cells. Cells growing at high density introduce large amounts of growth-promoting substances into the surrounding medium. These factors and metabolites diffuse out of cells and reach equilibrium with the surrounding medium sufficient to maintain adequate intracellular levels for biosynthesis and metabolism. However, at low population densities metabolite concentrations are too low to condition the surrounding medium effectively and factors inhibitory to cell growth cannot be neutralized (Eagle and Piez, 1962; Ham and McKeehan, 1979).

The first method to successfully establish clonal expansion of a single cell was the capillary tube technique developed by Sanford et al. (1948). Based on the assumption that single cells are unable to condition the surrounding medium sufficiently to promote cell division this method worked by reducing the volume of medium that a cell needed to adapt. Isolated cells were sucked from a Petri dish into a capillary tube about 100 μm in diameter that was then sealed. The low volume of medium inside the capillaries allowed conditioning effects to encourage cell growth. Once growth was established the capillary was opened and transferred to a culture dish into which the new colony of cells would eventually migrate.

Similar techniques developed based on the conditioning of small volumes of medium also included cloning of cells in small hanging droplets of medium (Rittenberg et al., 1986) and in droplets of medium under liquid paraffin (Wildy and Stoker, 1958). Initially, the limiting dilution method (see below) also encouraged growth by the provision of conditioning factors from a layer of irradiated feeder cells. Single cells were grown over layers of these feeder cells which, although non-dividing, continued to secrete growth factors to encourage clonal growth of isolated cells. With the development of better medium formulations it has become possible to isolate cells by this method without the use of feeder cells, however, this is largely dependent on cell type and for certain cell lines cloning efficiency can be quite low.

2.2 *Basic Cloning Techniques*

The cloning technique of choice is largely dependent on the individual properties of a particular cell line. Cloning of adherent cells is relatively easy and can be carried out in Petri dishes, multi-well plates or flasks as it is quite easy to discern individual colonies. Methods for attached cells include the spotting technique, cloning rings, and cloning on foil-bottomed petriperm dishes and subsequent excision of colonies (Clarke and Spier, 1980; Davis, 2002). For suspension cells some form of immobilization is generally required to retain cells as colonies. Cloning of cell suspensions can be carried out by seeding cells into gel, such as agar (Hamburger and Salmon, 1977) or agarose (Ayres, 1982) or a high viscosity solution such as methylcellulose (Davis, 1986). The viscosity ensures that cells are confined to the region where they originated and thus form a colony.

One of the most commonly used methods for clonal isolation is limiting dilution cloning (LDC). Although this method is labour intensive and low-throughput it is still the most commonly employed for selection of high producing clones due to its relative simplicity and low cost. LDC involves diluting a cell suspension to a very low density that when dispensed onto micro-well plates will give an average of >1 cell/well.

Wells are observed microscopically and those containing single cells are marked for later analysis. If the cell remains viable and proliferates then an isolated colony of clones will have been established in the well. Supernatants from these wells can be assayed for the desired product by methods such as ELISA; wells containing a high product titre are then selected for further growth. Following this a second round of cloning is generally recommended to ensure the isolation of a stable population of high-level antibody secreting cells, and to ensure the production of a true clone. This involves further testing of the culture medium for secreted product.

Similar to the method of LDC the Quixell system (Wewetzer and Seilheimer, 1995) is a semi-automated micromanipulator based cloning device. It is based on the same basic process; however, the Quixell system utilizes a robotic dispensing arm and microscopic visualization of the 'picked' cell to ensure efficiency of single cell plating. The system is made up of a motorized microscope stage holding both a donor and receiving culture dish and micropipette connected to an inverted phase-contrast microscope, the stage is operator-controlled by a joystick to position a cell of interest beneath the micropipette. The cell is gently drawn into and expelled from a glass micropipette by manipulation of the volume of air inside by Peltier cooling and heating. Captured cells are then transferred to a receiving dish. Since aspiration and ejection of the cell is carried out under microscopic control it can be guaranteed that each well is receiving a single clone. Although limited in the number of cells that can be picked in one session (approximately 25 cells in 20 min) the duration of cloning is reduced by about 70% (Wewetzer and Seilheimer, 1995) by the fact that each well is guaranteed to contain one cell, thus removing the need for microscopic marking and subsequent rounds of cloning. Cloning efficiency (defined as the percentage of seeded cells that survived single seeding and formed colonies) is similar to that of LDC but can be improved by the selective transfer of cells undergoing cell division (Wewetzer and Seilheimer, 1995). When used to clone cells co-expressing GFP the system showed a cloning efficiency averaging at over 40% and as high as 88% for a number of cell types. It was also capable of selecting rare cells present at frequencies as low as 1/100,000 from mixed populations (Caron et al., 2000).

2.3 *Drawbacks of Traditional Cloning Methods*

Even with the comparably high-throughput method of LDC the process of selecting a high-producing clone with subsequent product analysis can often run to over 8 months and cost millions of euros. Only a few hundred clones can realistically be characterized thereby increasing the chance of missing out on high producers due to the low number of cells screened. As a result of the poor cloning efficiency of mammalian cells a concession is often made by seeding more than 1 cell/well to encourage growth (Borth et al., 2000). Wells initially receiving more than one cell typically have a growth advantage due to the conditioning effects and inevitably cells with the highest growth rate will make up the majority of the population in the well.

Added to this is the fact that cells with a higher growth rate tend to have lower productivities as energy is redirected from recombinant protein synthesis towards proliferation and biomass production thus wells need to be screened and retested to find a high producer. With this in mind LDC is generally carried out two or three times to guarantee clonality of the chosen cell line.

One of the major problems with traditional methods (with the exception of Quixell) is that they make assumptions about the distribution of single cells without taking into account this growth advantage conferred by the presence of multiple cells (Marder et al., 1990). Although they are relatively simple and inexpensive there is no way to be entirely certain that the cell line generated is derived from a single cell, there is always the possibility of a second cell being present and thus cell lines can only be said to have a probability of being clonal. Underwood and Bean (1988) examined LDC for the selection of monoclonal hybridoma cells secreting anti-influenza antibodies from a mixed population. In spite of repeated rounds of LDC, when specificity of antibody from selected 'clonal' lines was analysed it could not be designated as monoclonal. Statistical analysis of the method shows that after two consecutive rounds of LDC a cell's clonality is still not guaranteed (Coller and Coller, 1986). At the same time the approach of direct cell micromanipulation (Quixell), although providing greater certainty of single cell deposition, is seriously prohibitive in terms of time and labour, and by extension cost.

All traditional cloning methods require downstream analysis of product levels to select high producing sub-clones. Protein secretion cannot be measured on an individual cell basis and thus these methods require outgrowth of subclones, analysis of generated lines for specific productivity and stability and subsequent selection. These methods are also constrained by the number of cells that can be feasibly screened. Because high producing cells often exist at very low levels within a population to find a cell line with favourable characteristics tens of thousands of clones need to be evaluated. For all traditional methods of selection this is not possible.

3 Flow Cytometric Methods

Flow cytometry and cell sorting has greatly enhanced the ability to select for high producing clones by considerably increasing screening capacity. Flow cytometry can screen several million cells in a short time and isolate sub-populations and single cells from heterogeneous populations, even rare cells present in populations at extremely low frequencies. Over the years flow cytometry has played a significant role in mammalian cell culture as a research tool (Al-Rubeai et al., 1991; Al-Rubeai and Emery, 1993) and is becoming increasingly important industrially. With the advent of online systems monitoring of cell cultures for the production of biopharmaceuticals can be carried out with much greater precision than by manual methods ultimately leading to improved processes and increased production (Kacmar and Srienc, 2005; Sitton et al., 2006; Sitton and Srienc, 2008; Zhao et al., 1999).

Flow cytometry facilitates the selection of high producing cells by allowing qualitative and quantitative analysis of protein expression. A number of methods have been developed to measure this expression on a per cell basis including measurement of cell surface expression of a protein of interest or reporter protein; measurement of intracellular fluorescence as from a reporter protein such as GFP (green fluorescent protein); or by retaining secreted protein in the vicinity of the cell for subsequent fluorescent staining and isolation such as gel microdrop and matrix based secretion assays. Below is a summary of some of these methods.

3.1 Cell Surface Expression

Flow cytometry has been a valuable tool in conjunction with hybridoma technology for the analysis of monoclonal antibody surface expression. Particularly in the isolation of hybridomas expressing specific antigen (Parks et al., 1979), and also in the isolation of bi-specific hybridomas (Karawajew et al., 1987) and isotype switch variants (Dangl et al., 1982). In terms of the selection of high producing cell lines a number of researchers have suggested a link between the level of cell surface expression of recombinant proteins and corresponding cellular productivity. In certain hybridoma lines a correlation can be seen between cell surface expression levels and levels of secreted antibody. Sen et al. (1990) and McKinney et al. (1995) demonstrated that surface antibody fluorescence patterns closely follow specific production for a number hybridoma cell lines. Marder et al. (1990) took this idea one step further by separating a hybridoma population into different groups based on fluorescence intensity corresponding to surface bound antibody and comparing fluorescence levels with specific antibody production of the ensuing clones. By separating cells into groups based on "dim, medium, and high" fluorescence they showed a correlation between surface expression and productivity – "dim" fluorescers showed no measurable antibody production, while the medium and high groups showed successive increases in production. However, not all hybridomas show this correlation (Meilhoc et al., 1989), and even different isotypes derived from the same line show variation in surface antibody staining (Marder et al., 1990). With some hybridomas a correlation existed at only certain phases of culture (Leno et al., 1991).

Selection based on a correlation between productivity and surface expression is not confined to hybridomas and has also been demonstrated in CHO cells stained with specific antibodies at low temperatures (Brezinsky et al., 2003). This method relies on the transient association of recombinant proteins with the cell surface during secretion. By carrying out the process at 4 C the aim was to extend the time that secreted protein remained associated with the cell membrane. Fluorescence of stained cells was found to be stable for over 1 hour. Twenty–three CHO cell lines were developed using this method, cells were stained with specific antibody and sorted based on highest fluorescence intensities. Increases were reported of 20–25-fold in specific production in the absence of gene amplification, and of 120-fold in conjunction with gene amplification. This method was also applied to hybridoma and NS0 cells with similar results.

Other selection methods based on cell surface expression have used co-expression of a cell surface reporter molecule along with the protein of interest to select for high producers (DeMaria et al., 2007). This method was described in CHO cells expressing various protein therapeutics including antibodies, a soluble receptor, and a glycoprotein hormone. Each was co-expressed from the same gene construct with CD20, a cell surface protein that is not normally expressed in CHO. Both genes are linked by an IRES (internal ribosome entry site) so that they are transcribed on the same mRNA but are translated separately. As they are transcribed on the same mRNA expression of the cell surface molecule correlates with recombinant protein expression and thus by staining cells with a fluorescent conjugated antibody to CD20 high producers can be selected based on relative fluorescence intensity. A major advantage of this method is that it doesn't rely on the availability of an antibody to the particular recombinant protein being expressed. And although a concern regarding this method would be a reduced level of expression of the therapeutic protein due to cellular resources being re-routed towards production of the co-expressed protein the lower translation efficiency of IRES-mediated translation relative to 5′ cap-mediated translation (Mizuguchi et al., 2000) means that reporter protein expression is much lower than that of the transgene and thus does not impose stress on the cell. Instability in protein expression from the resulting cell lines can also be monitored over time by flow cytometry quite easily without the need for secreted protein measurement by methods such as ELISA.

Another method for high producer isolation by co-expression of a cell surface capture molecule is the FASTR (Flow cytomerty-based autologous secretion trap) system developed by Regeneron ™ as a part of their VelociMab™ suite of technologies (www.regeneron.com/velocimab). With this method a membrane-bound cell surface capture molecule is inducibly expressed that binds the secreted protein of interest to the cell surface where it can be bound by a detection molecule and high producers isolated by sorting (Fig. 1). The use of a blocking molecule prevents "cross-talk" between cells expressing the protein of interest and non-expressing cells. The complex formed between the cell surface capture protein and secreted protein of interest is continuously internalized by the cell and replaced with new complexes, thus an equilibrium is established that removes the problem of saturation that is found with some surface display systems. This leads to a better correlation between fluorescence intensity and productivity. This system also allows for monitoring of expression stability over time.

3.2 Intracellular Reporter Proteins

In the absence of a suitable surface expression marker cells can be isolated based on levels of intracellular proteins using selectable markers such as GFP that are

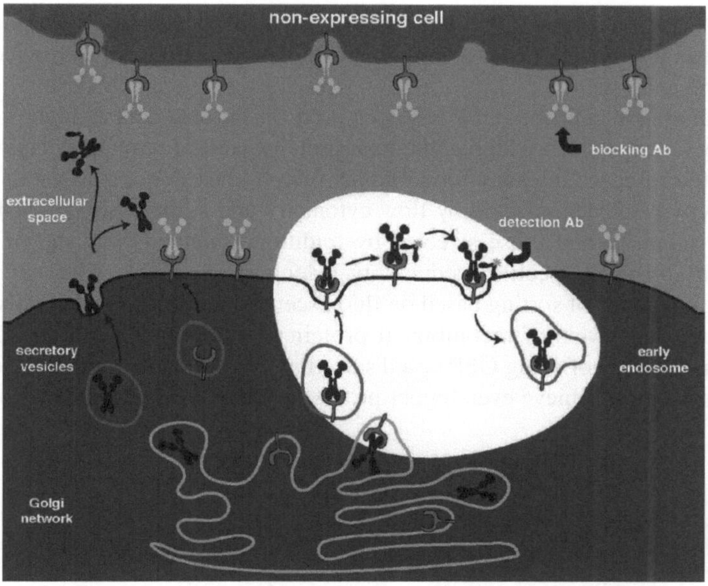

Fig. 1 FASTR (Flow cytometry-based autologous secretion trap) technology. Recombinant protein/antibody is co-expressed with a cell-surface capture protein which binds secreted protein and displays it on the cell surface for detection with a complementary detection antibody. The addition of a blocking antibody prevents binding of secreted protein to adjacent non-expressing cells (courtesy of James Fandl, Regeneron Inc)

co-expressed with the protein of interest or fluorescent detection molecules capable of permeating the cell surface without ensuing damage to the cell such as fluorescent methotrexate.

3.2.1 Green Fluorescent Protein

The green fluorescent protein (GFP) gene, from the jellyfish *Aequeorea victoria* has become an important reporter molecule of gene expression not just in mammalian cell culture, but also bacterial, fungal and insect systems and in transgenic animals and plants (Blumenthal et al., 1999; Chalfie et al., 1994; Plautz et al., 1996; Sarramegna et al., 2002; Suzuki et al., 2006). GFP is a naturally fluorescent protein that doesn't require any other enzymes, co-factors or substrates for its fluorescence and thus it has the advantage that expression can be measured in real-time in live cells.

A quantitative relationship has been confirmed between the level of GFP expression and its fluorescence intensity (Fig. 2c) (Meng et al., 2000; Subramanian and Srienc, 1996) thus making it feasible to select for high producers on the basis of high GFP fluorescence if the gene is co-expressed with a recombinant protein.

Also, as selection using GFP doesn't require any subsequent staining or manipulation of cells it may be a better alternative to antibody staining and subsequent flow cytometric analysis.

In mammalian cell lines GFP has been used for the selection of high producing clones by co-expression with recombinant proteins and subsequent selection based on GFP fluorescence intensity for a number of cell lines. A comparison of specific productivity of clones selected by flow cytometry and sorting based on GFP fluorescence intensity and those selected by traditional methods has shown up to a 6-fold difference in specific productivity (Meng et al., 2000). Most studies use consecutive rounds of sorting based on fluorescence intensity to yield pools of cells with increasing levels of recombinant protein production (Fig. 2a, b). There is also the possibility of using GFP based selection in conjunction with a gene amplification system to achieve even higher product yields. This has been demonstrated

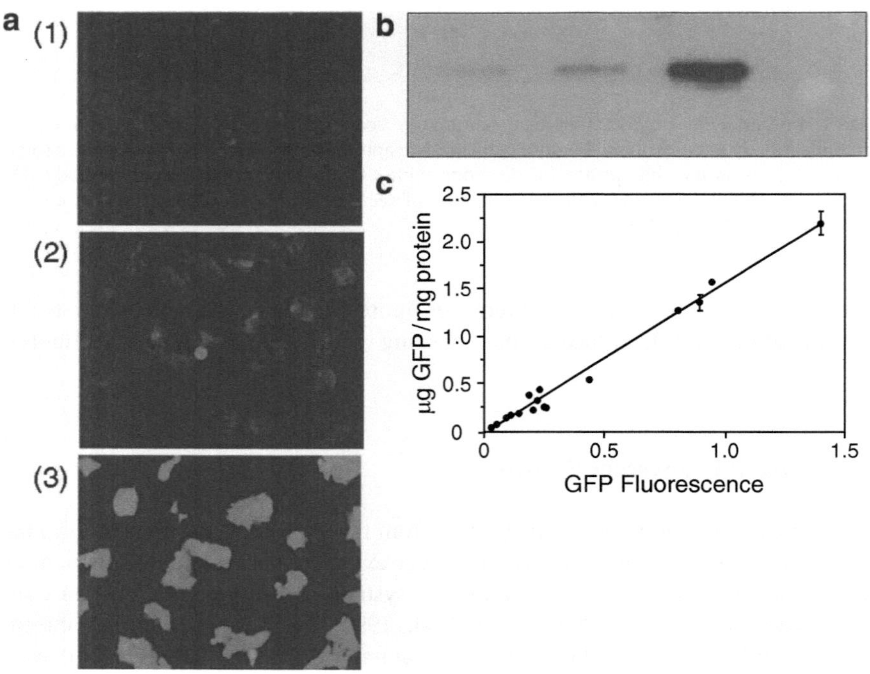

Fig. 2 Green fluorescent protein (GFP) selection of high-producing cells. (**a**) GFP fluorescence intensity monitored by fluorescence microscopy in human embryonic kidney (HEK) cells sorted based on GFP intensity, (1) unsorted cells, (2) following first GFP sort, (3) after final sort. (**b**) Corresponding recombinant protein expression measured by western blot (1) following transfection, (2) following antibiotic selection, (3) following GFP enrichment, (4) untransfected cells (adapted from Mancia et al., 2004). (**c**) Correlation between GFP expression and fluorescence intensity. GFP protein in cell lysate measured by ELISA correlated with GFP fluorescence intensity measured by flow cytometry in GFP expressing CHOK1 cells (from Meng et al., 2000)

in CHO cells using co-expression of GFP with dihydrofolate reductase (dhfr) (Meng et al., 2000) and metallothionein (Bailey et al., 2002) gene amplification systems.

A further development of this process focuses on cells expressing more than one recombinant protein, or separate fragments of a protein such as heavy and light chains of monoclonal antibodies, or multi-fragment proteins co-expressed with two different fluorescent proteins. This two colour fluorescence system incorporates GFP and another autofluorescent protein such as yellow fluorescent protein (YFP), a mutant form of GFP, or red fluorescent protein (RFP), derived from the coral *Discosoma*. Respective fluorescent genes are linked via IRES to genes for recombinant proteins or protein fragments, and only double positive cells showing a high level of fluorescence for both colours are sorted. In CHO cells a combination of GFP and YFP connected respectively to heavy and light chain genes of a recombinant antibody via an IRES in combination with metal amplification resulted in clones displaying a 30-fold increase in antibody production compared to the parental population (Sleiman et al., 2008). Another study by Assur et al. (2007) showed similar results using plasmids that linked GFP to heavy chain genes and RFP to light chain genes for a number of antibody Fab fragments in human embryonic kidney (HEK) cells (Fig. 3).

Co-expression with GFP or related fluorescent proteins is also useful in monitoring stability of protein production from cultures over time. However, both Assur et al. (2007) and Meng et al. (2000) report a fall off in GFP fluorescence intensity over time that is not necessarily linked to instability in recombinant protein production.

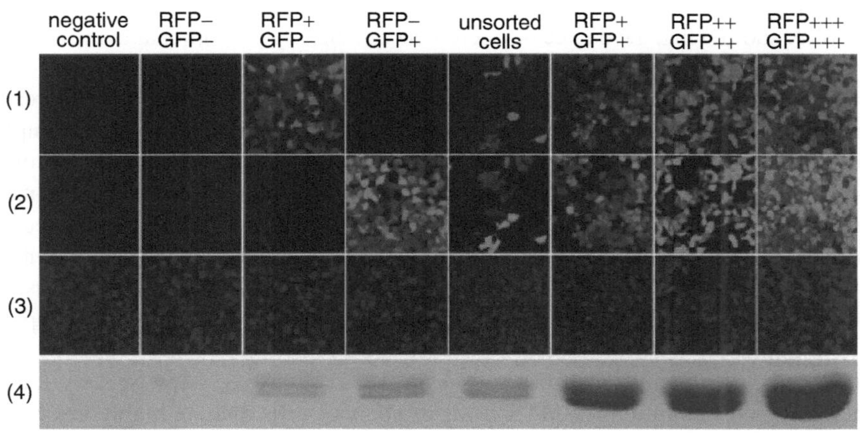

Fig. 3 Two-colour fluorescence selection. Cells co-expressing two recombinant proteins or protein fragments each linked to a fluorescent protein – GFP or RFP. Cells were sorted based on fluorescence intensity of each, or both, fluorescent markers: – no fluorescence, + low fluorescence, ++ medium fluorescence, +++ high fluorescence. Sorted cells were analyzed by fluorescence microscopy in the emission channels corresponding to (1) RFP, (2) GFP, (3) the nuclear stain TOTO-3. Corresponding Fab expression is shown in row (4) (from Assur et al., 2007)

Also, care must be taken to determine the optimum time-point for sorting based on GFP fluorescence and an analysis of the protein production kinetics of each producing cell line should be carried out. A study by Zeyda et al. (1999) has shown a "dilution" of fluorescence intensity in actively growing cultures. GFP concentration in growing mouse fibroblast cells was very low up until the time when cells reached confluency, at which point concentration increased significantly. Although GFP was constitutively expressed this dilution was attributed to the simultaneous production of biomass and only after biomass production has stopped could the intracellular GFP concentration increase. This feature was shown to be cell line specific, with liver carcinoma cells transfected with the same GFP plasmid showing a relatively constant level of GFP concentration as batch culture progressed.

Another drawback to the use of GFP as a marker of productivity is the relatively low fluorescence intensity exhibited, which can sometimes be indistinguishable from autofluorescence. However, the development of modified forms of GFP to improve expression efficiencies, increase fluorescence intensity and thermostability, and reduce photobleaching (Heim and Tsien, 1996; Siemering et al., 1996) may go some way to offset this issue.

3.2.2 Fluorescent Methotrexate

Methotrexate is a 4-amino analog of folic acid and cells that are resistant to this agent show increased levels of the enzyme dihydrofolate reductase (DHFR). The fluoresceinated-methotrexate (F-MTX) staining method was initially developed by Kaufman et al. (1978) in order to study the heterogeneity in DHFR levels in various murine cell lines. Staining of cell lines showing varying levels of resistance with a FITC-conjugate of MTX showed that the degree of fluorescence is directly proportional to the activity of dhfr in that cell line occurring as a result of amplified gene copy number.

The discovery that MTX resistance of cells correlates to a proportional increase in *dhfr* gene copy number eventually led to the development of the *dhfr* gene amplification expression system in *dhfr* deficient CHO cells. This system has become one of the most common expression systems for the production of recombinant protein from mammalian cells. Cells deficient in endogenous DHFR are transfected with an expression construct containing the *dhfr* gene and gene for the recombinant protein of interest. Rounds of selection are carried out in gradually increasing concentrations of MTX eventually yielding populations that can contain up to 1,000 copies of the transfected construct and greatly increasing yields of any co-transfected proteins (Kaufman and Sharp, 1982). However, stepwise selection in increasing concentrations of MTX leads to pools of cells that are highly heterogeneous with respect to productivity. These pools can often contain low and non-producing cells that have acquired resistance to MTX and thus some form of cloning and screening needs to be carried out to select for the highest producers.

As F-MTX can permeate the cell membrane without adverse effect to bind quantitatively to DHFR cells with the highest fluorescence intensity can be sorted to yield populations with high copy number *dhfr*. Yoshikawa et al. (2001) have shown this for dhfr deficient CHO cells producing human granulocyte macrophage colony stimulating (hGM-CSF) factor via a *dhfr* expression cassette. Cells were sorted into fractions based on fluorescence intensity and a strong correlation could be seen between F-MTX intensity and DHFR activity for all groups. Although a correlation between specific production of hGM-CSF and F-MTX intensity could be seen this was dependent on the chromosomal location of gene amplification. A correlation with F-MTX intensity was restricted to cells showing integration and gene amplification close to the telomeres. This group of cells also showed increased specific productivity and stability of production.

3.3 Methods Based on Cell Secretion Rate

As well as intracellular reporter detection some flow cytometric systems have been developed based on the amount of recombinant protein actively secreted from the cell. As flow cytometry cannot be used to measure protein that is not cell associated and secreted products quickly dissociate from the cells that produce them, the secreted protein needs to be somehow retained at the cell surface for measurement. The development of methods such as matrix-based secretion assays and gel microdrop technology have facilitated this.

3.3.1 Gel Microdrop Technology

Gel microdrop technology uses specialised equipment to isolate single cells within small droplets. Settings can be optimised to yield microscopic droplets ranging from 10 to 100 μm in diameter. Molten gels (such as agarose or alginate) are biotinylated, mixed with a low density cell suspension and then emulsified to generate droplets containing a matrix that can capture secreted antibody. This matrix is formed by the addition of a product-specific 'capture' antibody that binds to the biotinylated matrix via an avidin linker. Secreted protein is captured on the matrix and detected by a fluorescently conjugated antibody or specific antigen and high level secretors can be isolated by sorting based on fluorescence intensity. The porosity of the beads means that molecules of up to 500 kDa can easily and rapidly diffuse through it while specific product remains bound on the matrix (Weaver et al., 1997).

This method was first described by Weaver (1986) and was initially used to detect and quantify bacteria. Applied to mammalian systems it has been used to separate sub-populations of cells based on the production of cytokines (Atochina et al., 2004; Turcanu and Williams, 2001), for cytotoxicity assays (Bogen et al., 2001) and in populations of hybridomas to separate producers from non-producers (Powell and Weaver, 1990) or for the isolation of antigen specific hybridomas from mixed populations (Gray et al., 1995; Kenney et al., 1995).

Its use in the selection of high producing clones has resulted in a 2–5-fold increase in specific productivity of enriched subpopulations (Weaver et al., 1997).

The system has also been used to monitor stability of recombinant protein expression in CHO cells producing a recombinant antibody; here it was used to isolate emerging variant low/non-producing populations that had acquired a metabolic advantage over the course of long-term fermenter culture (Hammill et al., 2000).

Advantages of this system are the greater restriction on product diffusion and higher saturation levels than other cell surface staining or matrix based secretion assays – the surface area of the droplet means that many more capture sites can be generated than on the cell surface and the saturation limit is theoretically an order of magnitude higher than that of matrix-based assays (Frykman and Srienc, 1998) (see below). However, occupancy of the microdroplets is determined by Poisson statistics, thus there is no guarantee of single cell occupancy and in order to increase chances of droplets containing single cells a very low cell density cell suspension is used with the result that only about 10–15% of beads actually contain cells. Also, the process of encapsulation and subsequent removal from droplets may adversely affect cell viability.

3.3.2 Matrix-Based Secretion Assays

A similar approach is the matrix based secretion assay or "affinity capture surface display". Rather than enclosing cells in a droplet the capture matrix is generated on the cell surface. This method utilizes the naturally high affinity of biotin and avidin. Cells are labelled with biotin which readily binds at primary amines on cell surface residues. Biotinylated cells can then be bound directly with an avidinated capture antibody (Manz et al., 1995) or via an avidin bridge to a biotinylated capture antibody which is specific to the secreted product (Holmes and Al-Rubeai, 1999). The use of an avidin linker maximises the binding capacity of the matrix as avidins have four binding sites for biotin, thus increasing threefold the number of binding sites for the biotinylated capture antibody and, in turn, the surface area of the capture matrix. Also biotinylated ligands are generally more readily available on the market than avidinated ones.

Once the matrix has been generated, cells are incubated in a high-viscosity medium (containing gelatine or agar) that minimises diffusion of secreted protein – this ensures that secreted protein binds to the matrix of the cell it was secreted from and not that of a neighbouring cell. Bound product is subsequently labelled with a fluorescent tag and the highest producing cells, i.e. those with the most secreted product attached and consequently the highest levels of fluorescence can be separated by cell sorting (Fig. 4).

Manz et al. (1995) demonstrated the efficacy of this method by successfully separating heterogeneous pools of producers and non-producers for both hybridomas secreting IgM (~600 kDa) and activated T-lymphocytes secreting IFN-γ (~34 kDa active homodimer), showing the utility of the method for capturing secreted products over a wide size range. The affinity capture surface display (ACSD) method (Holmes and Al-Rubeai, 1999) applied to NSO cells producing a recombinant antibody simplified the assay by using commercially available reagents, increased the surface area of the capture matrix by using an avidin linker to generate more binding sites,

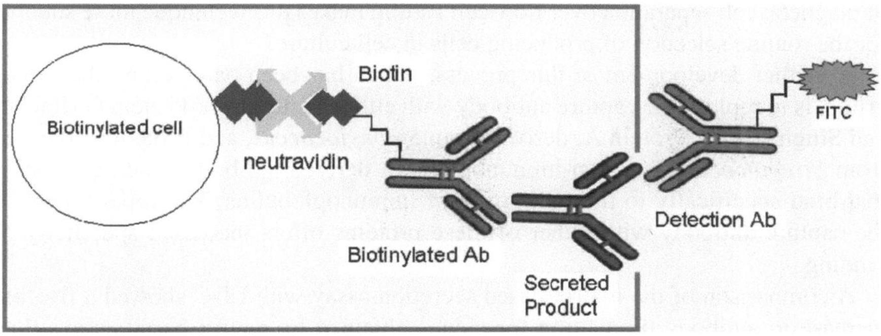

Fig. 4 Affinity capture surface display. A biotinylated "capture"-antibody specific to the secreted protein product is linked to biotinylated cells via a neutravidin bridge which increases the binding capacity of the matrix. This is carried out in a high viscosity medium (represented by the *red box*) to minimize "cross-talk" between cells. Secreted product is detected by a fluorescently labeled "detection"-antibody, in this case FITC (Carroll and Al-Rubeai, 2005)

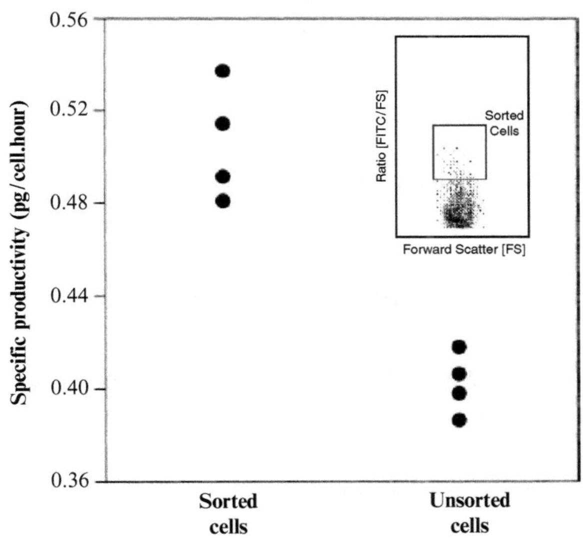

Fig. 5 Affinity capture surface display. The highest 10% of cells were sorted based on Fluorescence (FITC) to cell size (FS) ratio (*inset*) in a NS0 cell line expressing a chimeric antibody. Antibody expression of sorted clones showed a 30% increase compared to clones from the parental population as measured by ELISA (adapted from Holmes and Al-Rubeai, 1999)

and also reduced the chances of steric hindrance by using a modified biotin moiety that contained a 24 Å spacer arm. Sorting cells with the highest 10% ratio of fluorescence to forward scatter (i.e. taking into account the fact that larger cells will have a higher fluorescence due to increased surface area) yielded cells with 30% higher specific productivity than the original population (Fig. 5). Carroll and Al-Rubeai (2005) combined the ACSD method with MACS magnetic separation to isolate antibody producing cells from non producing cells in NSO culture. The advantages of magnetic cell separation over flow cell sorting makes this technique more suitable for the routine selection of producing cells in cell culture.

A further development of this process, which has been patented by the Lonza group, is to replace the capture antibody with either Protein A or Protein G (Racher and Singh, 2003). Protein A, derived from *Staphylococcus*, and Protein G, derived from *Streptococcus*, are non-immunoglobulin derived antibody binding proteins that bind specifically to the Fc portion of immunoglobulins. The replacement of the capture antibody with either of these proteins offers increased specificity of binding.

A comparison of the matrix based secretion assay with LDC showed a fivefold increase in antibody production for clones obtained by matrix based assays than and the highest secreting clone obtained by the limiting dilution method, indicating both higher efficiency (five 96-well plates were needed as opposed to 45 with LDC) and a decrease in the time required for screening (Borth et al., 2000).

As with gel microdrop assays matrix based secretion assays can be used to measure any product to which a complementary fluorochrome conjugated antibody is available. There is also the option of using fluorescently labelled specific antigen – although a limitation is the fact that conditions need to be optimised for each cell line analyzed. Also, for cell surface assays the matrices will only remain cell associated for a limited amount of time. However in spite of this and a lower saturation limit than that of the gel microdrop, this method has the advantage that all cells treated are labelled and thus the number of cells screened is increased, typically in the region of a few million cells. Also, cells are not exposed to encapsulation treatment and subsequent decapsulation of cells after the assay.

4 Fluorescent Methods and Automated Systems

4.1 HTRF (Homogeneous Time Resolved Fluorescence) Based MAb Assay

Fluorescence Resonance Energy Transfer (FRET) is a distance-dependent interaction between two fluorophores, a donor and an acceptor. In their excited state excitation is transferred from the donor to the acceptor if the two molecules are in close proximity leading to light emission from both. These interactions occur over distances in the region of 10–100 Å making them suitable to measure interactions at a molecular level. The application of FRET to molecular and cell biology has become widespread in recent years spanning a range of functions such as analysis of protein–protein or DNA–protein interactions, protein folding and assembly of protein complexes, signal transduction pathways and cellular localization and compartmentalization (Eidne et al., 2002; Stühmeier et al., 2000; Johnson, 2005; Aoki et al., 2008; Chen et al., 2003).

Homogeneous time-resolved fluorescence (HTRF) technology is a combination of fluorescence resonance energy transfer (FRET) and time-resolved fluorescence (TRF) whereby fluorescence is measured as a function of time after excitation; this coupling with TRF minimizes background and increases the sensitivity of the assay. This technology has been adapted into a 96-well plate ELISA format to measure antibody or Fc fragment production using a europium-cryptate donor and a modified allophycocyanin acceptor (Idusogie et al., 2008). When brought together some europium emission energy is released as light at 620 nm and this transfers energy to the allophycocyanin acceptor which emits fluorescence at 665 nm. Results are reported as the ratio of the 665 nm and 620 nm signals thus correcting for background interference.

To carry out the assay culture supernatants are mixed with HTRF reagent in a competitive binding assay. Allophycocyanin labeled Protein A binds antibody or Fc fragments in samples displacing the binding of a europium labeled antibody and thereby decreasing the FRET signal. Thus the fluorescent signal is inversely proportional to sample concentration. One issue, however, is that antibodies of a similar size don't necessarily display similar fluorescence profiles for relative concentrations, probably owing to differences in conformation and resulting accessibility to binding sites. While this is not an issue when simply ranking clones in terms of productivity it is important that an appropriate standard is available for accurate determination of protein concentration.

The HTRF assay is relatively simple requiring just the addition of reagents to wells followed by incubation and reading and the cost is comparable to that of ELISA. However, the added cost of manual input/high throughput systems for the required outgrowth and screening of a sufficient number of clones also needs to be taken into account, meaning this system remains on a par with similar established methods of cloning and supernatant screening by immunoassay.

4.2 *Laser-Enabled Analysis and Processing*

The Laser-Enabled Analysis and Processing (LEAP) system from Cyntellect is a high-throughput cell analysis system using laser-based negative selection of cells. Cells immobilized on a capture matrix in special culture dishes in the presence of Protein G are allowed to secrete protein which is fluorescently labeled. Cells of interest are identified based on the fluorescence intensity and a process of negative selection is employed to isolate them from heterogeneous populations by eliminating unwanted surrounding cells using laser ablation then allowing selected cells to proliferate. Image analysis software can quantify secreted protein surrounding each cell/colony and means clonality can be verified and automated selection can be based on pre-programmed criteria – for example colonies are selected not just on the basis of highest fluorescence but on fluorescence as a function of colony size. The system can be applied to both adherent and suspension cells as sample movement is minimized (Koller et al., 2004).

Applications of the LEAP system include manipulation of live cells and optoinjection of biomolecules such as siRNAs and plasmids (Clark et al., 2006; Szaniszlo et al., 2006). Applied to the selection of high-producers LEAP has been reported to routinely obtain clones with specific secretion rates of >50 pg/cell/day representing increases of 5–20-fold on parental populations and a decrease in heterogeneity in subsequent cell lines (Hanania et al., 2005). LEAP is particularly amenable to adherent cell lines which may be difficult to process by flow cytometry. A full account of this system is given in Chapter 6.

4.3 Automated Colony Picking

Based on a similar format to the LEAP system are automated colony pickers such as the ClonePix system from Genetix (www.genetix.com) and the CellCelector™ from Aviso (www.aviso-gmbh.de). Immobilized cells are stained and relative fluorescence of colonies is determined, although rather than destroying unwanted cells, cells and colonies of interest are "picked" and transferred to fresh wells. These systems allow picking of clonal colonies of interest, or in the case of the CellCelector single cells, based on a variety of criteria ranging from cell size, shape, and proximity to neighboring colonies, to quantitative protein secretion or specific protein production.

Initially used for the selection of bacterial or fungal colonies, the ClonePix has since been applied to the selection of mammalian cell lines. Using image analysis software cells can be monitored over time and then selected based on a wide range of criteria such as proliferation rate, cell size, expression of specific markers or other aspects of morphology. Fluorescent staining with antibody or specific antigen can be used for selection based on secretion of a specific protein, total protein secretion, or, membrane bound/intracellular fusion proteins. Antibody secreted by the colony is detected by a fluorescently labelled specific antigen, which can be added and allowed to diffuse through the semi-solid medium. The fluorescence can be viewed as a 'halo' around the colony which is imaged. Data from white light and fluorescent images are merged and fluorescence is normalised to colony size. Also, by merging data from white light and fluorescent images the system can exclude fluorescent colonies that are too close to non-fluorescent/non-producing colonies, and thus prevent contamination with non producing cells (Fig. 6).

4.4 Fully Automated Robotic Systems

Automation of previously manually performed procedures such as cell line maintenance and expansion and preparation of bioassay samples is becoming widespread with many high throughput systems available. The pinnacle in high-throughput systems for optimized cell line selection are fully automated systems such as the

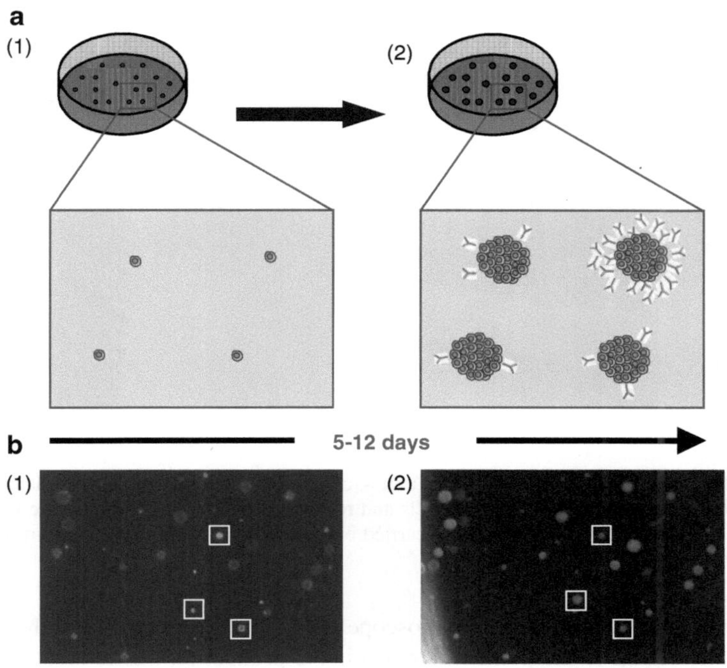

Fig. 6 ClonePix selection process. (**a**) (1) Cells are plated at low density in semi-solid medium and incubated to allow discrete colonies to form (2) Protein secreted from isolated colonies is detected using a fluorescently labeled detection antibody. (**b**) Fluorescent (1) and corresponding white light (2) image of fluorescently labeled colonies. High producers are selected based on the ratio of fluorescence intensity to colony size (courtesy of Chris Mann, Genetix)

Cello robotic cell culture system from The Automation Partnership (TAP) (http://automationpartnership.com) or the Cellerity™ system from TECAN (http://www.tecan.com/celerity). The function of these systems is really maintenance of cultures and sampling for downstream bioassays rather than direct identification of high producers. They can process hundreds of plates and multiple experiments in parallel and although they are not necessarily advancing the knowledge leading to identification and selection of highly productive cells they are vastly expanding the capacity of largely traditional screening formats.

The Cellerity™ system is a custom built system with different configurations depending on throughput and capacity requirements that can handle anywhere between 6 and 1,536 well plates or adapted culture flasks. This system processes adherent cells based on pre-programmed criteria with all liquid handling carried out in a laminar flow "clean bench".

The Cello system (Fig. 7) on the other hand is an entirely closed system contained within a negative pressure laminar flow cabinet. Bulk pools of transfectants are fed into the system and seeded automatically into plates. Clonality of plated

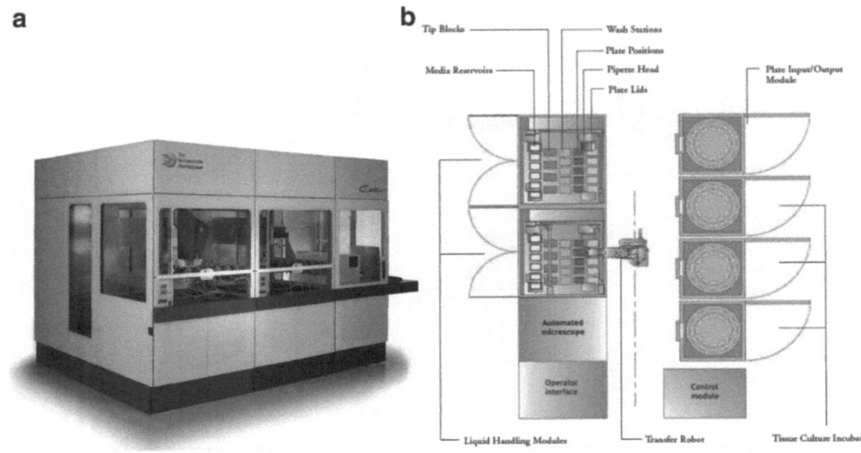

Fig. 7 The Cello system. (**a**) The Cello unit – contained in a negative pressure laminar flow cabinet. (**b**) Cello workflow – pools of cells and reagents are fed in and samples are fed out for analysis, all other cell culture actions are carried out robotically within the unit (courtesy of Tim Ward, The Automation Partnership)

cells is verified by an integrated microscope and all further processing is then carried out robotically up to the point where a suitable production line is identified. This includes medium handling, expansion of cells into fresh wells or plates, incubation of cells, measurement of cell growth; collection of cell/supernatant samples and removal for assays, or, collection of cell samples for banking; dilution sub-cloning of lines with favourable characteristics; and medium changes. It is applicable to both adherent and suspension cells lines and as with Cellerity™ can carry out enzymatic detachment of adherent cells. The Cello system also has the advantage of decision-making scheduling software that can interpret growth and assay data along with pre-programmed parameters to decide the next processing step without the need for operator intervention.

5 Concluding Remarks

The selection of high-producing cell lines is a significant economic factor in the industrial production of biopharmaceuticals, both in terms of development timelines and bioreactor capacity. Although there are many selection methods available each comes with a list of pros and cons and, with the exception of LDC, no method has established itself as a standard across the board. Traditional cloning and product analysis methods simply do not provide the scope necessary to select and screen the volume of cells required to find sub-clones with attractive characteristics in a reasonable time. The development of flow cytometric protocols has greatly increased the throughput of selection; however, these methods require optimization for individual cell lines and may not be suitable for less robust cell lines. Automated systems reduce the time required to select for cells with suitable characteristics while at the same time increasing the efficiency of the process significantly by increasing the number of clones that can be screened, however, although the capacity of screening is greatly increased and manual input is minimized these systems are still based on relatively simple screening methods.

As yet no method has been developed based on an understanding of the molecular basis of productivity. Much effort has recently been focused on transcriptomic and proteomic analysis of recombinant mammalian cell lines and an insightful review on this topic has been provided by Seth et al. (2007). Their conclusion that high productivity is not regulated by "master controllers" but rather by the accumulation of a number of subtle complementary changes in many pathways – not just protein secretion, but including energy metabolism, redox balance and control of cell death and proliferation, means it is unlikely that we will find a simple indicator or "bio-marker" of productivity. Thus the traditional format of cloning, outgrowth and screening will still be required although it can now be conducted in a highly automated manner. Although wasteful and expensive an easily implemented, simple and viable alternative has yet to present itself.

Acknowledgements We thank James C. Weaver, Division of Health Sciences and Technology, MIT; Tim Ward, The Automation Partnership; and James Fandl, Regeneron Pharmaceuticals for provision of information. We also thank Science Foundation Ireland (SFI) for funding.

References

Al-Rubeai M, Emery AN (1993) Flow cytometry in animal culture. *Biotechnology* 11:572–574

Al-Rubeai M, Emery AN, Chalder S (1991) Flow cytometric study of cultured mammalian cells. *J Biotechnol* 19:67–81

Aoki K, Kiyokawa E, Nakamura T, Matsuda M (2008) Visualization of growth signal transduction cascades in living cells with genetically encoded probes based on Förster resonance energy transfer. *Philos Trans R Soc Lond B Biol Sci* 363:2143–2151

Assur Z, Schieren I, Hendrickson WA, Mancia F (2007) Two-color selection for amplified co-production of proteins in mammalian cells. Protein Expr Purif 55:319–324

Atochina O, Mylvaganam R, Askelband Y, McGrath P (2004) Comparison of results using the gel microdrop cytokine secretion assay with ELISPOT and intracellular cytokine staining assay. Cytokine 27:120–128

Ayres KN (1982) High cloning efficiency of human lymphoid cells in agarose without feeder layers. J Natl Cancer Inst 68:919–923

Bailey CG, Tait AS, Sunstrom NA (2002) High-throughput clonal selection of recombinant CHO cells using a dominant selectable and amplifiable metallothionein-GFP fusion protein. Biotechnol Bioeng 80:670–676

Barnes LM, Bentley CM, Dickson AJ (2001) Characterization of the stability of recombinant protein production in the GS-NS0 expression system. Biotechnol Bioeng 73:261–270

Barnes LM, Moy N, Dickson AJ (2006) Phenotypic variation during cloning procedures: analysis of the growth behaviour of clonal cell lines. Biotechnol Bioeng 94:530–537

Blumenthal A, Kuznetzova L, Edelbaum O, Raskin V, Levy M, Sela I (1999) Measurement of green fluorescent protein in plants: quantification, correlation to expression, rapid screening and differential gene expression. Plant Sci 142:93–99

Bogen KT, Enns L, Hall LC, Keating GA, Weinfeld M, Murphy G, Wu RW, Panteleakos FN (2001) Gel microdrop flow cytometry assay for low-dose studies of chemical and radiation cytotoxicity. Toxicology 160:5–10
Borth N, Zeyda M, Kunert H, Katinger H (2000) Efficient selection of high-producing subclones during gene amplification of recombinant Chinese hamster ovary cells by flow cytometry and cell sorting. Biotechnol Bioeng 71:266–273
Brezinsky SC, Chiang GG, Szilvasi A, Mohan S, Shapiro RI, MacLean A, Sisk W, Thill G (2003) A simple method for enriching populations of transfected CHO cells for cells of higher specific productivity. J Immunol Methods 277:141–155
Butler M (2005) Animal cell cultures: recent achievements and perspectives in the production of biopharmaceuticals. Appl Microbiol Biotechnol 68:283–291
Caron AW, Massie B, Mosser DD (2000) Use of a micromanipulator for high-efficiency cloning of cells co-expressing fluorescent proteins. Methods Cell Sci 22:137–145
Carroll S, Al-Rubeai M (2005) ACSD labelling and magnetic cell separation: a rapid method of separating antibody secreting cells from non-secreting cells. J Immunol Methods 296:171–178
Chalfie M, Tu Y, Euskirchen G, Ward WW, Prasher DC (1994) Green fluorescent protein as a marker for gene expression. Science 263:802–805
Chen Y, Mills JD, Periasamy A (2003) Protein localization in living cells and tissues using FRET and FLIM. Differentiation 71:528–541
Clark IB, Hanania EG, Stevens J, Gallina M, Fieck A, Brandes R, Palsson BØ, Koller MR (2006) Optoinjection for efficient targeted delivery of a broad range of compounds and macromolecules into diverse cell types. J Biomed Opt 11:014034
Clarke JB, Spier RE (1980) Variation in the susceptibility of BHK populations and cloned cell lines to three strains of foot-and-mouth disease virus. Arch Virol 63:1–9
Coller HA, Coller BS (1986) Poisson statistical analysis of repetitive subcloning by the limiting dilution technique as a way of assessing hybridoma monoclonality. Methods Enzymol 121:412–417
Cox KM, Sterling JD, Regan JT, Gasdaska JR, Frantz KK, Peele CG, Black A, Passmore D, Moldovan-Loomis C, Srinivasan M, Cuison S, Cardarelli PM, Dickey LF (2006) Glycan optimization of a human monoclonal antibody in the aquatic plant *Lemna minor*. Nat Biotechnol 24:1591–1597
Dangl JL, Parks DR, Oi VT, Herzenberg LA (1982) Rapid isolation of cloned isotype switch variants using fluorescence activated cell sorting. Cytometry 2:395–401
Davis JM (1986) A single-step technique for selecting and cloning hybridomas for monoclonal antibody production. Methods Enzymol 121:307–322
Davis JM (2002) Basic cell culture – a practical approach, 2nd edn. Oxford University Press.
DeMaria CT, Cairns V, Schwarz C, Zhang J, Guerin M, Zuena E, Estes S, Karey KP (2007) Accelerated clone selection for recombinant CHO cells using a FACS-based high-throughput screen. Biotechnol Prog 23:465–472
Eagle H, Piez K (1962) The population-dependent requirement by cultured mammalian cells for metabolites which they can synthesize. J Exp Med 116:29–43
Eidne KA, Kroeger KM, Hanyaloglu AC (2002) Applications of novel resonance energy transfer techniques to study dynamic hormone receptor interactions in living cells. Trends Endocrinol Metab 13:415–421
Frykman S, Srienc F (1998) Quantitating secretion rates of individual cells: design of secretion assays. Biotechnol Bioeng 59:214–226
Gray F, Kenney JS, Dunne JF (1995) Secretion capture and report web: use of affinity derivatized agarose microdroplets for the selection of hybridoma cells. J Immunol Methods 182:155–163
Ham RG, McKeehan WL (1979) Media and growth requirements. Methods Enzymol 58:44–93
Hamburger A, Salmon SE (1977) Primary assay of human myeloma stem cells. J Clin Invest 60:846–854
Hamilton SR, Davidson RC, Sethuraman N, Nett JH, Jiang Y, Rios S, Bobrowicz P, Stadheim TA, Li H, Choi BK, Hopkins D, Wischnewski H, Roser J, Mitchell T, Strawbridge RR, Hoopes J,

Wildt S, Gerngross TU (2006) Humanization of yeast to produce complex terminally sialylated glycoproteins. Science 313:1441–1443

Hammill L, Welles J, Carson GR (2000) The gel microdrop secretion assay: identification of a low productivity subpopulation arising during the production of human antibody in CHO cells. Cytotechnology 34:27–37

Hanania EG, Fieck A, Stevens J, Bodzin LJ, Palsson BØ, Koller MR (2005) Automated in situ measurement of cell-specific antibody secretion and laser-mediated purification for rapid cloning of highly-secreting producers. Biotechnol Bioeng 91:872–876

Heim R, Tsien RY (1996) Engineering green fluorescent protein for improved brightness, longer wavelengths and fluorescence resonance energy transfer. Curr Biol 6:178–182

Hollister JR, Jarvis DL (2001) Engineering lepidopteran insect cells for sialoglycoprotein production by genetic transformation with mammalian β1, 4-galactosyltransferase and α2, 6-sialyltransferase genes. Glycobiology 11:1–9

Holmes P, Al-Rubeai M (1999) Improved cell line development by a high throughput affinity capture surface display technique to select for high secretors. J Immunol Methods 230:141–147

ICH Guideline (1996) Quality of biotechnological products: analysis of the expression construct in cells used for production of rDNA derived protein products. ICH harmonized tripartite guideline. http://www.fda.gov/cder/guidance/ichq5b.pdf

Idusogie EE, Castro JM, Casipit C, Sato A, Terasawa Y, Mulkerrin MG (2008) Development of an antibody screening assay for selection of production cell lines. BioProcess Int 6(4): 20–32

Johnson AE (2005) The co-translational folding and interactions of nascent protein chains: a new approach using fluorescence resonance energy transfer. FEBS Lett 579:916–920

Kacmar J, Srienc F (2005) Dynamics of single cell property distributions in Chinese hamster ovary cell cultures monitored and controlled with automated flow cytometry. J Biotechnol 120:410–420

Karawajew L, Micheel B, Behrsing O, Gaestel M (1987) Bi-specific antibody producing hybrid hybridomas selected by a fluorescence activated cell sorter. J Immunol Methods 96:265–270

Kaufman RJ, Sharp PA (1982) Amplification and expression of sequences cotransfected with a modular dihydrofolate reductase complementary DNA gene. J Mol Biol 159:601–621

Kaufman RJ, Bertino JR, Schimke RT (1978) Quantitation of dihydrofolate reductase in individual parental and methotrexate-resistant murine cells. Use of a fluorescence activated cell sorter. J Biol Chem 253:5852–5860

Kenney JS, Gray F, Ancel MH, Dunne JF (1995) Production of monoclonal antibodies using a secretion capture report web. Biotechnology (NY) 13:787–790

Kim NS, Kim SJ, Lee GM (1998) Clonal variability within dihydrofolate reductase-mediated gene amplified Chinese hamster ovary cells: stability in the absence of selective pressure. Biotechnol Bioeng 60:679–688

Kim NS, Byun TH, Lee GM (2001) Key determinants in the occurrence of clonal variation in humanized antibody expression of CHO cells during dihydrofolate reductase mediated gene amplification. Biotechnol Prog 17:69–75

Koller MR, Hanania EG, Stevens J, Eisfeld TM, Sasaki GC, Fieck A, Palsson BØ (2004) High-throughput laser-mediated in situ cell purification with high purity and yield. Cytometry A 61A:153–161

Leno M, Merten OW, Vuillier F, Hache J (1991) IgG production in hybridoma batch culture: kinetics of IgG mRNA, cytoplasmic-, secreted-, and membrane bound antibody levels. J Biotechnol 20:301–311

Mancia F, Patel SD, Rajala MW, Scherer PE, Nemes A, Schieren I, Hendrickson WA, Shapiro L (2004) Optimization of protein production in mammalian cells with a coexpressed fluorescent marker. Structure 12:1355–1360

Manz R, Assenmacher M, Pflüger E, Miltenyi S, Radbruch A (1995) Analysis and sorting of live cells according to secreted molecules, relocated to a cell-surface affinity matrix. Proc Natl Acad Sci USA 92:1921–1925

Marder P, Maciak RS, Fouts RL, Baker RS, Starling JJ (1990) Selective cloning of hybridoma cells for enhanced immunoglobulin production using flow cytometric cell sorting and automated laser nephelometry. Cytometry 11:498–505

McKinney KL, Dilwith R, Belfort G (1995) Optimizing antibody production in batch hybridoma culture. J Biotechnol 40:31–48

Meilhoc E, Wittrup KD, Bailey JE (1989) Application of flow cytometric measurement of surface IgG in kinetic analysis of monoclonal antibody synthesis and secretion by murine hybridoma cells. J Immunol Methods 121:167–174

Meng YG, Liang J, Wong WL, Chisolm V (2000) Green fluorescent protein as a second selectable marker for selection of high producing clones from transfected CHO cells. Gene 242:201–207

Mizuguchi H, Xu Z, Ishii-Watabe A, Uchida E, Hayakawa T (2000) IRES-dependent second gene expression is significantly lower than cap-mediated first gene expression in a bicistronic vector. Mol Ther 1:376–382

Parks DR, Bryan VM, Oi VT, Herzenberg LA (1979) Antigen-specific identification and cloning of hybridomas with a fluorescence activated cell sorter. Proc Natl Acad Sci USA 76:1962–1966

Plautz JD, Day RN, Dailey GM, Welsh SB, Hall JC, Halpain S, Kay SA (1996) Green fluorescent protein and its derivatives as versatile markers for gene expression in living *Drosophila melanogaster*, plant and mammalian cells. Gene 173:83–87

Powell KT, Weaver JC (1990) Gel microdroplets and flow cytometry: rapid determination of antibody secretion by individual cells within a cell population. Biotechnology (NY) 8:333–337

Racher A, Singh R (2003) Method for selecting antibody producing cells. Pub. No. WO/2003/012449.

Rittenberg MB, Buenafe A, Brown M (1986) A simple method for cloning hybridoma cells in 20-μl hanging drops. Methods Enzymol 121:327–331

Sanford KK, Earle WR, Likely GD (1948) The growth in vitro of single isolated tissue cells. J Natl Cancer Inst 9:229–246

Sanford KK, Covalesky AB, Dupree LT, Earle WR (1961) Cloning of mammalian cells by a simplified capillary technique. Exp Cell Res 23:361–372

Sarramegna V, Talmont F, Seree de Roch M, Milon A, Demange P (2002) Green fluorescent protein as a reporter of human μ-opioid receptor overexpression and localization in the methylotrophic yeast Pichia pastoris. J Biotechnol 99:23–29

Sato G, Fisher HW, Puck TT (1957) Molecular growth requirements of single mammalian cells. Science 126:961–964

Sen S, Hu WS, Srienc F (1990) Flow cytometric study of hybridoma cell culture: correlation between cell surface fluorescence and IgG production rate. Enzyme Microb Technol 12:571–576

Seth G, Charaniya S, Wlaschin KF, Hu WS (2007) In pursuit of a super producer – alternative pathways to high producing recombinant mammalian cells. Curr Opin Biotechnol 18:557–564

Sethuraman N, Stadheim TA (2006) Challenges in therapeutic glycoprotein production. Curr Opin Biotechnol 17:341–346

Siemering KR, Golbik R, Sever R, Haseloff J (1996) Mutations that suppress the thermosensitivity of green fluorescent protein. Curr Biol 6:1653–1663

Sitton G, Srienc F (2008) Mammalian cell culture scale-up and fed-batch control using automated flow cytometry. J Biotechnol 135:174–180

Sitton G, Hansgate A, Srienc F (2006) Transient gene expression in CHO cells monitored with automated flow cytometry. Cytotechnology 52:13–24

Sleiman RJ, Gray PP, McCall MN, Codamo J, Sunstrom NA (2008) Accelerated cell line development using two-color fluorescence activated cell sorting to select highly expressing antibody-producing clones. Biotechnol Bioeng 99:578–587

Stühmeier F, Hillisch A, Clegg RM, Diekmann S (2000) Fluorescence energy transfer analysis of DNA containing several bulges and their interaction with CAP. J Mol Biol 302:1081–1100

Subramanian S, Srienc F (1996) Quantitative analysis of transient gene expression in mammalian cells using the green fluorescent protein. J Biotechnol 49:137–151

Suzuki J, Fukuda M, Kawata S, Maruoka M, Kubo Y, Takeya T, Shishido T (2006) A rapid protein expression and purification system using Chinese hamster ovary cells expressing retrovirus receptor. J Biotechnol 126:463–474

Szaniszlo P, Rose WA, Wang N, Reece LM, Tsulaia TV, Hanania EG, Elferinck CJ, Leary JF (2006) Scanning cytometry with a LEAP: laser-enabled analysis and processing of live cells in situ. Cytometry A 69:641–651

Turcanu V, Williams NA (2001) Cell identification and isolation on the basis of cytokine secretion: a novel tool for investigating immune responses. Nat Med 7:373–376

Underwood PA, Bean PA (1988) Hazards of the limiting-dilution method of cloning hybridomas. J Immunol Methods 107:119–128

Walsh G (2006) Biopharmaceutical benchmarks 2006. Nat Biotechnol 24:769–776

Weaver JC (1986) Gel microdroplets for microbial measurement and screening: basic principles. Biotech Bioeng Symp 17:185–195

Weaver JC, McGrath P, Adams S (1997) Gel microdrop technology for rapid isolation of rare and high producer cells. Nat Med 3:583–585

Wewetzer K, Seilheimer B (1995) Establishment of a single-step hybridoma cloning protocol using an automated cell transfer system: comparison with limiting dilution. J Immunol Methods 179:71–76

Wildy P, Stoker M (1958) Multiplication of solitary HeLa cells. Nature 181:1407–1408

Yoshikawa T, Nakanishi F, Ogura Y, Oi D, Omasa T, Katakura Y, Kishimoto M, Suga K (2001) Flow cytometry: an improved method for the selection of highly productive gene-amplified CHO cells using flow cytometry. Biotechnol Bioeng 74:435–442

Zeyda M, Borth N, Kunert R, Katinger H (1999) Optimization of sorting conditions for the selection of stable, high-producing mammalian cell lines. Biotechnol Prog 15:953–957

Zhao R, Natarajan A, Srienc F (1999) A flow injection flow cytometry system for on-line monitoring of bioreactors. Biotechnol Bioeng 62:609–617i

Engineering Mammalian Cells for Recombinant Monoclonal Antibody Production

Sarah L. Davies and David C. James

Abstract Recombinant monoclonal antibodies represent one of the most important classes of new therapeutic entities. They are increasingly used to treat a variety of diseases, including several cancers. Large-scale commercial manufacture of recombinant antibodies is dependent upon the development of high-yielding production processes that are underpinned by expression of antibody genes in heterologous host cells – most frequently mammalian cells in culture. In this chapter we review recent advances in gene expression technologies that enable sustained high-level production of recombinant monoclonal antibodies in mammalian cells. We describe improvements in antibody expression vector design and recent attempts to improve the cellular process of antibody production through directed cell engineering.

1 Introduction

The production of recombinant proteins by the biopharmaceutical industry will serve a global market with a projected size of USD$70 billion by 2010 (Pavlou and Reichert, 2004; Walsh, 2003). The majority (approximately 70%) of recombinant therapeutic proteins are produced by mammalian cells in culture and this proportion is increasing. The sustained rapid growth of this sector is driven by the relatively high success rate of recombinant protein drugs in trials. For example, monoclonal antibodies (Mab's) are now the second largest category of biopharmaceutical products in development (Walsh, 2003) and approximately 18–29% of recombinant monoclonal antibodies (Mab's) in development succeed to market (Reichert et al., 2005). The forecast annual growth rate of the recombinant Mab sector is 21% (Pavlou and

S.L. Davies and D.C. James (✉)
Department of Chemical and Process Engineering, University of Sheffield,
Mappin Street, Sheffield, S1 3JD, UK
e-mail: d.c.james@sheffield.ac.uk

M. Al-Rubeai (ed.), *Cell Line Development*, Cell Engineering 6,
DOI 10.1007/978-90-481-2245-5_8, © Springer Science+Business Media B.V. 2009

Belsey, 2005), and it is anticipated that by 2015 biopharmaceuticals will constitute at least 30% of the total pharmaceutical market.

Chinese hamster ovary (CHO) cells are by far the most widely utilised mammalian cell type engineered to produce correctly folded, fully glycosylated recombinant Mab's (Walsh, 2006), although other mammalian host cells are also employed (e.g. murine myeloma NS0 and Sp2/0); correct glycosylation is crucial for Mab's requiring biological activities such as antibody dependent cell cytotoxicity and Complement mediated lysis (Jefferis, 2005). Other non-mammalian production systems are now in development that may ultimately compete with mammalian cell hosts, such as strains of *Pichia pastoris* and transgenic plants engineered to modify recombinant Mab's with mammalian-type N-glycans (Gomord et al., 2004; Li et al., 2006). Furthermore, full length correctly assembled yet aglycosylated Mab's have been produced in *Escherichia coli* (Simmons et al., 2002).

Typically, mammalian cell based Mab production systems capable of generating multi-kilogram quantities of product are required to support administration of relatively high doses (>100 mg) of these drugs in the clinic. This has placed significant demands on the biopharmaceutical industry to develop high-yielding production systems employing mammalian host cells. Accordingly, over the last fifteen years extensive empirical optimisation of mammalian cell based production systems has substantially increased both volumetric concentration of recombinant product and shortened cell line development time (Wurm, 2004). In the case of recombinant Mab's, volumetric productivities exceeding 5 g/L are now achievable in a drastically reduced development time (Birch and Racher, 2006).

Volumetric product yield is a function of two basic culture parameters, (i) cell specific production rate (qP) and (ii) the integral of viable cell concentration (commonly calculated as cell time per unit volume) during culture. For typical fed-batch production processes the ideal combination is a rapid accumulation of productive cellular biomass maintained at high viable cell concentration for as long as possible. Currently, volumetric Mab concentrations exceeding 5 g/L are not uncommon, although this has been achieved largely by systematic optimisation of media formulation and rational design of feeding regimes which primarily facilitate increased accumulation of cellular biomass in vitro and to a much lesser extent cell specific production rate (Wurm, 2004). The advent of rapid screening technologies able to identify and isolate productive transfectants has also increased the speed of cell line development (Browne and Al-Rubeai, 2007). In fact, the basic host cell (e.g. CHOK1 derivatives) and Mab gene expression systems used by industry (e.g. glutamine synthetase, dihydrofolate reductase selection/amplification systems (Bebbington et al., 1992; Page and Sydenham, 1991; Reff, 1993)) have not developed significantly for over ten years, with the specific productivity of proliferating cells in batch culture still typically 20–50 pg/cell/day. General developments in mammalian gene expression technology to increase recombinant gene copy number or transcriptional activity have occurred subsequently, e.g. use of more active promoters (Running Deer and Allison, 2004), targeting of recombinant DNA to transcriptionally active sites (Koduri et al., 2001), genomic DNA sequences (Fouser et al., 1992), chromatin opening elements (Antoniou et al., 2003), matrix attachment regions (Girod et al., 2005) etc. However, in practical terms, whilst these technologies may increase the proportion of stable transfectants with higher qP (Zahn-Zabal et al., 2001), step changes in cell specific production rate have not been achieved. Here we review the current status of gene expression and cell engineering technology aimed specifically at increasing recombinant monoclonal antibody production by mammalian cells.

2 General Considerations for Mab Expression Vector Design

Mab production requires the simultaneous expression of two genes, encoding both light chain (LC) and heavy chain (HC) polypeptides, plus a selectable marker. For industrial production in CHO or NS0 cells selectable marker systems most frequently utilise either recombinant glutamine synthetase (GS; (Cockett et al., 1990)) or dihydrofolate reductase (DHFR) in combination with their specific inhibitors methionine sulphoximine (MSX) and methotrexate (MTX) respectively, as the basis of selection/amplification systems. In the case of DHFR systems, mutant cell lines deficient in the DHFR activity have been isolated (e.g. DUKX-B11; (Urlaub and Chasin, 1980) and DG-44 (Urlaub et al., 1983)). The application of these systems for recombinant protein production in mammalian cells has been thoroughly reviewed elsewhere (Birch and Racher, 2006; Trill et al., 1995). Relevant to Mab production, Kim et al. (2001) report that whilst the use of the DHFR system permits gene amplification, effective amplification of both antibody chains, including the transgene not linked to the amplification marker may be quite rare. Out of 23 parental clones exhibiting a range of antibody production only one clone displayed an overall increase in production following stepwise increments in MTX levels. Lastly, in cell lines other than CHO or NS0, such as the human cell line PER.C6, antibiotic (neomycin) based selection has been employed to achieve reasonably high-level production using a single vector system (Jones et al., 2003). Other considerations for Mab expression vector design include general components such as the promoter, enhancer, signal sequences, polyadenylation motifs and the presence or absence of genomic introns (thoroughly reviewed by Makrides (1999)). Of note, although the human cytomegalovirus (CMV) promoter or its variants (CMV-IE) are utilised most prevalently (Boshart et al., 1985; Foecking and Hofstetter, 1986; Meier and Stinski, 1996), in some cases other mammalian derived elements have been utilised. For example Running Deer and Allison (2004) report the use of genomic human elongation factor 1α sequences for high level expression of proteins in CHO cells.

In general two types of transfection strategies have been employed, (i) both genes expressed under the control of individual promoters on the same vector (e.g. (Bebbington et al., 1992)) or (ii) co-transfection of two separate vectors encoding either HC or LC genes (e.g. Page and Sydenham, 1991). In the latter case different selectable markers may be associated with either HC or LC plasmids, for example HC expression selected by DHFR co-expression and LC via neomycin selection (Aldrich et al., 2003;

Brezinsky et al., 2003), or the use of two different antibiotic resistance genes on separate vectors (McLean et al., 2000), although various combinations of HC, LC and selectable markers have been reported. Other interesting variations in vector design for stable Mab production reported recently include "trans-complementing" vectors generated specifically for co-ordinated antibody production (Bianchi and McGrew, 2003). This system exploits the fact that the DHFR selectable marker can be divided into two individual fragments which can re-associate to form a fully functional molecule. Each fragment was encoded on separate a plasmid linked to either LC or HC genes by viral internal ribosome entry site (IRES) sequences. Stable transfection of both vectors was found to result in the simultaneous amplification of both antibody chains with a dramatic reduction in time required to select high antibody producing cell lines, supposedly without the need for cell cloning. Attenuated IRES elements have also been used recently to improve the selection of antibody producing clones by linking HC or LC expression directly to the expression of two different autofluorescent reporter proteins. Cell clones with high Mab production were selected on the basis of a two-colour cell sorting strategy within 12 weeks of transfection. Such clones were found to have a 38-fold increase in antibody production compared to clones isolated with a single round a cell sorting (Sleiman et al., 2008).

3 Transcriptional Enhancement

The variability of transgene expression experienced with stable transfection is often attributed to the number of transgene copies integrated, and to the particular site of integration within the host chromatin structure. Chromatin can be subdivided in two forms, the transcriptionally active decondensed euchromatin, and condensed heterochromatin which exists in a transcriptionally silent state. Considering heterochromatin forms a major proportion of the genome the probability that the transgene will integrate within this domain and consequently becomes repressed is high (Dillon and Festenstein, 2002), moreover it may be postulated that recombinant DNA is preferentially inserted into regions of genomic DNA which are susceptible to a high rate of genomic deletion (i.e. explaining its insertion there in the first place; Li et al., 2001). There are two principal methods currently employed to overcome these expression problems, (i) site specific integration – the transgene is directly integrated into a known transcriptionally active hotspot or (ii) flanking the transgene with genomic DNA elements that prevent heterochromatin repression and promote high transcriptional activity.

Targeted integration of transgenes into a pre-determined chromosomal site is a viable method to overcome variable expression levels between stable clones. Such site-specific recombination utilises recombinase enzymes such as the bacteriophage P1 Cre or the yeast Flp which mediate homologous recombination between a pair of target sequences specifically recognised by these enzymes. Site specific integration has been employed to engineer CHO cells for high level antibody production. Kito et al. (2002) used a two stage strategy whereby an integration plasmid was constructed that incorporated a target site into the CHO cell genome. The integration plasmid also carried a reporter gene which was used to identify gene expression levels. Clones were selected where integration was presumed to occur at transcriptionally active sites on the basis of high reporter levels. The selected clones were then co-transfected with a recombinase-expressing plasmid and a gene targeting plasmid, carrying the second target site and the antibody transgenes. Similar studies have been performed to generate a targeted integration system for antibody production (Huang et al., 2007; Wiberg et al., 2006). These studies could reproducibly generate high producing clones following gene targeting indicating reduced clonal variability, however Kito et al. (2002) noted that the efficiency of gene targeting was quite low presumably due to the fact that site specific recombination can potentially be reversible.

There are many possible trasngene flanking DNA elements which could be used to improve transgene expression and stability including locus control regions (LCRs), insulators, universal chromatin opening elements (UCOEs), matrix associated regions (MARs) and stabilising and antirepressor elements (STAR) (Kwaks and Otte, 2006; Kwaks et al., 2003). Zahn-Zabal et al. (2001) compared the effects of many DNA elements upon transgene expression in CHO cells. The only element that exerted a substantial positive effect was the chicken lysozyme 5′ MAR. Interestingly no effect was observed on transient expression elements emphasizing the role these elements play in chromatin remodelling. A further study revealed that the chicken lysozyme 5′ MAR increased both GFP and Mab expression levels when expressed either in *cis* (present on the same vector) or *trans* (cotransfected on a separate vector) configuration (Girod et al., 2005). Furthermore the addition of more than one MAR, in either *cis* or *trans* or a combination of both configurations, had a further positive effect. An overall shift in expression levels in both low and high producing cell clones was also observed with an appearance of a new very high producing sub-population. The mechanism behind this effect is not so easily explained. Suggestions include a highly protective role of MAR elements necessary even when transgenes are expressed in open permissive chromatin sites. An alternative is that MAR elements can increase transcription initiation or that these elements can increase the number of transgene copies integrated. Both Girod et al. (2005) and Kim et al. (2004) observed an increase in transgene copy number in the presence of a MAR element. Both STAR (Otte et al., 2007) and UCOEs (Benton et al., 2002) have also been shown to increase transgene expression levels in stable clones but to date no data on Mab production has been published.

Lastly, Morris et al. (1997) have also isolated an "Expression Augmenting Sequence Element" (EASE). This 5.7 kb genomic DNA element was identified from a cloned expression vector integration site derived from a CHO cell line producing high levels of recombinant protein from a single integration cassette. This element, when incorporated in Mab expression vector constructs was found to confer an increased frequency of high expression with low selective pressure. This group have reported a doubling in Mab expression levels from stable pools generated in the presence of a reduced EASE (3.7 kb; Aldrich et al., 2003).

4 Translational Control

The ability to control translation of recombinant Mab mRNA's is relevant to cell line development in two contexts. Firstly, we may be able to alter the efficiency of Mab mRNA utilisation. A variety of studies using clonally derived mammalian cell lines or hybridomas all demonstrate that at higher rates of Mab production there is no correlation between qMab and the corresponding cellular availability of HC or LC mRNA (Barnes et al., 2004; Flickinger et al., 1992; Kim et al., 1998a; Leno et al., 1992) – i.e. Mab mRNA translation rate may be determinant of cell specific production rate. Accordingly, Kalwy et al. (2006) have reported that optimisation of HC and LC Mab sequences by adjusting codon usage for CHO cells and removal of direct repeats, secondary structure elements and *cis*-acting elements such as cryptic splice sites increased levels of Mab production 1.5-fold. Furthermore, removal of introns in the coding sequence had no effect on Mab production. Carton et al. (2007) have reported that replacement of bacterial codon usage in the variable region of immunoglobulin genes with human codon usage yielded stable cell lines with a significantly higher Mab production.

Secondly, Mab production requires the coordinated expression, folding and assembly of both HC and LC polypeptides, and there is evidence that excess copies of LC polypeptide production are necessary for optimal rates of Mab assembly in the endoplasmic reticulum (ER; Jiang et al., 2006; Schlatter et al., 2005). An increase in Mab production efficiency may therefore be achieved via control of the relative rate of HC and LC mRNA translation.

For heteropolymers such as Mabs, internal ribosome entry sites (IRES's) permit the coordinated expression of multiple genes from a single plasmid (Fussenegger et al., 1998). An IRES element permits the efficient translation of mRNA devoid of either a cap structure or a free 5′ end, structures found to be necessary for the initiation of cap-dependent translation (Kozak, 1999). IRES's have been identified within the 5′ untranslated region (UTR) from many viruses, including piconaviruses and retroviruses (Houdebine and Attal, 1999), and certain cellular mRNA such as the immunoglobulin binding protein (BiP; Yang and Sarnow, 1997). Translation of such mRNAs is initiated without scanning of mRNA by the 40 S ribosomal subunit for a functional initiation codon. This is associated with a long 5′ UTR which is GC rich and highly structured, rendering the scanning process unable to occur. Instead, the IRES element is believed to direct the ribosomes straight to the initiation codon in a scanning-independent manner (Houdebine and Attal, 1999). Since their discovery many groups have utilised IRES's to construct bicistronic vectors to produce recombinant proteins with a linked selectable marker (Gurtu et al., 1996; Rees et al., 1996) and there are examples of the use of IRES-based bicistronic constructs to link Mab HC or LC expression to a selectable marker such as DHFR (Aldrich et al., 2003; Bianchi and McGrew, 2003; Brezinsky et al., 2003) or fluorescent proteins to facilitate Mab cell line selection (Sleiman et al., 2008) as mentioned above. Not much work has focused on the construction of tricistronic vectors which encode Mab polypeptides linked to a selectable marker on a single plasmid. Mielke et al. (2000) generated a tricistronic vector encoding a heterodimeric antibody fusion protein linked to puromycin. The tricistronic construct yielded higher stable antibody expression levels compared to co-transfection of individual vectors. One of the major issues relevant to the use of bi- or tricistronic IRES vectors for recombinant protein production is that cap-independent, IRES-mediated translation (second or third cistron) is less efficient than cap-dependent translation of the first cistron (Kaufman et al., 1991; Underhill et al., 2007). Utilising a range of EMCV IRES mutants differing in translational "strength" Li et al. (2007b) showed that optimal transient production of Mab in HEK293T cells was observed when the IRES-mediated translation efficiency of HC was 50% of the cap-dependent translation of LC, i.e. optimal Mab production was obtained when the ratio of HC to LC translational efficiency was 1:2. In another study the same group demonstrated that for stable Mab production in CHO cells, clones derived from bicistronic constructs produced similar Mab titres to those generated using monocistronic constructs (Li et al., 2007a). Bicistronic constructs with the LC gene in the first cistron were twice as effective as those with the HC gene in the first cistron, consistent with a model for Mab folding and assembly which requires a molar excess of LC production for optimal rates of Mab folding and assembly.

One expression tool that potentially overcomes the differential expression levels observed between cistons within a bi- or tri-cistronic IRES vector construct is the foot-and-mouth disease (FMDV)-derived 2A self-processing sequence. This particular sequence, consisting of as little of 19 amino acids, is able to cleave at its own C terminus allowing the generation of individual mature proteins from one single transcript (Ryan et al., 1991). This process is known to occur co-translationally and the "ribosome skipping" model for its mechanism of action has been proposed, whereby the nascent protein upstream of the 2A sequence is released and translation of the remainder of the transcript can be re-initiated (de Felipe et al., 2006). The 2A sequence has been used in many aspects of biotechnology including, plant engineering (Halpin et al., 1999), gene therapy (de Felipe et al., 1999) and recombinant Mab production (Fang et al., 2005). Fang et al. (2005) utilised a 24 amino acid variant of the FMDV 2A sequence to physically link the HC and LC of a IgG1 Mab cloned within an expression plasmid. Transient transfection in HEK293 cells resulted in a 16-fold increase in Mab production when compared to a control plasmid whereby the HC and LC were linked by an EMCV IRES. Characterisation of the antibody revealed full biological activity, although the HC (cistron upstream of 2A sequence) appeared to migrate slower on a western due to the remaining 23 amino acids of the 2A sequence following cleavage. Any potential adverse effects can be eliminated by the addition of a furin cleavage site sequence located immediately upstream of the 2A sequence. Western blot analysis demonstrated successful removal of the 23 amino acids. There also appeared to be an increase in Mab production following transient transfection compared to 2A plasmid without the cleavage sequence although this was not discussed by the authors. In vivo Mab production has also been successful with the application of the 2A antibody cassette introduced into a recombinant adeno-associated virus vector (Fang et al., 2007).

Due to the nature of FMDV 2A self processing sequence it is assumed to translate linked cistrons at a 1:1 ratio. However analysis of the self processing reaction within cell-free translation systems revealed an accumulation of proteins upstream of the 2A sequence compared to downstream proteins, assumed to be a cause of a termination of a subset of ribosomes at the self processing site (Donnelly et al., 2001). This appears to be an artefact of cell free translation systems, as cellular studies have not demonstrated such an accumulation (de Felipe et al., 2006). Although furin cleavage 2A sequence technology overcomes the limitations of IRES technology with respect to reduced translational efficiency across the transcript the issue of the crucial LC:HC ratio in antibody production is still not resolved. It remains unclear whether it can be used to generate the rare and important high Mab producing clones.

5 Modelling the Cellular Recombinant Monoclonal Antibody Production Process

Recombinant Mab production is a complex and dynamic process, cell specific recombinant Mab production rate (qMab) is a function of the relative rates of a diverse variety of cellular processes; recombinant gene transcription, mRNA decay, mRNA translation, co-translational glycosylation, nascent polypeptide folding, ER associated degradation and inter-vesicular transport of fully assembled protein. Crucially, we still do not *systematically* understand how the host cell coordinates and regulates the dynamic cellular processes that contribute to recombinant Mab synthesis during production processes (Fig. 1). Particular synthetic reaction steps, such as folding and assembly in the endoplasmic reticulum are likely to be rate limiting (Fig. 2). The first kinetic models of Mab synthesis in hybridomas by Bibila and Flickinger suggested that the rate-limiting step of Mab synthesis shifts from

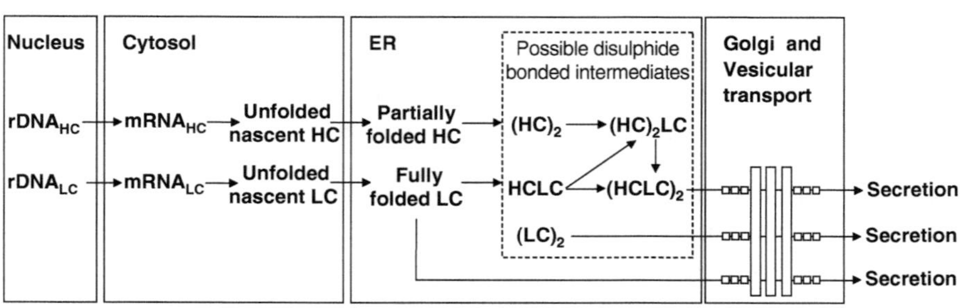

Fig. 1 Intracellular processes involved in recombinant monoclonal antibody biosynthesis and secretion. Mab production rate is a function of a wide variety of intracellular processes that are dynamically regulated during a production process

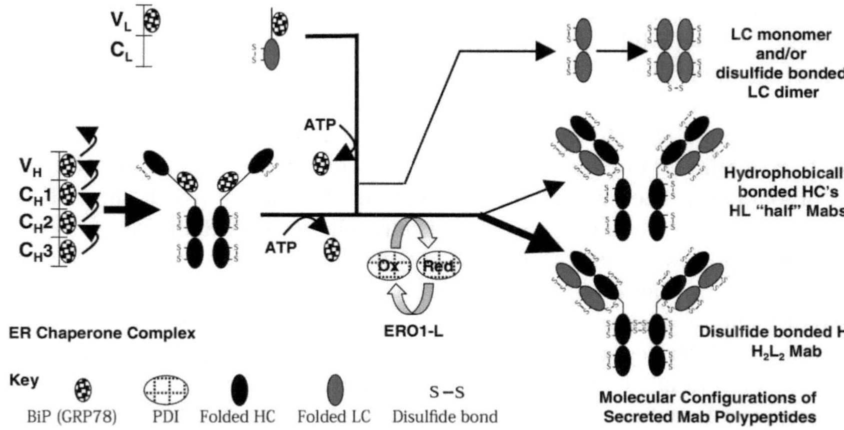

Fig. 2 Folding and assembly of IgG in the endoplasmic reticulum of mammalian cells. Immunoglobulin (Ig) polypeptides sequentially interact with a range of molecular chaperones, foldases and oxidoreductases present in a complex in the endoplasmic reticulum (Meunier et al., 2002). Whereas Ig LC polypeptide only has a comparatively transient association with BiP and (unlike HC polypeptide) can be secreted as both monomer and disulfide-bonded dimer, Ig heavy chain (HC) is known to interact extensively with BiP, and if expressed in the absence of LC will be retained in an unfolded state in the ER in persistent association (via the CHI domain) with BiP. In this case unassembled HC will eventually be trafficked to the proteasome for degradation (Fagioli et al., 2001). The LC both facilitates folding of the CHI domain and the release of HC from BiP (Lee et al., 1999; Vanhove et al., 2001). During assembly, ATP-dependent binding of unfolded polypeptides by BiP maintains them in a conformation in which cysteines are accessible to disulfide bond formation catalyzed by protein disulfide isomerase (PDI). Thick arrows indicate predominant processing pathways, thin arrows indicate alternative potential processing pathways yielding variant Mab molecular configurations, although these maybe IgG subclass specific (from Dinnis and James, 2005)

Mab assembly in fast-growing cells to mRNA translation in slow-growing cells (Bibila and Flickinger, 1991; Bibila and Flickinger, 1992). More recently, Gonzalez et al. (2001) have produced a simplified, theoretical metabolic control analysis of Mab synthesis. This model predicts shared control of flux to secreted antibody between cellular synthetic reactions, assembly processes and translocation events, where control varies with imposed constraints (precursor availability, recombinant gene expression levels etc). More recently, Ho et al. (2006) have modelled Mab production by GS-NSO cells. Using a global sensitivity analysis they report that specific transcription and translation rates for recombinant HC and LC genes/mRNA's remain important throughout a production process, although recombinant mRNA translation and half-life assume an increasing priority as culture progresses into stationary phase. These conclusions correlate with a recent study from this laboratory. Using a combination of biomolecular and proteome analyses Stansfield et al. (2007) demonstrated that although large changes in qMab occur during a fed-batch culture of GS-NSO cells, the cellular proteome remains remarkably constant, varying primarily with cell growth rate. However, during culture the cellular constraints on Mab production vary. Initially qMab correlates with increasing HC mRNA abundance, however from mid-exponential culture onwards this relationship does not hold, whilst HC and LC mRNA abundance and rate of polypeptide synthesis remain relatively high, qMab declines indicating a progressive restriction in cellular Mab production downstream of protein synthesis.

Other factors affecting the kinetics of Mab synthesis have also proved amenable to modelling. For example, based on a proteomic analysis of GS-NS0 cell lines producing a recombinant IgG_4 Mab in our laboratory which showed that stable transfectants had a molar excess of intracellular light chain (Smales et al., 2004), we hypothesised that excess LC polypeptide was necessary but not sufficient for efficient Mab folding and assembly in mammalian cells. Subsequently, we experimentally demonstrated more efficient use of Mab HC gene at excess LC production, and derived a mathematical model of Mab folding and assembly for prediction of optimal HC:LC gene ratio (Schlatter et al., 2005). Our conclusions have been confirmed in Sharfstein's laboratory (Jiang et al., 2006). These data suggest that forward folding and assembly reactions, specifically the intermolecular (HC–LC) disulphide bond formation catalysed by protein disulphide isomerase (PDI; see below) is promoted in the presence of excess LC, resulting in more rapid folding and clearance of HC polypeptide from the ER. We note that other CHO cell lines exhibiting high qMab have all exhibited an elevated ratio of intracellular LC to HC polypeptides (not published). These data therefore raise the question of how, when cells are transfected with a vector construct encoding a single copy of both LC and HC genes, specific transfectants can arise which produce a significant molar excess of LC (e.g. 10:1; Smales et al., 2004) to render intracellular Mab folding and assembly reactions efficient. We suggest that rearrangement or processing of recombinant genes post-transfection may be a significant factor with respect to the relative abundance of functional HC or LC genes (Barnes et al., 2007).

However, no models account for important parameters such as internal co-regulation of cellular processes at elevated recombinant gene expression levels. For example, it is important to note that mRNA translation rate, ER-associated degradation and ER folding *capacity* are linked via intracellular signaling pathways such as the unfolded protein response (UPR), and that as for antibody producing plasma cells in vivo, these are likely to be a functionally relevant interaction with respect to recombinant protein synthesis (Dinnis and James, 2005; Gass et al., 2002).

6 Engineering the Mammalian Cell Factory for Improved Mab Production Rate

Whilst intracellular recombinant mRNA level may be expected to be proportionately related to *q*P for monomeric proteins produced by amplified subclonal populations (Gu et al., 1992; Kaufman et al., 1985) and even different cell clones (Fann et al., 1999), the same does not hold true for recombinant Mab expression. Recombinant Mab production is limited at a post-transcriptional level. In vitro studies have shown that the rate of Mab folding in the ER is known to be relatively slow (Goto and Hamaguchi, 1981; Lilie et al., 1994). In the case of Mabs, which require the coordinated expression, folding and assembly of both HC and LC, a variety of studies using clonally derived mammalian cell lines or hybridomas all demonstrate that at higher volumetric Mab production there is no correlation between *q*Mab and the corresponding cellular availability of HC or LC mRNA (Barnes et al., 2004; Flickinger et al., 1992; Kim et al., 1998a; Leno et al., 1992). Although in some studies *q*Mab correlates more strongly with LC mRNA content than HC mRNA content (Borth et al., 1999; Merten et al., 1994; Strutzenberger et al., 1999).

Accordingly, the process of Mab folding and assembly in the ER has been a target for cell engineering for a number of years. During the folding and assembly process, immunoglobulin (Ig) polypeptides sequentially interact with a range of molecular chaperones, foldases and oxidoreductases present in a complex in the endoplasmic reticulum (Mayer et al., 2000; Melnick et al., 1994; Meunier et al., 2002). Most intensively studied are synergistic interactions with the molecular chaperone immunoglobulin binding protein (BiP; also known as GRP78) and PDI which catalyses intra- and intermolecular disulfide bond formation. Whereas Ig LC polypeptide only has a comparatively transient association with BiP and (unlike HC polypeptide) can be secreted as both monomer and disulfide-bonded dimer (Dul et al., 1996; Leitzgen et al., 1997), Ig HC is known to interact extensively with BiP, and if expressed in the absence of LC will be retained in an unfolded state in the ER in persistent association (via the C_H1 domain) with BiP (Vanhove et al., 2001). In this case unassembled HC will eventually be trafficked to the proteasome for degradation (Fagioli et al., 2001). The LC both facilitates folding of the C_H1 domain and the release of HC from BiP (Lee et al., 1999). During assembly, binding of unfolded polypeptides by BiP maintains them in a conformation in which cysteines are accessible to disulfide bond formation catalyzed by PDI. In an attempt to understand and optimize the kinetics of monoclonal antibody formation, mathematical models of Mab folding and assembly in vivo have been reported. These predict that increasing levels of BiP alone (Whiteley et al., 1997) or in combination with PDI (Gonzalez et al., 2002) will increase the rate of Mab formation. Interestingly, two independent proteomic investigations have revealed that PDI proteins and other ER chaperones known to interact with nascent Mab polypeptides such as endoplasmin are increased in abundance in NS0 cells with high Mab productivity (Borth et al., 2005; Seth et al. 2007; Smales et al., 2004).

Specifically in relation to Mab production, overexpression of discrete ER resident molecular chaperones has generally increased Mab production by eukaryotic cells. The best candidate for engineering is the redox reaction catalysed by PDI, which enables the formation of HC–LC and HC–HC disulphide bonds in the ER. This has been shown to facilitate expression of recombinant immunoglobulins by baculovirus-infected insect cells (Betenbaugh et al., 1996; Hsu et al., 1996) and Borth and co-workers have shown that over-expression of PDI in CHO cells increases qMab by 40% (Borth et al., 2005). PDI expression decreased intracellular accumulation of HC polypeptide, presumably in the ER. Conversely, over-expression of BiP alone, to increase cellular content by 48% resulted in a 34% reduction in qMab, with a concomitant accumulation of intracellular HC. Co-transfection of both PDI and BiP could not reverse the reduction in qMab observed on expression of BiP alone. A further study has employed tetracycline-repressible (Tet-Off) gene expression technology to investigate the effects of PDI over-expression. Using this system complete induction of PDI expression resulted in an increase in qMab of up to 27% (Mohan et al., 2007). Interestingly, PDI induction had no effect on the production of another recombinant protein, thrombopoietin, indicating the specificity of chaperone function in controlling the rate of protein folding and assembly. Lastly, we note that the rate of PDI catalyzed reactions also depends crucially on the activity of other ER-resident (and stress-inducible) oxidoreductases for re-oxidation such as the FAD-dependent Ero-1 proteins, which directly utilise molecular oxygen as the terminal electron acceptor (Tu and Weissman, 2004). Increasing Ero-1 levels in mammalian cells has been shown to increase the rate of PDI-dependent immunoglobulin oxidation, and expression of a mutant inactive form of an Ero-1 protein decreases Ig oxidation (Mezghrani et al., 2001).

Recently, a more global, genome scale cell engineering strategy has arisen from a new, mechanistic understanding of the cellular unfolded protein response (UPR) and its involvement in the process of B-cell differentiation into antibody producing plasma cells in vivo (Brewer and Hendershot, 2005; Dinnis and James, 2005). The UPR is an intracellular signalling system that coordinates the transcriptional upregulation of genes encoding for ER molecular chaperones, protein synthesis/degradation and apoptosis in response to an accumulation of unfolded proteins in the ER (Rutkowski and Kaufman, 2004). During the second phase of B-cell differentiation, components of the UPR coordinate protein expression such that maximal Mab production can be obtained (Ma and Hendershot, 2003). The key regulators in this process are transactivators such as X-box binding protein 1 (XBP-1) and activating transcription factors 6 and 4 (ATF6, ATF4) which induce the expression of a range of proteins whose overall function is to alleviate ER stress by increasing ER folding and assembly capacity, promote ER-associated degradation of unfolded proteins and stimulate amino acid sufficiency (Rutkowski and Kaufman, 2003; Rutkowski and Kaufman, 2004). Not surprisingly, these transactivators have recently been targeted as candidates for cell engineering and examples of whole organelle cell engineering based on this concept have been published recently. Expression of the bioactive form of the XBP-1 in CHO cells both expanded the endoplasmic reticulum and increased production of recombinant reporter proteins (secreted alkaline phosphatase and α-amylase), as well as human vascular endothelial growth factor (Tigges and Fussenegger, 2006). Recently Becker et al. (2008) stated that heterologous expression of XBP-1(s) could lead to an increase in ER content and antibody productivity in CHO suspension culture. However in this study whilst they demonstrated a 40% increase in overall antibody titre from fed-batch CHO cells they reported a difficulty in generating stable XBP-1(s) expressing monoclonal CHO cell lines, thus indicating a negative selection pressure imposed upon clonal cell populations expressing high levels of XBP-1(s). In contrast other studies have concluded that bioactive XBP-1 expression did not enhance stable Mab, interferon-γ or antithrombin III production by CHO cells (Ku et al., 2008;

Ohya et al., 2008). However, XBP-1 was effective when the existing ER folding/assembly pathway was saturated by high-level transient production of recombinant erythropoietin (Ku et al., 2008) and another UPR transactivator, ATF4, was able to significantly enhance AT-III production (Ohya et al., 2008).

7 The Effect of Cell Line Genetic Background on Cellular Productivity

A fundamental consideration highly relevant to cell line development is whether particular cells within a population of parental cells are more suited to act as Mab production vehicles than others, i.e. to what extent is clonal variation or genetic heterogeneity a determinant of functional competence? Clearly there can be heterogeneity in the functional state and organisation of introduced recombinant Mab genes (Barnes et al., 2007; Kim et al., 2001) and cell-to-cell variation arising from point mutations, epigenetic alteration or chromosomal aberrations within the host cell genome (Derouazi et al., 2006) can result in clone-specific differences in a variety of gross functional characteristics such as cell growth (Barnes et al., 2006), protein glycosylation and response to culture environment (Kim and Lee, 2007; Yoon et al., 2004). Accordingly, isolation of murine and rat cell lines has been shown to be a major determinant of genetic heterogeneity when the transcriptome of clonal derivatives was compared using oligonucleotide microarrays (Oh et al., 2003). In this experiment, clonal variation in gene expression exceeded that induced by recombinant protein induction. Therefore, it is perhaps unsurprising that recent –omic studies comparing engineered mammalian cells with different specific Mab productivity have identified significant clonal variation. However, the extent to which these differences have been correlated to cell line specific productivity (qMab) varies. Recent proteomic studies in our laboratory comparing GS-NS0 murine myeloma cell lines with different qMab have shown significant co-variation between qMab and cell line specific variation in the NS0 proteome. In initial studies comparing the relative abundance of single proteins, molecular chaperones known to be associated with Mab folding and assembly (notably BiP, PDI and endoplasmin) were significantly increased in abundance in the cell line with high qMab (Alete et al., 2005; Smales et al., 2004). A subsequent detailed statistical analysis of the proteomic data showed that six of nine functional groups of proteins were significantly increased in abundance in cell lines with elevated qMab: ER chaperone, non-ER chaperone, cytoskeletal, cell signaling, metabolic, and mitochondrial categories (Dinnis et al., 2006). Importantly, there was no evidence of an unfolded protein response derived induction of gene expression in high producing cell lines, indicating that the observed changes in the proteome were not themselves induced by the recombinant protein accumulation. These data imply that a number of cellular systems act in synergy to influence cell specific production rate (qP): qP is not just a function of cloned gene copy number in an otherwise "unemployed" host cell factory.

Other –omic studies on NS0 and CHO cells with varying qMab have identified a range of differentially expressed genes and proteins with altered relative abundance. Recently, a study by Seth et al. (2007) combined analyses of the transcriptome and proteome of eleven NS0 cell lines producing a recombinant Mab. Unfortunately no cell growth or specific productivity data were reported, although the "high-producing" cell lines were 2–11-fold more productive than low producing cell lines. A high degree of consistency between transcriptome and proteome data was reported, and a biological network analysis indicated that protein synthetic and cell growth/death pathways were altered in cells with high Mab productivity. In contrast, another transcriptomic study comparing non-transfected NS0 and a single GS-NS0 cell line producing a recombinant Mab reported that protein synthetic and ribosomal genes were more highly expressed in the non-producing wild-type cells (Khoo et al., 2007).

What conclusions can be drawn from these data? Despite the fact that recombinant protein production is not an objective function for CHO cells, it is likely that discrete cells within populations are more suited to the imposed burden of recombinant protein production than others. Genetic heterogeneity within the parental populations means that not all host cells are the same. It can be argued that all –omic studies indicate that cells with increased productivity have an elevated and coordinated "capacity for flux" in one or more pathways, and such cells are rare (Seth et al., 2007). Clearly in this case a cell line that harboured an elevated folding and assembly capacity would be more capable than other cells of combining an ability to produce the recombinant protein as well as its own biomass without invoking a cytostatic stress response to an accumulation of unfolded recombinant protein such as the unfolded protein response (Dinnis and James, 2005).

8 Stability of Mab Production During Extended Sub-Culture

Mammalian cell populations are inherently genetically heterogeneous, and the associated functional heterogeneity underpins the ability of some cells in a population to survive in synthetic growth environments, achieve high rates of cell proliferation in culture (Barnes et al., 2006) and efficiently manufacture recombinant proteins. Therefore, whilst exploitation of clonal genetic variation is necessary to generate productive cell lines capable of supporting biomanufacturing operations (e.g. "adaptation" to proliferation in suspension culture), spontaneously arising genetic heterogeneity (genetic instability) also contributes some undesirable side effects. Whilst in industry production cell lines have to exhibit stable production for at least 60 generations from the working cell bank (Brown et al., 1992), Clonally derived cell lines can exhibit unpredictable and uncontrollable behaviour in vitro such as loss of productivity (Barnes et al., 2003; Derouazi et al., 2006). This is particularly evident in the absence of selective pressure.

Few published reports studies have investigated the stability of Mab production in the presence or absence of selective pressure (Barnes et al., 2001; Barnes et al.,

2004; Kim et al., 1998a, b). Utilising DHFR selection/amplification Kim et al. (1998a) observed an overall decrease in Mab production in all clones tested following long term culture in the absence of MTX. It was reported that the decrease in Mab production was due to a decrease in both HC and LC gene copies. Similar results have been observed with the GS system. Barnes et al. (2004) reported an overall decrease in recombinant mRNA (HC, LC and GS) during long term culture in both "stable" and unstable Mab producing cell clones. However, the critical factor was found to be the absolute amount of mRNA expressed, in that clones exhibiting instability expressed lower levels of HC and LC at the start of extended culture than the stable clones. The authors postulated that a threshold level of recombinant mRNA was necessary to maintain qMab, and that this was not maintained in "unstable" clones. Although both Barnes et al. (2004) and Kim et al. (1998a) described a loss of both HC and LC during extended culture, other studies have demonstrated that an overall reduction in production stability can be directly related to the specific loss of HC alone (Couture and Heath, 1995) or the appearance of non-producing cells. In the case of the latter, if it is accepted that a non-producing sub-population (which does not have the burden of Mab production) has even a slight growth advantage over the producing population it may overgrow a culture within a limited number of generations (Lee et al., 1991; Kromenaker and Srienc, 1994).

Very little is known about the origins of genetic instability in engineered mammalian cell lines, as discussed in a review by Barnes et al. (2003), and genetic (e.g. mutations, deletions) as well as epigenetic (e.g. methylation) mechanisms may be responsible. More recently, CHO cell lines expressing a reporter GFP gene have recently been shown to exhibit instability in recombinant gene expression (Derouazi et al., 2006), although this was not apparently correlated with either the significant chromosomal instability observed or the number of recombinant gene copies. The latter is significant because it may be argued that maintenance of a low plasmid copy number avoids effects such as repeat-induced gene silencing (McBurney et al., 2002) which may be expected as recombinant genes tend to integrate at a single site (Derouazi et al., 2006; Wurm et al., 1991). These data do not support the hypothesis that loss of cell-specific production rate is a consequence of chromosomal instability *per se*.

References

Aldrich TL, Viaje A, Morris AE (2003) EASE vectors for rapid stable expression of recombinant antibodies. Biotechnol Prog 19:1433–1438

Alete DE, Racher AJ, Birch JR, Stansfield SH, James DC, Smales M (2005) Proteomic analysis of enriched microsomal fractions from GS-NS0 murine myeloma cells with varying secreted recombinant monoclonal antibody productivities. Proteomics 5:4689–4704

Antoniou M, Harland L, Mustoe T, Williams S, Holdstock J, Yague E, Mulcahy T, Griffiths M, Edwards S, Ioannou PA, Mountain A, Crombie R (2003) Transgenes encompassing dual-promoter CpG islands from the human TBP and HNRPA2B1 loci are resistant to heterochromatin-mediated silencing. Genomics 82:269–279

Barnes LM, Bently CM, Dickson AJ (2001) Characterization of the stability of recombinant protein production in the GS-NS0 expression system. Biotechnol Bioeng 73:261–270
Barnes LM, Bently CM, Dickson AJ (2003) Stability of protein production from recombinant mammalian cells. Biotechnol Bioeng 81:631–639
Barnes LM, Bentley CM, Dickson AJ (2004) Molecular definition of predictive indicators of stable protein expression in recombinant NS0 myeloma cells. Biotechnol Bioeng 85:115–121
Barnes LM, Moy N, Dickson AJ (2006) Phenotypic variation during cloning procedures: analysis of the growth behavior of clonal cell lines. Biotechnol Bioeng 94:530–537
Barnes LM, Bentley CM, Moy N, Dickson AJ (2007) Molecular analysis of successful cell line selection in transfected GS-NS0 myeloma cells. Biotechnol Bioeng 96:337–348
Bebbington CR, Renner G, Thomson S, King D, Abrams D, Yarranton GT (1992) High-level expression of a recombinant antibody from myeloma cells using a glutamine-synthetase gene as an amplifiable selectable marker. Biotechnology 10:169–175
Becker E, Florin L, Pfizenmaier K, Kaufmann H (2008) An XBP-1 dependent bottle-neck in production of IgG subtype antibodies in chemically defined serum-free Chinese hamster ovary (CHO) fed-batch processes. J Biotechnol 135:217–223
Benton T, Chen T, McEntee M, Fox B, King D, Crombie R, Thomas TC, Bebbington C (2002) The use of UCOE vectors in combination with a preadapted serum free suspension cell line allows for rapid production of large quantities of protein. Cytotechnology 38:43–46
Betenbaugh MJ, Ailor E, Whiteley E, Hinderliter P, Hsu TA (1996) Chaperone and foldase expression in the baculovirus-insect cell expression. Cytotechnology 20:149–159
Bianchi AA, McGrew JT (2003) High-level expression of full-length antibodies using trans-complementing expression vectors. Biotechnol Bioeng 84:439–444
Bibila T, Flickinger MC (1991) A structured model for monoclonal-antibody synthesis in exponentially growing and stationary phase hybridoma cells. Biotechnol Bioeng 37:210–226
Bibila T, Flickinger MC (1992) Use of a structured kinetic-model of antibody-synthesis and secretion for optimization of antibody-production systems 2. Transient analysis. Biotechnol Bioeng 37:262–272
Birch JR, Racher AJ (2006) Antibody production. Adv Drug Deliv Rev 58:671–685
Borth N, Strutzenberger K, Kunert R, Steinfellner W, Katinger H (1999) Analysis of changes during subclone development and ageing of human antibody-producing heterohybridoma cells by northern blot and flow cytometry. J Biotechnol 67:57–66
Borth N, Mattanovich D, Kunert R, Katinger H (2005) Effect of increased expression of protein disulfide isomerase and heavy chain binding protein on antibody secretion in a recombinant CHO cell line. Biotechnol Prog 21:106–111
Boshart M, Weber F, Jahn G, Dorsch-Hasler H, Fleckenstein B, Schaffner W (1985) A very strong enhancer is located upstream of an immediate early gene of human cytomegalovirus. Cell 41:521–530
Brewer JW, Hendershot LM (2005) Building an antibody factory: a job for the unfolded protein response. Nat Immunol 6:23–29
Brezinsky SC, Chiang GG, Szilvasi A, Mohan S, Shapiro RI, MacLean A, Sisk W, Thill G (2003) A simple method for enriching populations of transfected CHO cells for cells of higher specific productivity. J Immunol Methods 277:141–155
Brown ME, Renner G, Field RP, Hassell T (1992) Process development for the production of recombinant antibodies using the glutamine synthetase (GS) system. Cytotechnology 9:231–236
Browne SM, Al-Rubeai M (2007) Selection methods for high-producing mammalian cell lines. Trends Biotechnol 25:425–432
Carton JM, Sauerwald T, Hawley-Nelson P, Morse B, Peffer N, Beck H, Lu J, Cotty A, Amegadzie B, Sweet R (2007) Codon engineering for improved antibody expression in mammalian cells. Protein Expr Purif 55:279–286
Cockett MI, Bebbington CR, Yarranton GT (1990) High-level expression of tissue inhibitor of metalloproteinases in Chinese hamster ovary cells using glutamine-synthetase gene amplification. Biotechnology 8:662–667

Couture ML, Heath CA (1995) Relationship between loss of heavy chains and the appearance of nonproducing hybridomas. Biotechnol Bioeng 47:270–275

de Felipe P, Martin V, Cortes ML, Ryan M, Izquierdo M (1999) Use of the 2A sequence from foot-and-mouth disease virus in the generation of retroviral vectors for gene therapy. Gene Ther 6:198–208

de Felipe P, Luke GA, Hughes LE, Gani D, Halpin C, Ryan MD (2006) *E unum pluribus*: multiple proteins from a self-processing polyprotein. Trends Biotechnol 24:68–75

Derouazi M, Martinet D, Besuchet Schmutz N, Flaction R, Wicht M, Bertschinger M, Hacker DL, Beckmann JS, Wurm FM (2006) Genetic characterization of CHO production host DG44 and derivative recombinant cell lines. Biochem Biophys Res Commun 340:1069–1077

Dillon N, Festenstein R (2002) Unravelling heterochromatin: competition between positive and negative factors regulates accessibility. Trends Genet 18:252–258

Dinnis DM, James DC (2005) Engineering mammalian cell factories for improved recombinant monoclonal antibody production: lessons from nature? Biotechnol Bioeng 91:180–189

Dinnis DM, Stansfield SH, Schlatter S, Smales CM, Alete D, Birch JR, Racher AJ, Marshall CT, Nielsen LK, James DC (2006) Functional proteomic analysis of GS-NS0 murine myeloma cell lines with varying recombinant monoclonal antibody production rate. Biotechnol Bioeng 94:830–841

Donnelly ML, Luke G, Mehrotra A, Li X, Hughes LE, Gani D, Ryan MD (2001) Analysis of the aphthovirus 2A/2B polyprotein 'cleavage' mechanism indicates not a proteolytic reaction, but a novel translational effect: a putative ribosomal 'skip'. J Gen Virol 82:1013–1025

Dul JL, Aviel S, Melnick J, Argon Y (1996) Ig light chains are secreted predominantly as monomers. J Immunol 157:2969–2975

Fagioli C, Mezghrani A, Sitia R (2001) Reduction of interchain disulfide bonds precedes the dislocation of Ig-μ chains from the endoplasmic reticulum to the cytosol for proteasomal degradation. J Biol Chem 276:40962–40967

Fang J, Qian JJ, Yi S, Harding TC, Tu GH, VanRoey M, Jooss K (2005) Stable antibody expression at therapeutic levels using the 2A peptide. Nat Biotechnol 23:584–590

Fang J, Yi S, Simmons A, Tu GH, Nguyen M, Harding TC, VanRoey M, Jooss K (2007) An antibody delivery system for regulated expression of therapeutic levels of monoclonal antibodies in vivo. Mol Ther 15:1153–1159

Fann CH, Guarna MM, Kilburn DG, Piret JM (1999) Relationship between recombinant activated protein C secretion rates and mRNA levels in baby hamster kidney cells. Biotechnol Bioeng 63:464–472

Flickinger MC, Goebel NK, Bibila T, Boycejacino S (1992) Evidence for posttranscriptional stimulation of monoclonal-antibody secretion by L-glutamine during slow hybridoma growth. J Biotechnol 22:201–226

Foecking MK, Hofstetter H (1986) Powerful and versatile enhancer-promoter unit for mammalian expression vectors. Gene 45:101–105

Fouser LA, Swanberg SL, Lin BY, Benedict M, Kelleher K, Cumming DA, Riedel GE (1992) High-level expression of a chimeric antiganglioside Gd2 antibody – genomic kappa sequences improve expression in Cos and CHO Cells. Biotechnology 10:1121–1127

Fussenegger M, Moser S, Bailey JE (1998) pQuattro vectors allow one-step multigene metabolic engineering and auto-selection of quattrocistronic artificial mammalian operons. Cytotechnology 28:229–235

Gass JN, Gifford NM, Brewer JW (2002) Activation of an unfolded protein response during differentiation of antibody-secreting B cells. J Biol Chem 277:49047–49054

Girod PA, Zahn-Zabal M, Mermod N (2005) Use of the chicken lysozyme 5′ matrix attachment region to generate high producer CHO cell lines. Biotechnol Bioeng 91:1–11

Gomord V, Sourrouille C, Fitchette AC, Bardor M, Pagny S, Lerouge P, Faye L (2004) Production and glycosylation of plant-made pharmaceuticals: the antibodies as a challenge. Plant Biotechnol J 2:83–100

Gonzalez R, Andrews BA, Asenjo JA (2001) Metabolic control analysis of monoclonal antibody synthesis. Biotechnol Prog 17:217–226

Gonzalez R, Andrews BA, Asenjo JA (2002) Kinetic model of BiP- and PDI-mediated protein folding and assembly. J Theor Biol 214:529–537
Goto Y, Hamaguchi K (1981) Formation of the intrachain disulfide bond in the constant fragment of the immunoglobulin light chain. J Mol Biol 146:321–340
Gu MB, Kern JA, Todd P, Kompala DS (1992) Effect of Amplification of dhfr and lac Z genes on growth and beta-galactosidase expression in suspension-cultures of recombinant CHO cells. Cytotechnology 9:237–245
Gurtu V, Yan G, Zhang G (1996) IRES bicistronic expression vectors for efficient creation of stable mammalian cell line. Biochem Biophys Res Commun 229:295–298
Halpin C, Cooke SE, Barakate A, El Amrani A, Ryan MD (1999) Self-processing 2A polyproteins – a system for co-ordinates expression of multiple proteins in transgenic plants. Plant J 17:453–459
Ho Y, Varley J, Mantalaris A (2006) Development and analysis of a mathematical model for antibody-producing GS-NS0 cells under normal and hyperosmotic culture conditions. Biotechnol Prog 22:1560–1569
Houdebine LM, Attal J (1999) Internal ribosome entry sites (IRESs): reality and use. Transgenic Res 8:157–177
Hsu TA, Watson S, Eiden JJ, Betenbaugh MJ (1996) Rescue of immunoglobulins from insolubility is facilitated by PDI in the baculovirus expression system. Protein Expr Purif 7:281–288
Huang Y, Li Y, Wang YG, Gu X, Wang Y, Shen BF (2007) An efficient and targeted gene integration for high-level antibody expression. J Immunol Methods 322:28–39
Jefferis R (2005) Glycosylation of recombinant antibody therapeutics. Biotechnol Prog 21:11–16
Jiang Z, Huang Y, Sharfstein ST (2006) Regulation of recombinant monoclonal antibody production in chinese hamster ovary cells: a comparative study of gene copy number, mRNA level, and protein expression. Biotechnol Prog 22:313–318
Jones D, Kroos N, Anema R, Van Montfort B, Vooys A, Van Der Kraats S, Van Der Helm E, Smits S, Schouten J, Brouwer K, Lagerwerf F, van Berkel P, Opstelten DJ, Logtenberg T, Bout A (2003) High-level expression of recombinant IgG in the human cell line per.c6. Biotechnol Prog 19:163–168
Kalwy S, Rance J, Young R (2006) Toward more efficient protein expression: keep the message simple. Mol Biotechnol 34:151–156
Kaufman RJ, Wasley LC, Spiliotes AJ, Gossels SD, Latt SA, Larsen GR, Kay RM (1985) Coamplification and coexpression of human tissue-type plasminogen activator and murine dihydrofolate reductase sequences in Chinese hamster ovary cells. Mol Cell Biol 5:1750–1759
Kaufman RJ, Davies MV, Wasley LC, Michnick D (1991) Improved vectors for stable expression of foreign genes in mammalian cells by use of the untranslated leader sequence from EMC virus. Nucleic Acids Res 19:4485–4490
Khoo SH, Falciani F, Al-Rubeai M (2007) A genome-wide transcriptional analysis of producer and non-producer NS0 myeloma cell lines. Biotechnol Appl Biochem 47:85–95
Kim HS, Lee GM (2007) Differences in optimal pH and temperature for cell growth and antibody production between two Chinese hamster ovary clones derived from the same parental clone. J Microbiol Biotechnol 17:712–720
Kim NS, Kim SJ, Lee GM (1998a) Clonal variability within dihydrofolate reductase-mediated gene amplified Chinese hamster ovary cells: stability in the absence of selective pressure. Biotechnol Bioeng 60:679–688
Kim SJ, Kim NS, Ryu CJ, Hong HJ, Lee GM (1998b) Characterization of chimeric antibody producing CHO cells in the course of dihydrofolate reductase-mediated gene amplification and their stability in the absence of selective pressure. Biotechnol Bioeng 58:73–84
Kim NS, Byun TH, Lee GM (2001) Key determinants in the occurrence of clonal variation in humanized antibody expression of cho cells during dihydrofolate reductase mediated gene amplification. Biotechnol Prog 17:69–75
Kim JM, Kim JS, Park DH, Kang HS, Yoon J, Baek K, Yoon Y (2004) Improved recombinant gene expression in CHO cells using matrix attachment regions. J Biotechnol 170:95–105

Kito M, Itami S, Fukano Y, Yamana K, Shibui T (2002) Construction of engineered CHO strains for high-level production of recombinant proteins. Appl Microbiol Biotechnol 60:442–448
Koduri RK, Miller JT, Thammana P (2001) An efficient homologous recombination vector pTV(I) contains a hot spot for increased recombinant protein expression in Chinese hamster ovary cells. Gene 280:87–95
Kozak M (1999) Initiation of translation in prokaryotes and eurkaryotes. Gene 234:187–208
Kromenaker SJ, Srienc F (1994) Stability of producer hybridoma cell lines after cell sorting: a case study. Biotechnol Prog 10:299–307
Ku SC, Ng DT, Yap MG, Chao SH (2008) Effects of overexpression of X-box binding protein 1 on recombinant protein production in Chinese hamster ovary and NS0 myeloma cells. Biotechnol Bioeng 99:155–164
Kwaks TH, Otte AP (2006) Employing epigenetics to augment the expression of therapeutic proteins in mammalian cells. Trends Biotechnol 24:137–142
Kwaks TH, Barnett P, Hemrika W, Siersma T, Sewalt RG, Satijn DP, Brons JF, van Blokland R, Kwakman P, Kruckeberg AL, Kelder A, Otte AP (2003) Identification of anti-repressor elements that confer high and stable protein production in mammalian cells. Nat Biotechnol 21:553–558
Lee GM, Varma A, Palsson BO (1991) Application of population balance model to the loss of hybridoma antibody productivity. Biotechnol Prog 7:72–75
Lee YK, Brewer JW, Hellman R, Hendershot LM (1999) BiP and immunoglobulin light chain cooperate to control the folding of heavy chain and ensure the fidelity of immunoglobulin assembly. Mol Biol Cell 10:2209–2219
Leitzgen K, Knittler MR, Haas IG (1997) Assembly of immunoglobulin light chains as a prerequisite for secretion. A model for oligomerization-dependent subunit folding. J Biol Chem 272:3117–3123
Leno M, Merten OW, Hache J (1992) Kinetic-analysis of hybridoma growth and monoclonal-antibody production in semicontinuous culture. Biotechnol Bioeng 39:596–606
Li CY, Little JB, Hu K, Zhang W, Zhang L, Dewhirst MW, Huang Q (2001) Persistent genetic instability in cancer cells induced by non-DNA-damaging stress exposures. Cancer Res 61:428–432
Li H, Sethuraman N, Stadheim TA, Zha D, Prinz B, Ballew N, Bobrowicz P, Choi BK, Cook WJ, Cukan M, Houston-Cummings NR, Davidson R, Gong B, Hamilton SR, Hoopes JP, Jiang Y, Kim N, Mansfield R, Nett JH, Rios S, Strawbridge R, Wildt S, Gerngross TU (2006) Optimization of humanized IgG's in glycoengineered *Pichia pastoris*. Nat Biotechnol 24:210–215
Li J, Menzel C, Meier D, Zhang C, Dubel S, Jostock T (2007a) A comparative study of different vector designs for the mammalian expression of recombinant IgG antibodies. J Immunol Methods 318:113–124
Li J, Zhang C, Jostock T, Dubel S (2007b) Analysis of IgG heavy chain to light chain ratio with mutant Encephalomyocarditis virus internal ribosome entry site. Protein Eng Des Sel 20:491–496
Lilie H, McLaughlin S, Freedman R, Buchner J (1994) Influence of protein disulfide-isomerase (Pdi) on antibody folding in-vitro. J Biol Chem 269:14290–14296
Ma Y, Hendershot LM (2003) The stressful road to antibody secretion. Nat Immunol 4:310–311
Makrides SC (1999) Components of vectors for gene transfer and expression in mammalian cell. Protein Expr Purif 17:183–202
Mayer M, Kies U, Kammermeier R, Buchner J (2000) BiP and PDI cooperate in the oxidative folding of antibodies in vitro. J Biol Chem 275:29421–29425
McBurney MW, Mai T, Yang X, Jardine K (2002) Evidence for repeat-induced gene silencing in culture Mammalian cells: inactivation of tandem repeats of transfected genes. Exp Cell Res 274:1–8
McLean GR, Nakouzi A, Casadevall A, Green NS (2000) Human and murine immunoglobulin expression vector cassettes. Mol Immunol 37:837–845
Meier JL, Stinski MF (1996) Regulation of human cytomegalovirus immediate-early gene expression. Intervirology 39:331–342

Melnick JH, Dul JL, Argon Y (1994) Sequential interaction of the chaperones BiP and GRP94 with immunoglobulin chains in the endoplasmic reticulum. Nature 370:373–375

Merten OW, Moeurs D, Keller H, Leno M, Palfi GE, Cabanie L, Couve E (1994) Modified monoclonal-antibody production kinetics, kappa/gamma mRNA levels, and metabolic-activities in a murine hybridoma selected by continuous culture. Biotechnol Bioeng 44:753–764

Meunier L, Usherwood YK, Chung KT, Hendershot LM (2002) A subset of chaperones and folding enzymes form multiprotein complexes in endoplasmic reticulum to bind nascent proteins. Mol Biol Cell 13:4456–4469

Mezghrani A, Fassio A, Benham A, Simmen T, Braakman I, Sitia R (2001) Manipulation of oxidative protein folding and PDI redox state in mammalian cells. EMBO J 20:6288–6296

Mielke C, Tummler M, Schubeler D, von Hoegen I, Hauser H (2000) Stabilized, long-term expression of heterodimeric proteins from tricistronic mRNA. Gene 254:1–8

Mohan C, Park SH, Chung JY, Lee GM (2007) Effect of doxycycline-regulated protein disulfide isomerase expression on the specific productivity of recombinant CHO cells: thrombopoietin and antibody. Biotechnol Bioeng 98:611–615

Morris AE, Lee CC, Hodges K, Aldrich TL, Krantz C, Smidt PS, Thomas JN (1997) Expression augmenting sequence elements (EASE) isolated from Chinese hamster ovary cells. In: Carrondo MJT, Griffiths B, Moreira JLP (eds) Animal cell technology. Kluwer Academic, The Netherlands, pp 529–534

Oh MK, Scoles DR, Haipek C, Strand AD, Gutmann DH, Olson JM, Pulst SM (2003) Genetic heterogeneity of stably transfected cell lines revealed by expression profiling with oligonucleotide microarrays. J Cell Biochem 90:1068–1078

Ohya T, Hayashi T, Kiyama E, Nishii H, Miki H, Kobayashi K, Honda K, Omasa T, Ohtake H (2008) Improved production of recombinant human antithrombin III in Chinese hamster ovary cells by ATF4 overexpression. Biotechnol Bioeng 100:317–324

Otte AP, Kwaks TH, van Blokland RJ, Sewalt RG, Verhees J, Klaren VN, Siersma TK, Korse HW, Teunissen NC, Botschuijver S, van Mer C, Man SY (2007) Various expression-augmenting DNA elements benefit from STAR-Select, a novel high stringency selection system for protein expression. Biotechnol Prog 23:801–807

Page MJ, Sydenham MA (1991) High-level expression of the humanized monoclonal-antibody campath-1 h in Chinese-hamster ovary cells. Biotechnology 9:64–68

Pavlou AK, Belsey MJ (2005) The therapeutic antibodies market to 2008. Eur J Pharm Biopharm 59:389–396

Pavlou AK, Reichert JM (2004) Recombinant protein therapeutics-success rates, market trends and values to 2010. Nat Biotechnol 22:1513–1519

Rees S, Coote J, Stables J, Goodson S, Harris S, Lee MG (1996) Bicistronic vector for the creation of stable mammalian cell lines that predisposes all antibiotic resistant cells to express recombinant protein. Biotechniques 20:102–110

Reff ME (1993) High-level production of recombinant immunoglobulins in mammalian cells. Curr Opin Biotechnol 4:573–576

Reichert JM, Rosensweig CJ, Faden LB, Dewitz MC (2005) Monoclonal antibody successes in the clinic. Nat Biotechnol 23:1073–1078

Running Deer J, Allison DS (2004) High-level expression of proteins in mammalian cells using transcription regulatory sequences from the Chinese hamster EF-1alpha gene. Biotechnol Prog 20:880–889

Rutkowski DT, Kaufman RJ (2003) All roads lead to ATF4. Dev Cell 4:442–444

Rutkowski DT, Kaufman RJ (2004) A trip to the ER: coping with stress. Trends Cell Biol 14:20–28

Ryan MD, King AM, Thomas GP (1991) Cleavage of foot-and-mouth disease virus polyprotein is mediated by residues located within a 19 amino acid sequence. J Gen Virol 72:2727–2732

Schlatter S, Stansfield SH, Dinnis DM, Racher AJ, Birch JR, James DC (2005) On the optimal ratio of heavy to light chain genes for efficient recombinant antibody production by CHO cells. Biotechnol Prog 21:122–133

Seth G, Philp RJ, Lau A, Jiun KY, Yap M, Hu WS (2007) Molecular portrait of high productivity in recombinant NS0 cells. Biotechnol Bioeng 97:933–951

Simmons LC, Reilly D, Klimowski L, Raju TS, Meng G, Sims P, Hong K, Shields RL, Damico LA, Rancatore P, Yansura DG (2002) Expression of full-length immunoglobulins in *Escherichia coli*: rapid and efficient production of aglycosylated antibodies. J Immunol Methods 263:133–147

Sleiman RJ, Gray PP, McCall MN, Codamo J, Sunstrom NA (2008) Accelerated cell line development using two-color fluorescence activated cell sorting to select highly expressing antibody-producing clones. Biotechnol Bioeng 99:578–587

Smales CM, Dinnis DM, Stansfield SH, Alete D, Sage EA, Birch JR, Racher AJ, Marshall CT, James DC (2004) Comparative proteomic analysis of GS-NS0 murine myeloma cell lines with varying recombinant monoclonal antibody production rate. Biotechnol Bioeng 88:474–488

Stansfield SH, Allen EE, Dinnis DM, Racher AJ, Birch JR, James DC (2007) Dynamic analysis of GS-NS0 cells producing a recombinant monoclonal antibody during fed-batch culture. Biotechnol Bioeng 97:410–424

Strutzenberger K, Borth N, Kunert R, Steinfellner W, Katinger H (1999) Changes during subclone development and ageing of human antibody-producing recombinant CHO cells. J Biotechnol 69:215–226

Tigges M, Fussenegger M (2006) Xbp1-based engineering of secretory capacity enhances the productivity of Chinese hamster ovary cells. Metab Eng 8:264–272

Trill JJ, Shatzman AR, Ganguly S (1995) Production of monoclonal antibodies in COS and CHO Cells. Curr Opin Biotechnol 6:553–560

Tu BP, Weissman JS (2004) Oxidative protein folding in eukaryotes: mechanisms and consequences. J Cell Biol 164:341–346

Underhill MF, Smales CM, Naylor LH, Birch JR, James DC (2007) Transient gene expression levels from multigene expression vectors. Biotechnol Prog 23:435–443

Urlaub G, Chasin LA (1980) Isolation of Chinese hamster cell mutants deficient in dihydrofolate reductase activity. Proc Natl Acad Sci USA 77:4216–4220

Urlaub G, Kas E, Carothers AM, Chasin LA (1983) Deletion of the diploid dihydrofolate reductase locus from cultured mammalian cells. Cell 33:405–412

Vanhove M, Usherwood YK, Hendershot LM (2001) Unassembled Ig heavy chains do not cycle from BiP in vivo but require light chains to trigger their release. Immunity 15:105–114

Walsh G (2003) Biopharmaceutical benchmarks 2003. Nat Biotechnol 21:865–870

Walsh G (2006) Biopharmaceutical benchmarks 2006. Nat Biotechnol 24:769–776

Whiteley EM, Hsu TA, Betenbaugh MJ (1997) Modeling assembly, aggregation and chaperoning of immunoglobulin g production in insect cells. Biotechnol Bioeng 56:106–116

Wiberg FC, Rasmussen SK, Frandsen TP, Rasmussen LK, Tengbjerg K, Coljee VW, Sharon J, Yang CY, Bregenholt S, Nielsen LS, Haurum JS, Tolstrup AB (2006) Production of target-specific recombinant human polyclonal antibodies in mammalian cells. Biotechnol Bioeng 94:396–405

Wurm FM (2004) Production of recombinant protein therapeutics in cultivated mammalian cells. Nat Biotechnol 22:1393–1398

Wurm FM, Pallavicini MG, Arathoon R (1991) Integration and stability of CHO amplicons containing plasmid sequences. Dev Biol Stand 76:69–82

Yang Q, Sarnow P (1997) Location of the internal ribosome entry site in the 5′ non-coding region of the immunoglobulin heavy-chain binding protein (BiP) mRNA: evidence for specific RNA-protein interactions. Nucleic Acids Res 25:2800–2807

Yoon SK, Hwang SO, Lee GM (2004) Enhancing effect of low temperature on specific antibody productivity of recombinant Chinese hamster ovary cell: clonal variation. Biotechnol Prog 20:1683–1688

Zahn-Zabal M, Kobr M, Girod PA, Imhof M, Chatellard P, de Jesus M, Wurm F, Mermod N (2001) Development of stable cell lines for production or regulated expression using matrix attachment regions. J Biotechnol 87:29–42

Engineering Cell Function by RNA Interference

Joseph A. Gredell, Hemant K. Kini, and S. Patrick Walton

Abstract RNA interference (RNAi) is a recently discovered technique for the directed inhibition of gene expression at the level of the mRNA. Its specificity and potency make it a promising method for manipulating cellular phenotypes in eukaryotic cell lines and primary cells. RNAi is already being explored for therapeutic and bioprocessing applications, and it is expected that this will only expand as understanding of the mechanism increases. In this chapter, the mechanism of RNAi is described, followed by a description of some of the technical challenges that remain to realize the full benefits of RNAi. The progress to date on RNAi-based applications will also be detailed.

1 Introduction

A variety of methods are available to manipulate cellular function to achieve a desired response or phenotype. These include gross methods such as manipulation of the culture media, the adhesion matrix, or other factors in the extracellular environment. Those of greater interest, and, hence, those that are more frequently applied, are highly-tunable techniques for specific manipulation of cells, in particular at the level of gene expression. For upregulation of gene expression, transient or stable plasmid transfection or viral transduction are now standard tools for eukaryotic and prokaryotic expression manipulation (Chen et al., 2003). Fine control of the expression of the gene inserted by these techniques is now a relatively easy task through the use of promoter engineering (Alper et al., 2005). In prokaryotes, a recently discovered gene expression control modality, riboswitches (Mandal and Breaker, 2004), which alter gene expression in response to changes in metabolite concentrations, function in both the upregulation and downregulation of gene expression (Cheah et al., 2007). Combinations of these high affinity ligand binding riboswitches have been used to perturb specific metabolic pathways (Liao, 2004).

J.A. Gredell, H.K. Kini, and S.P. Walton (✉)
Department of Chemical Engineering and Materials Science,
Michigan State University, USA e-mail: spwalton@egr.msu.edu

M. Al-Rubeai (ed.), *Cell Line Development*, Cell Engineering 6,
DOI 10.1007/978-90-481-2245-5_9, © Springer Science+Business Media B.V. 2009

Technologies for downregulation of byproduct/undesired pathways, such as antisense oligonucleotides (Walton et al., 2000) and knockouts (Gallitaliadoros et al., 1995), have been used in metabolic engineering for manipulating gene functions and targeting metabolic pathways since the early 1990s (Warner, 1999). Antisense oligonucleotides showed significant promise for highly-specific downregulation of mRNA levels of target genes through activation of RNase H (Walton et al., 2000). However, the relatively high concentrations of antisense oligonucleotides required to achieve the desired effect were often a challenge, in particular for in vivo applications, due to immunogenicity of natural and chemically-modified oligonucleotides (Crooke, 2004). Nonetheless, transient gene expression control on established cell lines provided a powerful tool for the engineering of cell function.

While searching for the mechanism of antisense RNA regulation of gene expression in *Caenorhabditis elegans*, it was determined that double-stranded RNA (dsRNA) was far more active in inhibiting gene expression than the antisense single-stranded RNA (ssRNA) (Fire et al., 1998). This dsRNA regulatory mechanism was coined RNA interference (RNAi), and it has since become a widespread tool for manipulation of cellular function for metabolic engineering, protein expression control, and therapeutic applications (Kim et al., 2007). This pathway, which is functional only in eukaryotes, is hypothesized to have arisen as a means to defend the host organism against viral infection and other events that result in alteration of the host genome. Given the extraordinary specificity of RNAi, one of its earliest applications has been in functional genomics. Phenotypic profiles for individual knockdowns of 98% of the predicted genes in *C. elegans* were characterized using RNAi (Kamath et al., 2003; Sonnichsen et al., 2005). The specificity of RNAi is perhaps best demonstrated by the effective and specific silencing of mutant mRNAs that differ at only a single nucleotide from the wild-type alleles, such as in amyotrophic lateral sclerosis and Huntington's disease (Maxwell et al., 2004; Schwarz et al., 2006). Continued expansion of the use of RNAi as both a research and therapeutic tool will depend on enhanced understanding of the mechanistic details of the process and solving critical challenges that even now impact the straightforward application of RNAi.

2 Mechanism of RNAi

RNAi was first characterized in *C. elegans* after dsRNAs were observed to silence the expression of a gene containing a sequence complementary to one strand of the dsRNA (Fire et al., 1998). It was shown that exogenously introduced 80–100 base pair (bp) dsRNAs triggered RNAi and silenced the gene of interest with the effect being potent enough to pass from mother to daughter cells (Hammond et al., 2001; Hannon, 2002). In the cytoplasm, these long dsRNAs are then cleaved by the RNase III family enzyme Dicer (Bernstein et al., 2001) to yield ~21-mer short interfering RNAs (siRNAs) with 5′ phosphates and 3′ dinucleotide overhangs

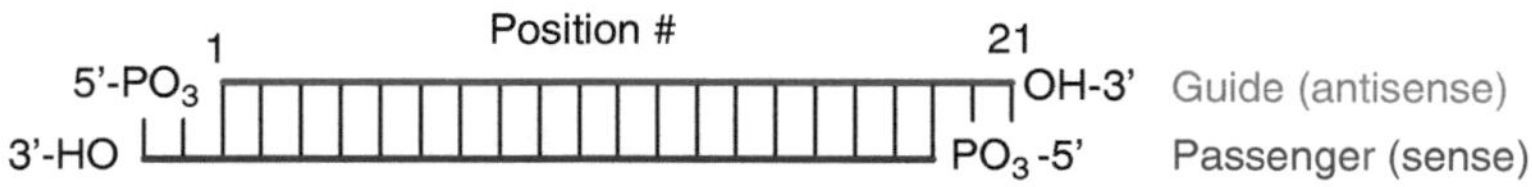

Fig. 1 Structure of an siRNA. siRNAs have 19 base pairs in their core with 2 nt overhangs at each 3′-end, making each strand 21 nt in length. If not already phosphorylated, the 5′-ends are phosphorylated upon entry into the cell cytoplasm

(Elbashir et al., 2001b) (Fig. 1). After cleavage by Dicer, additional proteins are recruited to the siRNAs to form the RNA-induced silencing complex (RISC) and its predecessor, the RISC loading complex (RLC). In humans, the proteins required to assemble the minimal forms of the RLC and RISC are Dicer, TAR RNA Binding Protein (TRBP), and Argonaute 2 (Ago2) (MacRae et al., 2008). After assembly of the complex, RISC becomes activated when Ago2 cleaves the passenger (non-targeting) strand of the siRNA (Fig. 1), leaving the remaining strand to function as a guide to the complementary target mRNA. Active RISC then binds to the target by hybridization between the guide strand and the mRNA allowing Ago2 to cleave the mRNA 10 nt from the 5′-end of the guide strand (Elbashir et al., 2001a), resulting in inhibition of gene expression.

In parallel to characterizing the initiation of RNAi by exogenous dsRNAs, it was shown that RNAi is a constitutive pathway in eukaryotes, activated by endogenously-expressed small RNAs, termed micro RNAs (miRNAs) (Bartel, 2004). miRNAs are expressed as hairpin transcripts that are processed by the RNases Drosha and then Dicer to their final form (Lee et al., 2003), which is essentially identical to an siRNA. Unlike siRNAs, natural miRNAs do not exhibit perfect complementarity between the two miRNA strands or between the guide strand and the target mRNA (Bartel, 2004). Nonetheless, miRNAs are thought to regulate at least one-third of mRNAs expressed in humans, being involved in cellular processes such as proliferation, metabolism, and differentiation (Bartel, 2004; Cui et al., 2007). The final step in each pathway represents the major difference between siRNA-mediated and miRNA-mediated expression control. At this step, miRNA-guided RISC does not cleave the target but rather remains bound to the target to inhibit translation by a steric mechanism (Bartel, 2004). Moreover, humans have eight Argonaute family proteins, of which Ago2 is the only member capable of guide strand directed mRNA cleavage (Liu et al., 2004), with others involved in miRNA-mediated RNAi (Carmell et al., 2002). Exogenous siRNAs, then, take advantage of the presence of the components for the constitutively-active, endogenous miRNA pathway. The remainder of this chapter will focus on siRNA-mediated RNAi (Fig. 2), as this is the primary mode for exogenous manipulation of cell function currently in use.

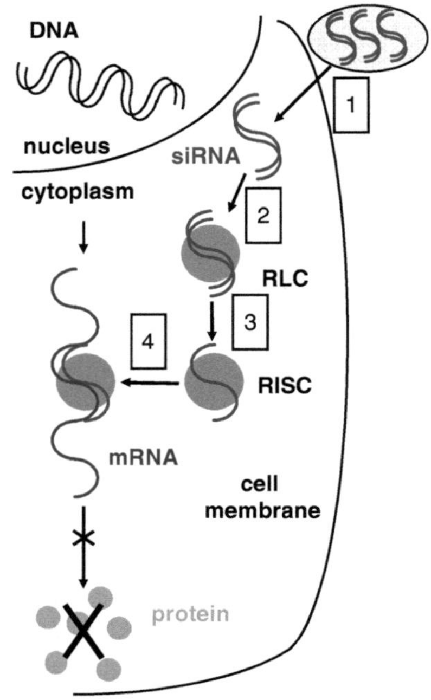

Fig. 2 siRNA-mediated RNAi pathway. Key steps in the process include: access of the siRNA to the cytoplasm (1), binding of the siRNA to proteins to form the RLC and RISC (2), removal of the passenger strand leaving only the proper strand to guide RISC (3), and hybridization to the target region on the mRNA (4)

3 Current Challenges for RNAi

For readily accessible cells, such as in culture, the current mechanistic understanding of the RNAi pathway underscores that silencing cannot occur without: (i) access of the siRNAs to the cytoplasm, (ii) binding of siRNAs by RNAi pathway proteins, (iii) selection of the appropriate strand to guide RISC cleavage, and (iv) binding of active RISC to the intended target mRNA (Fig. 2). To design the most active siRNAs, it would be necessary to optimize each of these processes, with these optimization choices often being incompatible. Yet, even these are only a subset of the design choices that must be made when using RNAi.

The selection of a target gene is typically the first design decision in an RNAi application. Subsequently, the choice of an siRNA sequence to target the intended gene also impacts the incidence of other secondary effects such as off-target silencing and the stimulation of immune responses. Moreover, for in vivo applications particularly, successful delivery of the designed siRNAs to the intended cells or tissues prior to degradation or renal clearance is critical. Together, siRNA design and delivery are the two most significant current challenges to the effective and widespread application of RNAi. It is therefore essential to analyze these processes using quantitative and mechanistic approaches to supplement cell engineering efforts that rely on this endogenous pathway.

3.1 siRNA Sequence Design

siRNAs possess unique structural characteristics that are essential for their silencing activity and yet are independent of their sequence. siRNAs are ~21 nt long, with 2 nt overhangs on either 3′-end (Fig. 1) (Bernstein et al., 2001; Elbashir et al., 2001b). Overhangs of this kind are characteristic of RNase III cleavage of a dsRNA, in this case by Dicer. When cleaved by Dicer from a long dsRNA template, siRNAs always possess phosphates on the 5′-ends. In contrast, chemically-synthesized siRNAs typically have hydroxyl groups in these positions, which would limit their activity in RNAi (Weitzer and Martinez, 2007). However, when exogenous siRNAs are delivered to cells, if either 5′-end lacks a phosphate, one is added by the kinase Clp in the cytoplasm (Weitzer and Martinez, 2007). These defining characteristics, length, overhang structure, and 5′-phosphorylation, are significant for incorporation of the siRNA into RISC, although the exact contributions of each of the factors remain unknown. That notwithstanding, knowledge of the general structural requirements of siRNAs allows direct chemical synthesis of active agents whose activities differ only due to changes in their sequence and, by extension, their target location on the mRNA.

Early sequence design was not based on understanding of the silencing mechanism and instead relied more on empirical rules. When it was determined that exogenous, chemically-synthesized siRNAs could initiate RNAi (Elbashir et al., 2001a), sequence-based rules were established to maximize the efficiency of siRNA synthesis, including elimination of candidate sequences based on excessive GC content and stretches of greater than four consecutive identical bases (e.g., GGGG) (Elbashir et al., 2002). Other positional base preferences (e.g., an A at position 19 of the sense strand) have been correlated from large data sets (Jagla et al., 2005; Khvorova et al., 2003; Reynolds et al., 2004; Ui-Tei et al., 2004). These preferences have been the foundation for elaborate computational algorithms that use ten or more parameters to identify candidate siRNAs (Ge et al., 2005; Lu and Mathews, 2007; Shah et al., 2007; Vert et al., 2006).

Subsequent identification of additional mechanistic requirements has led to other design rules. One example is derived from analysis of the relative stability of hybridization at one end (~5 bp) of the siRNA duplex relative to the other. Naïvely, it would appear that either strand of the siRNA may be incorporated into active RISC with approximately equal probability. Half of the activated RISCs would then target the complementary sequence of the guide strand on the intended mRNA while the other half would target any mRNAs with sequences complementary to the passenger strand. Though the likelihood of there being naturally occurring, perfectly complementary targets for both the guide and passenger strands is quite small, even partial complementarity between an mRNA and the passenger strand can lead to some silencing of transcripts other than the target, often called off-target silencing (Jackson and Linsley, 2004; Svoboda, 2007). Even if no significant off-target silencing occurs, loading of RISCs with the passenger strand increases the concentration of siRNA required for effective silencing by occupying RISCs that would otherwise contain the guide strand and be active against the intended target (Matranga et al., 2005). Active siRNAs, often defined as those that reduce the expression of the target gene by at least 75% at a concentration of ≤30 nM for cell culture experiments, tend to exhibit a bias in their internal stability that leads to the strand whose 5′-end is less stably hybridized within the siRNA duplex becoming preferentially incorporated into active RISC (Khvorova et al., 2003; Schwarz et al., 2003; Tomari et al., 2004). This is typically referred to as siRNA duplex asymmetry. The result is that a higher proportion of active RISCs contain the guide strand (for the desired target), leading to more active target silencing at any siRNA concentration. Many of the positional base preferences identified as being useful for selecting active siRNAs tend to yield the desired differential stability between the two ends (Jagla et al., 2005; Reynolds et al., 2004; Ui-Tei et al., 2004).

Until recently, siRNA selection guidelines have not included the possible impact of the mRNA sequence and structure on silencing efficiency. siRNAs can be equally effective when targeting inside the coding region of the mRNA or in the 5′- and 3′-untranslated regions (e.g., Yoshinari et al., 2004). Whereas siRNA sequences are only ~21 nt long and are double-stranded most of the time, single-stranded mRNAs can be several thousand nucleotides long and therefore tend to possess significant intramolecular base-pairing, termed secondary structure. This intramolecular secondary structure can impair the ability of RISC to form intermolecular interactions with its target mRNA (Ameres et al., 2007; Bohula et al., 2003; Brown et al., 2005; Far and Sczakiel, 2003; Gredell et al., 2008; Overhoff et al., 2005; Schubert et al., 2005; Shao et al., 2007; Vickers et al., 2003; Westerhout and Berkhout, 2007). Other work has also recently shown that target accessibility is one of the most important factors in defining siRNA efficacy (Tafer et al., 2008). Furthermore, it was shown that the guide strand of the siRNA can also form limited secondary structure, and that this effect can significantly impact the silencing efficiency of the siRNA (Patzel et al., 2005). Our computational analyses on a large set of siRNAs supported this finding, suggesting that guide strand structure does indeed limit the interaction between the guide strand and the complementary mRNA target site (Gredell et al., 2008). Results such as these are now being included in siRNA selection algorithms (Heale et al., 2005; Lu and Mathews, 2007; Shao et al., 2007) using programs that predict RNA secondary structure such as UNAfold (a newer version of mfold) (Markham and Zuker, 2008), Sfold (Ding et al., 2004), and the Vienna RNA package (Hofacker, 2003), that can be coupled with other sequence filtering algorithms.

3.2 *siRNA Structure Design*

With an active siRNA sequence identified, it is still necessary to design the siRNA construct to meet the demands of the application for which it is intended. This may be achieved through chemical and structural manipulations. For chemical modifications, all three functional units of each nucleotide, the phosphate backbone, the ribose sugar, and the nucleotide base, can be targeted for modification. However, to maintain the activity of the siRNA, these modifications must not appreciably alter the A-form helical structure of the dsRNA as this is critical for recognition by the proteins of the RLC and RISC (Amarzguioui et al., 2003; Chiu and Rana, 2002; Haley and Zamore, 2004). It has generally been found that the 5′-end of the guide strand should contain a free hydroxyl (which can be phosphorylated upon cell entry) or a phosphate group, and, therefore, this location cannot be otherwise modified (Rana, 2007).

One significant concern for siRNA design is off-target silencing, as alluded to above. It has been shown that off-target effects could be reduced by replacing the 5′-hydroxyl on the passenger strand with a 5′-methoxyl modification (Chen et al., 2008). This prevents phosphorylation and subsequent incorporation of the passenger strand into active RISC. Thus, any active RISCs must contain unmodified guide strands and therefore will be directed only at the intended target mRNA. It was also shown that substituting DNA nucleotides everywhere in the 5′-third of the siRNA duplex, on both the guide and passenger strands, eliminated off-target effects without substantial reduction in siRNA activity (Ui-Tei et al., 2008). Presumably, RNA at the 3′-end of the guide strand is necessary for interactions with TRBP or Ago2, either for formation of a stable RLC and RISC or for stabilizing hybridization to the target mRNA; conversely, the DNA:RNA hybrid formed at the 3′-end of the passenger strand appears not to permit the interactions that are essential for RNAi.

Other modifications can be made to the siRNA structure with the goal of increasing the longevity and specificity of the siRNA in the cellular environment. These strategically placed modifications can improve resistance to RNases, enhance the biodistribution of the molecules in vivo, as well as facilitate cellular uptake and localization (Corey, 2007). Phosphorothioate (PS) linkages are a particularly common backbone modification that specifically enhances the resistance of the backbone to cleavage by RNases. Unfortunately, substantial PS modifications result in increased cytotoxicity (Corey, 2007). Boranophosphate (BO) linkages, while having been studied less frequently and being limited in scale by synthesis techniques, appear to offer similar benefits (Corey, 2007). An alternative is to modify the ribose sugar, specifically at the 2′-position, using bulky groups that interfere with hydrolysis. Such groups include, but are certainly not limited to, 2′-OCH_3, 2′-F, and locked nucleic acids (LNA) (Corey, 2007; Rana, 2007). In some cases, they can even enhance activity relative to unmodified sequences (Elmen et al., 2005). Other modifications that were originally devised to improve the biodistribution of antisense oligonucleotides, such as direct conjugation to cholesterol, receptor ligands, and transport peptides, can potentially be applied for siRNAs as well, provided they do not prevent recognition of the modified siRNA by RNAi proteins (recently reviewed in de Fougerolles et al., 2007).

3.3 siRNA Delivery Methods

While the improvement of algorithms that effectively identify highly functional siRNAs is a point of emphasis for in vitro applications of RNAi, the main limitation for in vivo use, and consequently use in a human clinical setting, is delivery of siRNAs in sufficient concentrations to the tissues and cells of interest. In some lower organisms, such as worms, long dsRNAs can be eaten or absorbed, resulting in highly efficient dsRNA delivery to many tissues (Fire et al., 1998; Hannon, 2002). However, in mammals, systemic administration of long dsRNAs results in immune stimulation, and naked siRNAs (siRNA alone with no delivery agent) do not diffuse directly through cellular membranes in sufficient quantities to initiate RNAi in most tissues (de Fougerolles et al., 2007). This constraint on systemic siRNA administration is greatly compounded by the rapid degradation of siRNAs by nucleases (<1 h half-life in serum), binding interactions between the siRNAs and blood components, and clearance by the kidneys (Zhang et al., 2007). Although viral delivery methods have the ability to overcome these limitations by stable incorporation of silencing constructs into the genomes of infected cells, viral gene therapy methods still suffer from a number of safety issues that have caused multiple gene therapy clinical trials to be abandoned or curtailed (e.g., http://www.fda.gov/bbs/topics/NEWS/2007/NEW01672.html). Thus, much effort is being spent developing non-viral delivery methods for siRNAs for therapeutic and other applications.

In some instances, cellular uptake can be forced through physical methods, including intravascular injection, ultrasound, electroporation, and gene guns (Wolff and Rozema, 2008). In other circumstances, delivery can be accomplished by aerosolizing the siRNA for administration to the lungs, or by application of siRNA in a topical cream for dermal delivery (de Fougerolles et al., 2007). The siRNAs currently in clinical trials are delivered either by injection into the eye (for age-related macular degeneration, AMD) or inhalation to the lungs (for respiratory syncytial virus, RSV) (de Fougerolles, 2008). Unfortunately, these methods are not viable for delivery to deep tissues or solid tumors, targets of significant interest for siRNA therapeutics. As an alternative, both lipid and polymer based reagents are being pursued as delivery vehicles that can be administered systemically while retaining the potential for localized targeting. These approaches hold promise for delivery to even the most refractory of tissues, including transport through the blood-brain-barrier for treatment of neurological diseases (Pardridge, 2007).

Due to their phosphate backbone, all nucleic acids are highly negatively charged. Accordingly, delivery vehicles for siRNAs, plasmids, antisense oligonucleotides, ribozymes, or aptamers typically utilize electrostatic interactions between the nucleic acid backbone and either cationic lipids (lipoplexes) or cationic polymers (polyplexes). Condensing the siRNA in the vehicle can be achieved electrostatically through simple mixing of the siRNA and vehicle in an aqueous buffer and allowing them to self-assemble or by more complex techniques such as spray drying (Takashima et al., 2007). The resulting complexes, which can range in size from ~50–300 nm, have been applied in vitro and in vivo. For several of the complexation strategies studied to date, the final complex charge tends to be slightly positive overall, which facilitates contact with the negative charge of the cellular membrane without causing substantial interactions with blood components that would lead to complement activation (Bartlett and Davis, 2007). For systemic delivery, while there is no universally optimal size, it is generally considered that complexes must be large enough to protect the siRNA from nuclease digestion and renal clearance

(>10 nm) but also small enough (<70–100 nm) to allow access to cells such as hepatocytes through capillary circulation (Bartlett and Davis, 2007; Wolff and Rozema, 2008). After association with the extracellular membrane, the complexes are then endocytosed (Zuhorn et al., 2007). The siRNAs must then escape the endocytotic vesicles into the cytosol to initiate RNAi.

The traditional structure of lipid-based delivery reagents is that of a fatty acid with a cationic head group and a non-polar hydrocarbon tail. Use of this type of transfection was first demonstrated using DOTMA with plasmids (Felgner et al., 1987). Early RNAi transfection was performed using lipids developed for delivery of either plasmids or antisense oligonucleotides. Since the discovery of siRNA-mediated RNAi, a variety of proprietary lipid formulations have been developed that are specifically designed for siRNA transfection to cultured cells, such as Lipofectamine RNAiMAX (Invitrogen) or siPORT NeoFX (Applied Biosystems – Ambion). These reagents can be very effective at delivering siRNAs to diverse cell lines, but, unfortunately, they can be toxic at concentrations only moderately higher than concentrations required for effective siRNA delivery; in vivo experiments have also demonstrated similar toxicity concerns (Zhang et al., 2007). While generally not an issue for cultured cell applications, this small therapeutic window limits the prospects for use of these types of delivery agents for clinical applications. Recently, a combinatorial approach was used to test lipid-like compounds termed "lipidoids" to deliver siRNAs, finding that a diverse subset caused extensive silencing in conditions ranging from cell culture to non-human primate animals (Akinc et al., 2008). In many cases, these molecules were also found to have acceptable toxicity profiles in vivo. These readily synthesized materials provide an alternative to traditional lipids, expand the set of available transfection reagents, and provide insight as to the physical and chemical characteristics necessary for the most effective lipid vehicles. However, due to the relatively limited chemical and structural diversity in lipid and lipid-like vehicles, considerably more effort at creating in vivo delivery vehicles has been invested in using polymeric vehicles for nucleic acid delivery, in general, and for delivery of siRNAs, specifically.

Polymeric vehicles have strong potential for siRNA delivery applications because they provide protection from nucleases and other serum components and facilitate endocytosis of the siRNA and its release into the cytosol. Of these, polyethylenimine (PEI) is perhaps the most commonly used and well-studied (Kircheis et al., 2001). At physiologic pH, the amine groups of PEI are protonated, providing the positive charge necessary to condense the nucleic acid (Clamme et al., 2003). Both linear PEI (LPEI) and branched PEI (BPEI) have shown encouraging results in silencing applications, with some molecular weights also demonstrating delivery of active siRNAs in vivo (Urban-Klein et al., 2005). In several instances, modifications, such as biomolecule addition (e.g., cholesterol, antibodies/peptides, or aptamers) or crosslinking have been made to minimize the toxicity associated with the larger molecular weight PEIs and to improve and target delivery (Swami et al., 2007a, b). A particularly common modification is the addition of polyethylene glycol (PEG), which partially shields the charges on the polymer and siRNA, improves membrane interactions, increases product release, and decreases toxicity

(Wolff and Rozema, 2008). Interestingly, there appears to be an optimal amount and length of PEG per polymer, although the combinations studied have not led to a consensus set of guidelines (Brus et al., 2004; Kunath et al., 2002; Mao et al., 2006).

Although PEI is perhaps the most commonly studied polymer, two other considerably more complex systems have shown intriguing in vivo results. The first is called a "Dynamic PolyConjugate" (Rozema et al., 2003; Rozema et al., 2007; Wakefield et al., 2005). This vehicle contains multiple functional entities to address delivery in a stepwise manner. In a manner akin to a multi-stage rocket, groups are released from the conjugate once they have served their purpose, thus preventing interruption of downstream events. Starting with a membrane-active poly butyl amino vinyl ether (PBAVE) backbone, PEG and N-acetylgalactosamine (for hepatocyte targeting) were attached via maleamate linkages. The siRNA is also covalently linked to the backbone via a disulfide bond. The maleamate linkages are reduced in the endosomal vesicles, exposing the PBAVE, which lyses the endosomal vesicles. The PBAVE-siRNA disulfide bond is reduced once the complex enters the cytoplasm, thereby protecting the siRNA from degradation until it is released in the compartment where it can access the RNAi machinery. Two different siRNAs delivered using this vehicle silenced their target genes by ~60–80% (Rozema et al., 2007)

A second system is built by self-assembly of an siRNA with a cyclodextrin-modified PEI, PEG, and transferrin (for targeting to tumor cells) (Hu-Lieskovan et al., 2005; Pun et al., 2004). This complex shows low toxicity and effective silencing of the target gene in a mouse model and in non-human primates. Together, these two polymeric systems highlight the utility of incorporating factors for targeted delivery and endosomal release. However, they also illustrate the potential complexity of the polymers required to address the many limitations that restrict delivery of active siRNAs in vivo.

4 Applications of RNAi in Biotechnology and Biomedicine

Metabolic flux analysis and genetic screens have identified cellular pathways influencing protein synthesis and cell survival (Srivastava et al., 2007; Tewari and Vidal, 2003). Those pathways identified as being too active to achieve the desired phenotype can be specifically targeted and their activities reduced using RNAi. One specific application in which this has been demonstrated is in the expression of proteins by mammalian cells, which grow more slowly than bacterial cells, but provide proper folding and post-translational modification of the expressed proteins (Wurm, 2004).

Since the approval of the tissue plasminogen activator as the first recombinantly expressed therapeutic protein from Chinese Hamster Ovary (CHO) cells, the CHO cell line and its variations have become the dominant mammalian system for recombinant protein synthesis (Kumar et al., 2008). Maximal protein production from CHO cells requires high viability and initial rapid proliferation of the cells. Apoptosis from stress or other signals reduces the yield and purity of the recombinant protein product. As cells lacking both the proapoptotic proteins BAK and BAX exhibit considerably higher viability compared to cells expressing both these proteins (Wei et al., 2001), siRNAs were used to target these proteins in recombinant human IFN-γ expressing CHO cells. With BAK and BAX mRNAs reduced to 10% of their normal level, the CHO cultures showed 30–50% greater viable cell density, and 35% higher IFN-γ production relative to control cells (Lim et al., 2006).

Cellular pathways in plants can also be manipulated using RNAi to engineer and over-express mammalian antibodies and plant specific proteins. Plants have emerged as a safe and economical alternative to microbial and mammalian cell lines for expressing vaccines and therapeutic antibodies (Floss et al., 2007), in part because they have minimal risk of contamination from potential human pathogens and can be easily scaled up for mass production without the need for extensive purification steps (Daniell et al., 2001). For example, the aquatic plant *Lemna minor* has been developed for producing high yields of therapeutic recombinant proteins (Neuenschwander et al., 1991). These proteins are, however, susceptible to plant specific glycosylation by the genes α-1,3-fucosyltransferase and β-1,2-xylosyltransferase (Gomord et al., 2005). After silencing these enzymes by vector expressed shRNAs, their activities were reduced to the levels of negative control. When used to produce monoclonal antibodies against human CD30, a human cell surface receptor that is specifically upregulated in certain tumor cells such as Hodgkin and Reed-Sternberg cells, adult T cell leukemia, and embryonal carcinoma of the testis (Dong et al., 2003), the antibodies did not have any detectable plant specific glycans and were more potent than antibodies expressed in mammalian cell lines (Cox et al., 2006).

Interestingly, storage organs such as potato tubers can also serve as bioreactors for mass production of human proteins and vaccines, as they provide superb environments for maintaining protein stability (Arntzen et al., 2005; Farran et al., 2002). A major problem for proteins expressed in tubers is patatin contamination (Arntzen et al., 2005). Patatins are a family of glycoproteins that constitute nearly 40% of the soluble protein in potato tubers (Prat et al., 1990). shRNA vectors have been used to target the highly conserved gene (*pat3-k1*) of the patatin family, thereby reducing patatin expression by nearly 99% (Kim et al., 2008). The expressed glycoproteins did not then require purification steps to remove patatin, resulting in significant improvement in protein yield.

In cases where the native plant proteins are detrimental, RNAi has been shown to be active in reducing expression of the undesired protein. Peanut allergy is one of the most severe food allergies, affecting nearly 1% of the US population (Sicherer et al., 1999), with 86% of cases resulting from reactions to the protein Ara h 2 (Koppelman et al., 2004). Knocking down the expression of Ara h 2 using plasmid-expressed shRNAs resulted in hypoallergenic seeds showing significant reduction in allergenic potency, as measured by IgE binding capacity, compared to wild-type peanuts (Dodo et al., 2008).

RNAi can also be coupled with metabolic engineering to provide several advantages over classical plant breeding techniques, such as control of spatial and temporal expression of genes of interest (Tang et al., 2007). These new methods of breeding are being employed to improve the nutritional value of food, offset the loss of agricultural land, and help satisfy the food demand of the growing world population (Doos, 2002). Cotton tissues express the protein Gossypol, which acts as a defense against insects and pathogens (Sunilkumar et al., 2006). However, Gossypol's toxicity to humans prevents the use of cotton seeds as a source of food in developing countries (Townsend and Llewellyn, 2007). shRNAs targeting the enzyme δ-cadinene, which is crucial for Gossypol synthesis, reduced the Gossypol level in transgenic seeds by 99% compared to wild-type seeds, with the transgenic plant showing normal growth and development (Sunilkumar et al., 2006).

A similar approach was used to enhance the plant production of lysine, an essential amino acid important for human nutrition and also livestock growth. Lysine is found in limiting amounts in corn and other cereal grains (Singh et al., 2001), as it negatively regulates the activity of dihydropicolinate synthase (DHPS), the first enzyme in the lysine biosynthesis pathway (Tang et al., 2007). Plants engineered to express DHPS mutants insensitive to lysine showed increased lysine synthesis in all plant organs (Frankard et al., 1992). Unfortunately, elevated lysine levels caused abnormal tissue and flower development, which in turn reduced seed yield (Frankard et al., 1992). The quality of the seeds was also inferior due to defective post-germination lysine catabolism (Zhu and Galili, 2004). Lys-ketoglutarate reductase (LKR) and saccharopine dehydrogenase (SDH) are key enzymes in the lysine catabolism pathway (Zhu and Galili, 2004). In Arabidopsis plants, both the lysine content and seed quality were improved by seed-specific lysine over-expression using both the DHPS mutant and temporal shRNA silencing of the LKR and SDH enzymes during seed development (Zhu and Galili, 2004). These seeds had about 25 mol% higher lysine content than wild-type seeds and also grew faster than seeds from plants over-expressing lysine without LKR and SDH knockdown (Zhu and Galili, 2004).

Alternatively, RNAi modified plants have been shown to be effective for therapeutic drug production. Morphinan alkaloids such as morphine, codeine, oripavine, and thebaine have direct therapeutic use and serve as intermediates for manufacture of synthetic analgesics used to treat opiate addiction (Allen et al., 2008). These alkaloids are only expressed in plants of the genus *Papaver* (Allen et al., 2008). Upregulating the flux through the alkaloid synthesis pathway by over-expression of the enzymes salutaridinol 7-O-acetyltransferase (SalAT) or codeinone reductase (COR) results in increased production of the alkaloids (Allen et al., 2004; Grothe et al., 2001). Morphine, however, remains a significant product of the pathway, impairing the expression and purification of thebaine and codeine, products of significantly greater therapeutic value. Targeting the COR transcript using shRNAs leads to almost complete inhibition of its enzymatic activity, resulting in the upstream accumulation of (S)-reticuline in the alkaloid biosynthesis pathway (Allen et al., 2004). (S)-reticuline is a valuable pharmaceutical intermediate that can easily be converted to salutaridine and then to thebaine (Page, 2005).

Engineering poppy plants in this manner is therefore a means to generate plants with significant medicinal value, while also possibly limiting their use in the production of illicit opioid drugs.

Use of the RNAi pathway to treat human disease is another popular goal. siRNAs are already being tested in vivo, and several are entering, or are currently in, clinical trials (http://clinicaltrials.gov). These first generation therapeutic siRNAs are typically delivered locally by physical means and target well-studied pathways. In particular, two siRNAs in development, siRNA-027 by Sirna Therapeutics (http://www.sirna.com) and Cand5 by Acuity Pharmaceuticals (http://acuitypharma.com), are entering Phase II and III clinical trials, respectively. Both of these target the vascular endothelial growth factor (VEGF) pathway for treatment of the wet form of age-related macular degeneration (AMD). As they target a disease of the eye, both are administered by intravitreal injection, which is also the mode of administration for the only antisense oligonucleotide-based therapeutic, Vitravene, that has gained FDA approval (http://www.isispharm.com/vitravene.html). Similarly, Alnylam Pharmaceuticals (http://www.alnylam.com) has completed a Phase II study targeting the respiratory syncytial virus (RSV) with an siRNA delivered to the lungs as a nasal spray. The treatment was well-tolerated and studies to assess the silencing ability of the siRNA can be expected (DeVincenzo et al., 2008). Furthermore, a second trial is underway with the same molecule for delivery to lung transplant patients infected with RSV.

Several other new siRNAs are being explored at the Phase I stage. Quark Pharmaceuticals (http://www.quarkpharma.com) is investigating siRNA I5NP/Akli-5 for its protective effects on acute kidney failure following major cardiovascular surgery. They have also been granted approval to test DGFi, an siRNA targeting the p53 gene, shown in preclinical studies to protect the kidney during cold ischemia, a lack of oxygen that can occur when organs are stored in low temperatures prior to transplantation (http://clinicaltrials.gov). It is expected that this drug might be useful for delayed graft function (DGF), which is a complication commonly seen following transplantation. In these cases, the siRNAs are being delivered systemically, but they are chemically modified to enhance their stability for such an approach.

In a slightly different approach, Duke University is sponsoring a Phase I trial using an siRNA to reduce dendritic cell expression of the immunoproteasome (http://clinicaltrials.gov). Cells derived from a patient's monocytes are to be treated and then injected into melanoma tumors as a means to induce immune responses to combat tumor growth. This study is unique in that the siRNA itself is not the therapeutic but rather acts to tune the treated cells for maximum immunogenicity. This approach highlights the potential use of siRNAs for indirect therapeutic benefit or as adjuvants for existing therapeutics.

Another intriguing study aims to use local siRNA injection to treat the feet of patients with the rare autosomal dominant disorder, pachyonychia congenital (http://clinicaltrials.gov). This siRNA is designed to target a specific mutant form of the keratin K6a gene. As there are only six known patients with the particular mutation targeted by this siRNA, the application of this particular siRNA cannot be expanded beyond the trial stage. However, successful silencing would validate use of this delivery strategy for RNAi on easily accessible tissues and would provide valuable biological information about the clinical feasibility of treating single-nucleotide mutant genes.

An interesting development is the start of a Phase I trial to establish the safety associated with an siRNA delivered by means of a stabilized polymeric nanoparticle (Calando Pharmaceuticals; http://www.insertt.com). As mentioned above, the cyclodextrin-PEI nanoparticle is intended to protect the siRNA from nuclease degradation thus improving its delivery (Bartlett et al., 2007). Moreover, the nanoparticle contains transferrin as a targeting molecule for tumor cells, which is intended to enhance levels of siRNA locally within the tumor. The encapsulated siRNA targets the M2 subunit of ribonucleotide reductase. This enzyme, which participates in deoxynucleotide synthesis, is therefore essential for cell division. Thus, this treatment potentially provides a means of generally inhibiting the rate of growth of any tumor. This trial represents one of the first to examine the feasibility of delivering siRNAs by polymeric vehicles designed specifically for in vivo applications.

In each of these clinical trials using an siRNA as the therapeutic agent, the intention is to reduce the expression of a *specific* gene. As with any therapeutic, the expected mechanism of action may not always be the actual mechanism. In one study, the mode of action of VEGF and VEGF receptor targeting siRNAs was studied. The tested siRNAs each resulted in the desired reduction in choroidal neovascularization (CNV) due to wet AMD (Kleinman et al., 2008). However, a similar reduction in CNV was achieved using non-targeting siRNAs. Further analysis showed that essentially all of the tested siRNAs were functioning through activation of Toll-like Receptor 3 (TLR3), a cell-surface protein that binds dsRNA and initiates an immune response. The immune stimulation, as characterized by the induction of IFN-γ and IL-12, was therefore principally responsible for the suppression of CNV, independent of the RNAi pathway. Thus, the specificity of RNAi does not completely preclude the possibility of side-effects.

Another complication has been encountered of late when using shRNAs for stable induction of RNAi in mice. shRNAs delivered via viral vectors were found to be lethal depending on dose but independent of the vector length and shRNA sequence (Grimm et al., 2006). The shRNAs appeared to be competing with endogenous miRNAs, possibly causing a saturation of some of the RNAi pathway proteins, particularly Exportin-5, which is responsible for transport of miRNAs from the nucleus to the cytoplasm. A second study reached similar conclusions except that a reduction of dosage only reduced silencing without a concomitant improvement in toxicity (McBride et al., 2008). Instead, toxicity was linked to high levels of guide strand RNA. This work, however, did demonstrate that an artificial system expressing a hairpin that looks more like a miRNA seemed to utilize the endogenous miRNA processing pathway more naturally, leading to a reduction in guide strand RNA levels and toxicity, while also maintaining high levels of silencing. All of these results from shRNA mediated silencing further support investigation of the use of siRNA-based RNAi, as siRNAs are not trafficked and processed in the same manner as shRNAs and miRNAs. Nonetheless, understanding strategies for mitigating the side-effects of stable in vivo silencing will be useful for development of treatments for chronic diseases as well as for establishing cell lines with constitutively modified gene expression. Taken together, these studies underscore the complexity of the various overlapping pathways that can be initiated when using dsRNA and that care must be taken when eventually progressing to in vivo applications.

Acknowledgements Financial support for this work was provided in part by Michigan State University, the National Science Foundation (#0425821), and the National Institutes of Health (#CA126136, #GM079688, #RR024439).

References

Akinc A, Zumbuehl A, Goldberg M, Leshchiner ES, Busini V, Hossain N, Bacallado SA, Nguyen DN, Fuller J, Alvarez R et al (2008) A combinatorial library of lipid-like materials for delivery of RNAi therapeutics. Nat Biotech 26:561–569

Allen RS, Millgate AG, Chitty JA, Thisleton J, Miller JA, Fist AJ, Gerlach WL, Larkin PJ (2004) RNAi-mediated replacement of morphine with the nonnarcotic alkaloid reticuline in opium poppy. Nat Biotechnol 22:1559–1566

Allen RS, Miller JA, Chitty JA, Fist AJ, Gerlach WL, Larkin PJ (2008) Metabolic engineering of morphinan alkaloids by over-expression and RNAi suppression of salutaridinol 7-O-acetyltransferase in opium poppy. Plant Biotechnol J 6:22–30

Alper H, Fischer C, Nevoigt E, Stephanopoulos G (2005) Tuning genetic control through promoter engineering. Proc Natl Acad Sci U S A 102:12678–12683

Amarzguioui M, Holen T, Babaie E, Prydz H (2003) Tolerance for mutations and chemical modifications in a siRNA. Nucleic Acids Res 31:589–595

Ameres SL, Martinez J, Schroeder R (2007) Molecular basis for target RNA recognition and cleavage by human RISC. Cell 130:101–112

Arntzen C, Plotkin S, Dodet B (2005) Plant-derived vaccines and antibodies: potential and limitations. Vaccine 23:1753–1756

Bartel DP (2004) MicroRNAs: genomics, biogenesis, mechanism, and function. Cell 116:281–297

Bartlett DW, Davis ME (2007) Physicochemical and biological characterization of targeted, nucleic acid-containing nanoparticles. Bioconjug Chem 18:456–468

Bartlett DW, Su H, Hildebrandt IJ, Weber WA, Davis ME (2007) Impact of tumor-specific targeting on the biodistribution and efficacy of siRNA nanoparticles measured by multimodality in vivo imaging. Proc Natl Acad Sci U S A 104:15549–15554

Bernstein E, Caudy AA, Hammond SM, Hannon GJ (2001) Role for a bidentate ribonuclease in the initiation step of RNA interference. Nature 409:363–366

Bohula EA, Salisbury AJ, Sohail M, Playford MP, Riedemann J, Southern EM, Macaulay VM (2003) The efficacy of small interfering RNAs targeted to the type 1 insulin-like growth factor receptor (IGF1R) is influenced by secondary structure in the IGF1R transcript. J Biol Chem 278:15991–15997

Brown KM, Chu CY, Rana TM (2005) Target accessibility dictates the potency of human RISC. Nat Struct Mol Biol 12:469–470

Brus C, Petersen H, Aigner A, Czubayko F, Kissel T (2004) Physicochemical and biological characterization of polyethylenimine-graft-poly(ethylene glycol) block copolymers as a delivery system for oligonucleotides and ribozymes. Bioconjug Chem 15:677–684

Carmell MA, Xuan Z, Zhang MQ, Hannon GJ (2002) The Argonaute family: tentacles that reach into RNAi, developmental control, stem cell maintenance, and tumorigenesis. Genes Dev 16:2733–2742

Cheah MT, Wachter A, Sudarsan N, Breaker RR (2007) Control of alternative RNA splicing and gene expression by eukaryotic riboswitches. Nature 447:497–500
Chen ZY, He CY, Ehrhardt A, Kay MA (2003) Minicircle DNA vectors devoid of bacterial DNA result in persistent and high-level transgene expression in vivo. Mol Ther 8:495–500
Chen PY, Weinmann L, Gaidatzis D, Pei Y, Zavolan M, Tuschl T, Meister G (2008) Strand-specific 5′-O-methylation of siRNA duplexes controls guide strand selection and targeting specificity. RNA 14:263–274
Chiu Y, Rana TM (2002) RNAi in human cells: basic structural and functional features of small interfering RNA. Mol Cell 10:549–561
Clamme JP, Azoulay J, Mely Y (2003) Monitoring of the formation and dissociation of polyethylenimine/DNA complexes by two photon fluorescence correlation spectroscopy. Biophys J 84:1960–1968
Corey DR (2007) Chemical modification: the key to clinical application of RNA interference? J Clin Invest 117:3615–3622
Cox KM, Sterling JD, Regan JT, Gasdaska JR, Frantz KK, Peele CG, Black A, Passmore D, Moldovan-Loomis C, Srinivasan M et al (2006) Glycan optimization of a human monoclonal antibody in the aquatic plant Lemna minor. Nat Biotechnol 24:1591–1597
Crooke ST (2004) Progress in antisense technology. Annu Rev Med 55:61–95
Cui Q, Yu Z, Pan Y, Purisima EO, Wang E (2007) MicroRNAs preferentially target the genes with high transcriptional regulation complexity. Biochem Biophys Res Commun 352:733–738
Daniell H, Streatfield SJ, Wycoff K (2001) Medical molecular farming: production of antibodies, biopharmaceuticals and edible vaccines in plants. Trends Plant Sci 6:219–226
de Fougerolles A, Vornlocher HP, Maraganore J, Lieberman J (2007) Interfering with disease: a progress report on siRNA-based therapeutics. Nat Rev Drug Discov 6:443–453
de Fougerolles AR (2008) Delivery vehicles for small interfering RNA in vivo Hum Gene Ther 19:125–132
DeVincenzo J, Cehelsky JE, Alvarez R, Elbashir S, Harborth J, Toudjarska I, Nechev L, Murugaiah V, Van Vliet A, Vaishnaw AK, Meyers R (2008) Evaluation of the safety, tolerability and pharmacokinetics of ALN-RSV01, a novel RNAi antiviral therapeutic directed against respiratory syncytial virus (RSV). Antiviral Res 77:225–231
Ding Y, Chan CY, Lawrence CE (2004) Sfold web server for statistical folding and rational design of nucleic acids. Nucleic Acids Res 32:W135–W141
Dodo HW, Konan KN, Chen FC, Egnin M, Viquez OM (2008) Alleviating peanut allergy using genetic engineering: the silencing of the immunodominant allergen Ara h 2 leads to its significant reduction and a decrease in peanut allergenicity. Plant Biotechnol J 6:135–145
Dong L, Hulsmeyer M, Durkop H, Hansen HP, Schneider-Mergener J, Ziegler A, Uchanska-Ziegler B (2003) Human CD30: structural implications from epitope mapping and modeling studies. J Mol Recognit 16:28–36
Doos BR (2002) Population growth and loss of arable land. Global Environ Chang 12:303–311
Elbashir SM, Lendeckel W, Tuschl T (2001a) RNA interference is mediated by 21- and 22-nucleotide RNAs. Genes Dev 15:188–200
Elbashir SM, Martinez J, Patkaniowska A, Lendeckel W, Tuschl T (2001b) Functional anatomy of siRNAs for mediating efficient RNAi in *Drosophila melanogaster* embryo lysate EMBO J 20:6877–6888
Elbashir SM, Harborth J, Weber K, Tuschl T (2002) Analysis of gene function in somatic mammalian cells using small interfering RNAs. Methods 26:199–213
Elmen J, Thonberg H, Ljungberg K, Frieden M, Westergaard M, Xu Y, Wahren B, Liang Z, Orum H, Koch T, Wahlestedt C (2005) Locked nucleic acid (LNA) mediated improvements in siRNA stability and functionality. Nucleic Acids Res 33:439–447
Far R, Sczakiel G (2003) The activity of siRNA in mammalian cells is related to structural target accessibility: a comparison with antisense oligonucleotides. Nucleic Acids Res 31: 4417–4424
Farran I, Sanchez-Serrano JJ, Medina JF, Prieto J, Mingo-Castel AM (2002) Targeted expression of human serum albumin to potato tubers. Transgenic Res 11:337–346

Felgner PL, Gadek TR, Holm M, Roman R, Chan HW, Wenz M, Northrop JP, Ringold GM, Danielsen M (1987) Lipofection: a highly efficient, lipid-mediated DNA-transfection procedure. Proc Natl Acad Sci U S A 84:7413–7417

Fire A, Xu S, Montgomery MK, Kostas SA, Driver SE, Mello CC (1998) Potent and specific genetic interference by double-stranded RNA in *Caenorhabditis elegans*. Nature 391:806–811

Floss DM, Falkenburg D, Conrad U (2007) Production of vaccines and therapeutic antibodies for veterinary applications in transgenic plants: an overview. Trangenic Res 16:315–332

Frankard V, Ghislain M, Jacobs M (1992) Two feedback-insensitive enzymes of the aspartate pathway in nicotiana sylvestris. Plant Physiol 99:1285–1293

Gallitaliadoros LA, Sedgwick JD, Wood SA, Korner H (1995) Gene knock-out technology – a methodological overview for the interested novice. J Immunol Methods 181:1–15

Ge G, Wong GW, Luo B (2005) Prediction of siRNA knockdown efficiency using artificial neural network models. Biochem Biophys Res Commun 336: 723–728

Gomord V, Chamberlain P, Jefferis R, Faye L (2005) Biopharmaceutical production in plants: problems, solutions and opportunities. Trends Biotechnol 23:559–565

Gredell JA, Berger AK, Walton SP (2008) Impact of target mRNA structure on siRNA silencing efficiency: a large-scale study. Biotechnol Bioeng 100:744–755

Grimm D, Streetz KL, Jopling CL, Storm TA, Pandey K, Davis CR, Marion P, Salazar F, Kay MA (2006) Fatality in mice due to oversaturation of cellular microRNA/short hairpin RNA pathways. Nature 441:537–541

Grothe T, Lenz R, Kutchan TM (2001) Molecular characterization of the salutaridinol 7-O-acetyltransferase involved in morphine biosynthesis in opium poppy Papaver somniferum. J Biol Chem 276:30717–30723

Haley B, Zamore PD (2004) Kinetic analysis of the RNAi enzyme complex. Nat Struct Mol Biol 11:599–606

Hammond SM, Caudy AA, Hannon GJ (2001) Post-transcriptional gene silencing by double-stranded RNA. Nat Rev Genet 2:110–119

Hannon GJ (2002) RNA interference. Nature 418:244–251

Heale BS, Soifer HS, Bowers C, Rossi JJ (2005) siRNA target site secondary structure predictions using local stable substructures. Nucleic Acids Res 33:e30

Hofacker IL (2003) Vienna RNA secondary structure server. Nucleic Acids Res 31:3429–3431

Hu-Lieskovan S, Heidel JD, Bartlett DW, Davis ME, Triche TJ (2005) Sequence-specific knockdown of EWS-FLI1 by targeted, nonviral delivery of small interfering RNA inhibits tumor growth in a murine model of metastatic Ewing's sarcoma. Cancer Res 65:8984–8992

Jackson AL, Linsley PS (2004) Noise amidst the silence: off-target effects of siRNAs? Trends Genet 20:521–524

Jagla B, Aulner N, Kelly PD, Song DA, Volchuk A, Zatorski A, Shum D, Mayer T, De Angelis DA, Ouerfelli O et al (2005) Sequence characteristics of functional siRNAs. RNA 11:864–872

Kamath RS, Fraser AG, Dong Y, Poulin G, Durbin R, Gotta M, Kanapin A, Le Bot N, Moreno S, Sohrmann M et al (2003) Systematic functional analysis of the *Caenorhabditis elegans* genome using RNAi. Nature 421:231–237

Khvorova A, Reynolds A, Jayasena SD (2003) Functional siRNAs and miRNAs exhibit strand bias. Cell 115:209–216

Kim EJ, Kramer SF, Hebert CG, Valdes JJ, Bentley WE (2007) Metabolic engineering of the baculovirus-expression system via inverse "shotgun" genomic analysis and RNA interference (dsRNA) increases product yield and cell longevity. Biotechnol Bioeng 98:645–654

Kim YS, Lee YH, Kim HS, Kim MS, Hahn KW, Ko JH, Joung H, Jeon JH (2008) Development of patatin knockdown patato tubers using RNA interference (RNAi) technology, for the production of human-therapeutic glycoproteins. BMC Biotechnol 8:36

Kircheis R, Wightman L, Wagner E (2001) Design and gene delivery activity of modified polyethylenimines. Adv Drug Deliv Rev 53:341–358

Kleinman ME, Yamada K, Takeda A, Chandrasekaran V, Nozaki M, Baffi JZ, Albuquerque RJC, Yamasaki S, Itaya M, Pan Y et al (2008) Sequence- and target-independent angiogenesis suppression by siRNA via TLR3. Nature 452:591–597

Koppelman SJ, Wensing M, Ertmann M, Knulst AC, Knol EF (2004) Relevance of Ara h1, Ara h2 and Ara h3 in peanut-allergic patients, as determined by immunoglobulin E Western blotting, basophil-histamine release and intracutaneous testing: Ara h2 is the most important peanut allergen. Clin Exp Allergy 34: 583–590

Kumar N, Gammell P, Meleady P, Henry M, Clynes M (2008) Differential protein expression following low temperature culture of suspension CHO-K1 cells. BMC Biotechnol 8:42

Kunath K, von Harpe A, Petersen H, Fischer D, Voigt K, Kissel T, Bickel U (2002) The structure of PEG-modified poly(ethylene imines) influences biodistribution and pharmacokinetics of their complexes with NF-kappaB decoy in mice. Pharm Res 19:810–817

Lee Y, Ahn C, Han J, Choi H, Kim J, Yim J, Lee J, Provost P, Radmark O, Kim S, Kim VN (2003) The nuclear RNase III Drosha initiates microRNA processing. Nature 425:415–419

Liao JC (2004) Custom design of metabolism. Nat Biotechnol 22:823–824

Lim SF, Chuan KH, Liu S, Loh SO, Chung BY, Ong CC, Song Z (2006) RNAi suppression of Bax and Bak enhances viability in fed-batch cultures of CHO cells. Metab Eng 8: 509–522

Liu J, Carmell MA, Rivas FV, Marsden CG, Thomson JM, Song JJ, Hammond SM, Joshua-Tor L, Hannon GJ (2004) Argonaute2 is the catalytic engine of mammalian RNAi. Science 305:1437–1441

Lu ZJ, Mathews DH (2007) Efficient siRNA selection using hybridization thermodynamics. Nucleic Acids Res 36: 640–647

MacRae IJ, Ma E, Zhou M, Robinson CV, Doudna JA (2008) In vitro reconstitution of the human RISC-loading complex. Proc Natl Acad Sci U S A 105:512–517

Mandal M, Breaker RR (2004) Gene regulation by riboswitches. Nat Rev Mol Cell Biol 5:451–463

Mao S, Neu M, Germershaus O, Merkel O, Sitterberg J, Bakowsky U, Kissel T (2006) Influence of polyethylene glycol chain length on the physicochemical and biological properties of poly(ethylene imine)-*graft*-poly(ethylene glycol) block copolymer/SiRNA polyplexes. Bioconjug Chem 17:1209–1218

Markham NR, Zuker M (2008) UNAFold: software for nucleic acid folding and hybridization. Methods Mol Biol 453:3–31

Matranga C, Tomari Y, Shin C, Bartel DP, Zamore PD (2005) Passenger-strand cleavage facilitates assembly of siRNA into Ago2-containing RNAi enzyme complexes. Cell 123:607–620

Maxwell MM, Pasinelli P, Kazantsev AG, Brown RH Jr (2004) RNA interference-mediated silencing of mutant superoxide dismutase rescues cyclosporin A-induced death in cultured neuroblastoma cells. Proc Natl Acad Sci U S A 101:3178–3183

McBride JL, Boudreau RL, Harper SQ, Staber PD, Monteys AM, Martins I, Gilmore BL, Burstein H, Peluso RW, Polisky B et al (2008) Artificial miRNAs mitigate shRNA-mediated toxicity in the brain: implications for the therapeutic development of RNAi. Proc Natl Acad Sci U S A 105:5868–5873

Neuenschwander U, Suter M, Brunold C (1991) Regulation of sulfate assimilation by light and O-acetyl-l-serine in lemna minor L. Plant Physiol 97:253–258

Overhoff M, Alken M, Far RK, Lemaitre M, Lebleu B, Sczakiel G, Robbins I (2005) Local RNA target structure influences siRNA efficacy: a systematic global analysis. J Mol Biol 348:871–881

Page JE (2005) Silencing nature's narcotics: metabolic engineering of the opium poppy. Trends Biotechnol 23:331–333

Pardridge WM (2007) shRNA and siRNA delivery to the brain. Adv Drug Deliv Rev 59:141–152

Patzel V, Rutz S, Dietrich I, Koberle C, Scheffold A, Kaufmann SH (2005) Design of siRNAs producing unstructured guide-RNAs results in improved RNA interference efficiency. Nat Biotechnol 23:1440–1444

Prat S, Frommer WB, Hofgen R, Keil M, Kossmann J, Koster-Topfer M, Liu XJ, Muller B, Pena-Cortes H, Rocha-Sosa M et al (1990) Gene expression during tuber development in potato plants. FEBS Lett 268:334–338

Pun SH, Bellocq NC, Liu A, Jensen G, Machemer T, Quijano E, Schluep T, Wen S, Engler H, Heidel J, Davis ME (2004). Cyclodextrin-modified polyethylenimine polymers for gene delivery. Bioconjug Chem 15:831–840

Rana TM (2007) Illuminating the silence: understanding the structure and function of small RNAs. Nat Rev Mol Cell Biol 8:23–36

Reynolds A, Leake D, Boese Q, Scaringe S, Marshall WS, Khvorova A (2004) Rational siRNA design for RNA interference. Nat Biotechnol 22:326–330

Rozema DB, Ekena K, Lewis DL, Loomis AG, Wolff JA (2003) Endosomolysis by masking of a membrane-active agent (EMMA) for cytoplasmic release of macromolecules. Bioconjug Chem 14:51–57

Rozema DB, Lewis DL, Wakefield DH, Wong SC, Klein JJ, Roesch PL, Bertin SL, Reppen TW, Chu Q, Blokhin AV et al (2007) Dynamic PolyConjugates for targeted in vivo delivery of siRNA to hepatocytes. Proc Natl Acad Sci U S A 104:12982–12987

Schubert S, Grunweller A, Erdmann VA, Kurreck J (2005) Local RNA target structure influences siRNA efficacy: systematic analysis of intentionally designed binding regions. J Mol Biol 348:883–893

Schwarz DS, Hutvagner G, Du T, Xu Z, Aronin N, Zamore PD (2003) Asymmetry in the assembly of the RNAi enzyme complex. Cell 115:199–208

Schwarz DS, Ding H, Kennington L, Moore JT, Schelter J, Burchard J, Linsley PS, Aronin N, Xu Z, Zamore PD (2006) Designing siRNA that distinguish between genes that differ by a single nucleotide. PLoS Genet 2:e140

Shah JK, Garner HR, White MA, Shames DS, Minna JD (2007) sIR: siRNA information resource, a web-based tool for siRNA sequence design and analysis and an open access siRNA database. BMC Bioinformatics 8:178

Shao Y, Chan CY, Maliyekkel A, Lawrence CE, Roninson IB, Ding Y (2007) Effect of target secondary structure on RNAi efficiency. RNA 13:1–10

Sicherer SH, Munoz-Furlong A, Burks AW, Sampson HA (1999) Prevalence of peanut and tree nut allergy in the US determined by a random digit dial telephone survey. J Allergy Clin Immunol 103:559–562

Singh J, Sharp PJ, Skerritt JH (2001) A new candidate protein for high lysine content in wheat grain. J Sci Food Agr 81:216–226

Sonnichsen B, Koski LB, Walsh A, Marschall P, Neumann B, Brehm M, Alleaume AM, Artelt J, Bettencourt P, Cassin E et al (2005) Full-genome RNAi profiling of early embryogenesis in *Caenorhabditis elegans*. Nature 434:462–469

Srivastava S, Li Z, Yang X, Yedwabnick M, Shaw S, Chan C (2007) Identification of genes that regulate multiple cellular processes/responses in the context of lipotoxicity to hepatoma cells. BMC Genomics 8:364

Sunilkumar G, Campbell LM, Puckhaber L, Stipanovic RD, Rathore KS (2006) Engineering cottonseed for use in human nutrition by tissue-specific reduction of toxic gossypol. Proc Natl Acad Sci U S A 103:18054–18059

Svoboda P (2007) Off-targeting and other non-specific effects of RNAi experiments in mammalian cells. Curr Opin Mol Ther 9:248–257

Swami A, Aggarwal A, Pathak A, Patnaik S, Kumar P, Singh Y, Gupta KC (2007a). Imidazolyl-PEI modified nanoparticles for enhanced gene delivery. Int J Pharm 335:180–192.

Swami A, Kurupati RK, Pathak A, Singh Y, Kumar P, Gupta KC (2007) A unique and highly efficient non-viral DNA/siRNA delivery system based on PEI-bisepoxide nanoparticles. Biochem Biophys Res Commun 362:835–841

Tafer H, Ameres SL, Obernosterer G, Gebeshuber CA, Schroeder R, Martinez J, Hofacker IL (2008) The impact of target site accessibility on the design of effective siRNAs. Nat Biotechnol 26:578–583

Takashima Y, Saito R, Nakajima A, Oda M, Kimura A, Kanazawa T, Okada H (2007) Spray-drying preparation of microparticles containing cationic PLGA nanospheres as gene carriers for avoiding aggregation of nanospheres. Int J Pharm 343:262–269

Tang GL, Galili G, Zhuang X (2007) RNAi and microRNA: breakthrough technologies for the improvement of plant nutritional value and metabolic engineering. Metabolomics 3:357–369

Tewari M, Vidal M (2003) RNAi on the apoptosis TRAIL: the mammalian cell genetic screen comes of age. Dev Cell 5:534–535

Tomari Y, Matranga C, Haley B, Martinez N, Zamore PD (2004) A protein sensor for siRNA asymmetry. Science 306:1377–1380

Townsend BJ, Llewellyn DJ (2007) Reduced terpene levels in cottonseed add food to fiber. Trends Biotechnol 25:239–241

Ui-Tei K, Naito Y, Takahashi F, Haraguchi T, Ohki-Hamazaki H, Juni A, Ueda R, Saigo K (2004) Guidelines for the selection of highly effective siRNA sequences for mammalian and chick RNA interference. Nucleic Acids Res 32:936–948

Ui-Tei K, Naito Y, Zenno S, Nishi K, Yamato K, Takahashi F, Juni A, Saigo K (2008) Functional dissection of siRNA sequence by systematic DNA substitution: modified siRNA with a DNA seed arm is a powerful tool for mammalian gene silencing with significantly reduced off-target effect. Nucleic Acids Res 36:2136–2151

Urban-Klein B, Werth S, Abuharbeid S, Czubayko F, Aigner A (2005) RNAi-mediated gene-targeting through systemic application of polyethylenimine (PEI)-complexed siRNA in vivo. Gene Ther 12:461–466

Vert JP, Foveau N, Lajaunie C, Vandenbrouck Y (2006) An accurate and interpretable model for siRNA efficacy prediction. BMC Bioinformatics 7:520.

Vickers TA, Koo S, Bennett CF, Crooke ST, Dean NM, Baker BF (2003) Efficient reduction of target RNAs by small interfering RNA and RNase H-dependent antisense agents. A comparative analysis. J Biol Chem 278:7108–7118

Wakefield DH, Klein JJ, Wolff JA, Rozema DB (2005) Membrane activity and transfection ability of amphipathic polycations as a function of alkyl group size. Bioconjug Chem 16:1204–1208

Walton SP, Roth CM, Yarmush ML (2000) Antisense technology. In: The biomedical engineering handbook. CRC Press LLC, Boca Raton, FL, pp 103–119.

Warner TG (1999) Enhancing therapeutic glycoprotein production in Chinese hamster ovary cells by metabolic engineering endogenous gene control with antisense DNA and gene targeting. Glycobiology 9:841–850

Wei MC, Zong WX, Cheng EH, Lindsten T, Panoutsakopoulou V, Ross AJ, Roth KA, MacGregor GR, Thompson CB, Korsmeyer SJ (2001) Proapoptotic BAX and BAK: a requisite gateway to mitochondrial dysfunction and death. Science 292:727–730.

Weitzer S, Martinez J (2007) The human RNA kinase hClp1 is active on 3′ transfer RNA exons and short interfering RNAs. Nature 447:222–226

Westerhout EM, Berkhout B (2007) A systematic analysis of the effect of target RNA structure on RNA interference. Nucleic Acids Res 35:4322–4330

Wolff JA, Rozema DB (2008) Breaking the bonds: non-viral vectors become chemically dynamic. Mol Ther 16:8–15

Wurm FM (2004) Production of recombinant protein therapeutics in cultivated mammalian cells. Nat Biotechnol 22:1393–1398

Yoshinari K, Miyagishi M, Taira K (2004) Effects on RNAi of the tight structure, sequence and position of the targeted region. Nucleic Acids Res 32:691–699

Zhang S, Zhao B, Jiang H, Wang B, Ma B (2007) Cationic lipids and polymers mediated vectors for delivery of siRNA. J Control Release 123:1–10

Zhu X, Galili G (2004) Lysine metabolism is concurrently regulated by synthesis and catabolism in both reproductive and vegetative tissues. Plant Physiol 135:129–136

Zuhorn IS, Engberts JB, Hoekstra D (2007) Gene delivery by cationic lipid vectors: overcoming cellular barriers. Eur Biophys J 36:349–362

Apoptosis and Autophagy Cell Engineering

Chaya Mohan, Yeon-Gu Kim, and Gyun Min Lee

Abstract Programmed cell death (apoptosis and autophagy) in cell cultures is considered an important problem to be dealt with as it affects the viable cell concentration and the product quality. This chapter describes various strategies employed to confront and prevent programmed cell death in biotechnologically important mammalian cell lines, mainly Chinese Hamster Ovary (CHO) cells, with special importance to the genetic manipulation of cells for anti-apoptosis and anti-autophagic engineering.

1 Introduction

Of all the strategies implemented for cell culture optimization, increasing the cell viability is of paramount importance. Viability is one main parameter that decides the success of the bioprocess, and in turn, viability can be affected by a number of factors, such as cell line instability, nutrient limitation, chemical and physical factors, and accumulation of waste byproducts. Hence, tackling cell death in bioreactors has been pursued with great interest.

Although a variety of expression systems exists, such as, microbial, insect, transgenic animals and plants, mammalian cells still remain the most sought-after expression system for commercial therapeutic protein production. Of the various mammalian cells, immortalized Chinese hamster ovary (CHO) cells, and other cell lines like NS0, Baby hamster kidney (BHK), and human embryo kidney (HEK-293) cells have gained FDA approval for large-scale commercial production of therapeutic proteins. CHO cells have been a popular host for the production of therapeutics for diverse reasons such as the biosimilar glycosylation pattern of the proteins produced

C. Mohan, Y-G. Kim, and G.M. Lee (✉)
Department of Biological Sciences, KAIST, 335 Gwahangno, Yuseong-Gu,
Daejeon, 305-701, Republic of Korea
e-mil: gyunminlee@kaist.ac.kr

M. Al-Rubeai (ed.), *Cell Line Development*, Cell Engineering 6,
DOI 10.1007/978-90-481-2245-5_10, © Springer Science+Business Media B.V. 2009

in them, powerful and easily amplifiable genetics, and ease with which they can be adapted to growth in the regulatory-friendly serum-free suspension condition.

Large-scale culture requires cells to be grown in bioreactors at high cell densities under rigorous optimization schemes to be able to meet the demands of the ever-increasing highly competitive market (Al-Rubeai and Singh, 1998). Cells grow and proliferate in the controlled environment of the bioreactor but succumb to even slightest perturbations. The disturbances that can stress a cell in culture include nutrient limitation and accumulation of acidic metabolic by-products during the later stages of the culture, elevated osmolality, and the hydrodynamic environment of the bioreactor, including agitation and aeration. The way a cell responds to such perturbations can vary from a temporary cessation of the cell cycle letting the cells cope up, to a more serious state of leading the severely stressed cells to the path of cell death. Death comes to cells in a bioreactor in either of two forms – necrosis or programmed cell death (PCD). Since necrosis is a sudden, passive, form of cell death, not much research has been of help in alleviating this occurrence (Bowen and Lockshin, 1981). On the other hand, programmed cell death, as the name suggests, is an active, well co-ordinated, genetically controlled death process, where the fate of the cell depends on the relative abundance of survival and death proteins.

This chapter begins with a brief overview of the forms of cell death in mammalian cell cultures, mainly focusing on the two types of programmed cell death, namely, apoptosis and autophagy, followed by a discussion of cell engineering strategies that have been employed to overcome programmed cell death in mammalian cells.

2 Cell Death in Bioreactors

2.1 Necrosis

This form of cell death occurs when cells are subjected to sudden, severe stress like exposure to high level of toxins, sharp changes in pH and high agitation rates (Singh et al., 1994). At such extreme stress levels cells die of necrosis, essentially because they have no time to respond to the stimulus and hence die instantly. Morphologically, a cell dying of necrosis can be differentiated by the features it displays, such as, cell swelling, disruption of plasma membrane and organelle membranes. At later stages, extensive DNA hydrolysis, vacuolation of the endoplasmic reticulum, organelle breakdown, and cell lysis occurs (Buja et al., 1993; Trump et al., 1981; Wyllie et al., 1980). The enzymes released by the necrotic cell lysis results in the destruction of the neighboring cells thus culminating in total metabolic collapse of the cell and the neighboring population. Since in this type of cell death, the cell is at the absolute mercy of its environment and has no control over its fate, technological intervention to prevent or alleviate it has been of little assistance. Figure 1 illustrates the forms of cell death.

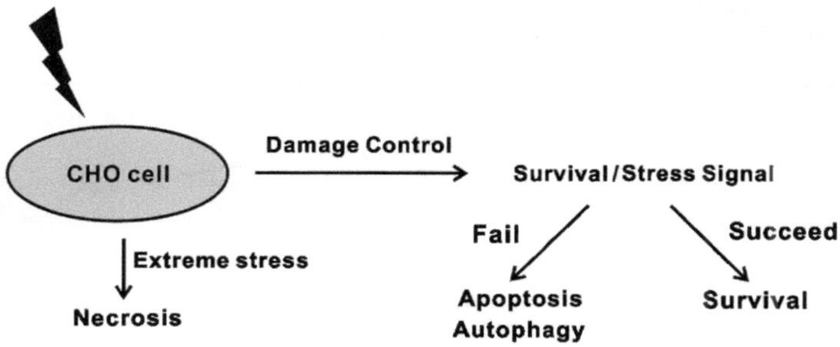

Fig. 1 Forms of cell death commonly experienced by mammalian cells in culture. Extremely severe stress drives a cell to an irreparable state and finally necrosis whereas mild stress activates a damage control process and finally death by apoptosis/autophagy

2.2 *Programmed Cell Death*

The term programmed cell death was introduced in 1964, proposing that cell death during development is not of accidental nature but follows a sequence of controlled steps leading to locally and temporally defined self-destruction (Lockshin and Williams, 1964). Programmed cell death can be predominantly of two types- apoptosis and autophagy. Apoptosis has been known to have tremendous implication in the medical field, specifically relating to cancer, viral diseases, neurodegeneration, heart diseases and autoimmunity and its importance in the field of animal cell technology has been gaining interest in the past two decades. Autophagy, on the other hand, had been known as a survival mechanism (Lum et al., 2005) but recently with regard to animal cell culture, it has been identified as a form of cell death the cell undertakes while it encounters stressful conditions such as nutrient deprivation (Hwang and Lee, 2008b).

2.2.1 Apoptosis

One of the widely studied forms of programmed cell death, apoptosis derives its name from the Greek word *apo-* from, *ptosis-* falling off, meaning 'falling/dropping off'. This process signifies cell suicide with the full cooperation of the cellular machinery when a cell is damaged beyond repair. The studies of Kerr et al. in 1972 are often cited as an important milestone in the study of apoptosis (Kerr et al., 1972).

Two major signaling pathways of apoptosis, depicted in Fig. 2, have been recognized and the pathways amply elucidated - the intrinsic and the extrinsic pathway.

Intrinsic Pathway: As the name suggests, intrinsic pathway is initiated from within the cell, usually in response to cellular signals resulting from DNA damage, a defective cell cycle, detachment from the extracellular matrix, hypoxia, loss of cell survival factors, or other types of cell stress (Coultas and Strasser, 2003; Fulda

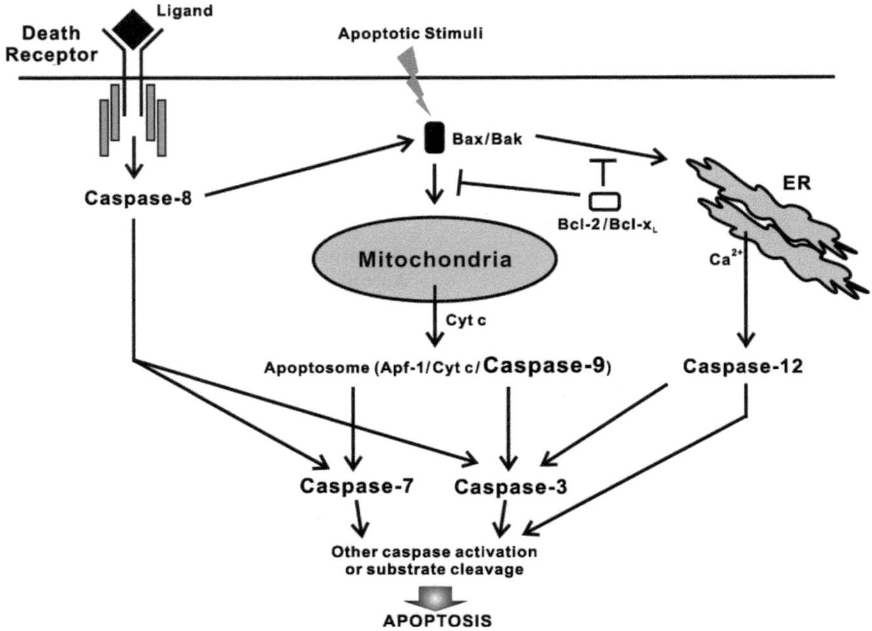

Fig. 2 Common pathways of apoptosis

and Debatin, 2006). The intrinsic apoptotic pathway hinges on the balance of activity between pro- and anti-apoptotic members of the Bcl-2 family of proteins which act to regulate the permeability of the mitochondrial membrane. This pathway involves the release of pro-apoptotic proteins from the mitochondria that activate caspase enzymes, which ultimately trigger apoptosis, hence also referred to as the mitochondrial apoptotic pathway (Thornberry and Lazebnik, 1998).

Extrinsic Pathway: The extrinsic pathway begins outside the cell through the activation of specific pro-apoptotic receptors on the cell surface by specific molecules known as pro-apoptotic ligands. These ligands include Apo2L/TRAIL (receptors DR4, DR5), and CD95L/FasL (receptor CD95/Fas). Once activated by extracellular ligand binding, the intracellular domains of these receptors, known as the 'death domains' (hence also called the death receptor pathway), bind to the adaptor protein Fas-associated death domain (FADD), leading to the assembly of the death-inducing signaling complex (DISC). The recruitment and assembly of initiator caspases and the subsequent activation of the effector caspases results in the convergence of this pathway on the intrinsic pathway (Arden and Betenbaugh, 2004).

Apoptosis has gained importance in the field of animal cell culture for the large scale production of therapeutics over the last two decades. Studies on the occurrence of apoptosis in hybridoma cultures (Al-Rubeai et al., 1995; Ishaque and Al-Rubeai, 1998; Singh et al., 1994; Singh et al., 1997) provided knowledge of this phenomenon with respect to large scale bioreactor cultures. Apoptosis has been found to be a universal death mechanism opted by various cell types under irreparable stress (Nivitchanyong et al., 2007; Singh et al., 1994; Tey et al., 2000a). Specifically, with respect to CHO cells, a majority of the reports observed cultures undergoing apoptotic cell death and suggested remedial measures (Arden et al., 2007a; Goswami et al., 1999; Lee and Lee, 2003; Moore et al., 1995; Tey et al., 2000b).

Apoptosis limits the maximum viable cell density, promotes the release of toxic metabolites from dead cells, and potentially decreases heterologous protein yield and quality (Chiang and Sisk, 2005; Figueroa et al., 2004; Figueroa et al., 2001; Figueroa et al., 2003; Goldman et al., 1997; Mastrangelo et al., 2000a,b; Mercille and Massie, 1999; Munzert et al., 1996; Teige et al., 1994). Hence, applying all available resources and finding new ones, for the alleviation or prevention of apoptosis seem worthwhile for obvious reasons.

2.2.2 Autophagy

Autophagy, the term originated from Greek 'to eat oneself', is an evolutionarily conserved, catabolic process through the lysosomal-mediated degradation pathway (Levine, 2005). Upon various autophagic stresses including limitation of nutrients, autophagosomes, a hallmark of autophagy, fuse with lysosomes to form autophagolysosomes, in which intracellular components are degraded (Fig. 3). Degraded cytoplasmic contents are recycled for macromolecular synthesis and ATP generation. It is regulated by an autophagic-related (Atg) gene that encodes for essential proteins in autophagy machinery (Levine and Klionsky, 2004, Yorimitsu and Klionsky, 2005). Autophagy is an evolutionarily conserved process among all eukaryotes and the molecular basis of autophagy has been extensively studied, mainly in yeast, by investigating the autophagy-defective mutants to identify the responsible genes (Tsukada and Ohsumi, 1993). When cells are subjected to nutrient-limiting conditions, they simultaneously decrease overall protein synthesis and increase rates of protein degradation by an autophagic pathway (Mortimore and Pösö, 1988). Thus, prolonged survival of cells for a short time is achieved under

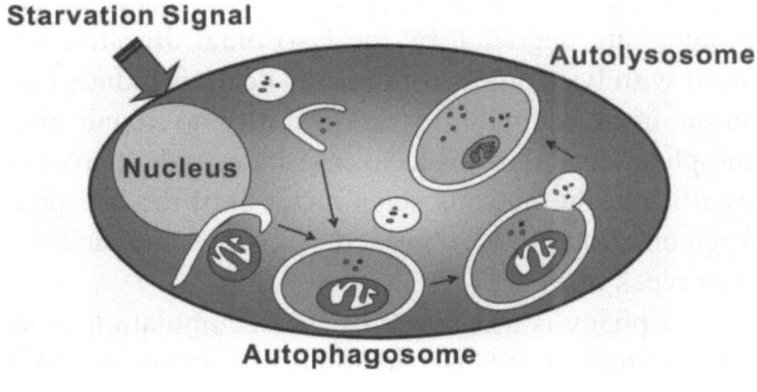

Fig. 3 Formation of an autophagolysosome by the fusion of an autophagic cell to a lysosome

starvation conditions. In addition to its role in normal degradation, autophagy may promote type II PCD that is distinguished from apoptosis (Bursch, 2001; Gozuacik and Kimchi, 2004; Levine and Yuan, 2005; Yorimitsu and Klionsky, 2005).

In mammalian cells, like apoptosis, autophagy can also be regulated through multiple signaling pathways (Levine and Klionsky, 2004; Meijer and Codogno, 2004). The major component of autophagy is the mammalian target of rapamycin (mTOR), evolutionarily conserved Ser/Thr kinase, which acts as a central regulator of several starvation responses (Lum et al., 2005; Miron and Sonenberg, 2001; Oldham and Hafen, 2003). Treatment of cells with rapamycin suppresses mTOR activity and stimulates autophagy (Blommaart et al., 1995). Several additives, including nutrients (amino acids), energy (ATP), and growth factors regulate mTOR activation through the action of a class I phosphoinositide 3-kinase (PI3-K) and protein kinase B (PKB/Akt). (Brown and Schreiber, 1996; Hay and Sonenberg, 2004; Schmelzle and Hall, 2000). PI3-K activation determines the induction of the proteins of the growth signaling pathway such as the p70 kDa S6 kinase (S6K) (Alvarez et al., 2003). Activation of mTOR leads to the phosphorylation of S6K which coincides with the inhibition of autophagy (Raught et al., 2001; Tokunaga et al., 2004). Numerous studies have shown that protein kinase B (PKB/Akt) functions downstream of PI3-K. Akt can activate a number of downstream targets including S6K and TOR (Scott et al., 1998).

2.2.3 Characteristic Features of Apoptosis and Autophagy

The morphologies of apoptosis and autophagy differ from each other noticeably.

Apoptotic cells undergo highly characteristic condensation and fragmentation of chromatin and cell shrinkage. At a later stage, they undergo intensive blebbing of the plasma membrane during which numerous protrusions form at the surface of the cell. These protrusions eventually break away from the plasma membrane, forming intact cytoplasmic packages called apoptotic bodies. Associated with the diminution in cell volume is condensation of nuclear chromatin and, in some types, this is followed by extensive nuclear fragmentation into an oligo nucleosomal DNA ladder, multimers of about 180 bp.

The morphology of autophagy is characterized by the appearance of double- or multi-membrane cytosolic vesicles engulfing cytoplasmic organelles such as mitochondria and endoplasmic reticulum (autophagosomes). Autophagosomes and their contents are degraded by the lysosomal digestive system of the same cell, when fused with lysosomes. Autophagic death is induced when the nutrient limiting condition lasts beyond the cell's ability to repair the damage. Unlike apoptosis, autophagic cell death is a caspase-independent process and DNA degradation is not significant, the nucleus stays intact until the late phases of cell death and cellular fragmentation is not observed. Table 1 summarizes the characteristic features of both types of PCD.

Autophagy is monitored by the accumulation of microtubule-associated protein 1 light chain 3 (LC3, a mammalian orthologue of yeast Atg8) by Western blot

Table 1 Comparison of the two types of programmed cell death (adapted with permission from Gozuacik and Kimchi, 2004)

	Type I Apoptotic	Type II Autophagic
Nucleus	Chromatin condensation	Partial chromatin condensation
	Pyknosis of nucleus	Sometimes pyknosis of nucleus
	DNA laddering and nuclear fragmentation	Nucleus intact until late stages
		No DNA laddering
Cytoplasm	Cytoplasmic condensation	Increased autophagic vesicle number
	Ribosome loss from RER	Increased autolysosome number
	Fragmentation to apoptotic bodies	Increased lysosomal activity
	Lysosomal protease release to cytosol may be involved	Enlarged Golgi, sometimes dilatation of ER
	Mitochondrial permeability transition is often involved	Mitochondrial permeability transition may be involved
	Caspases are active	Caspase-independent
Cell membrane	Blebbing	Blebbing
Corpse clearance	Heterophagy by other cells	Late and occasional heterophagy by other cells
Detection methods	Electron microscopy	Electron microscopy
	Nuclear/cellular fragmentation detection	Test of increased long-lived protein degradation
	Caspase activation tests	Tests of increased lysosomal activity (MDC, acridine orange or lysotracker staining, etc.)
	Caspase substrate cleavage tests	Test of increased cytoplasmic sequestration (LDH or sucrose sequestration tests)
	DNA laddering detection	Detection of LC3 recruitment to autophagic membranes (protein band shift or change in intracellular localization)
	TUNEL staining	
	Increase in sub G1 cell population assessed by FACS analysis	
	Annexin V staining	

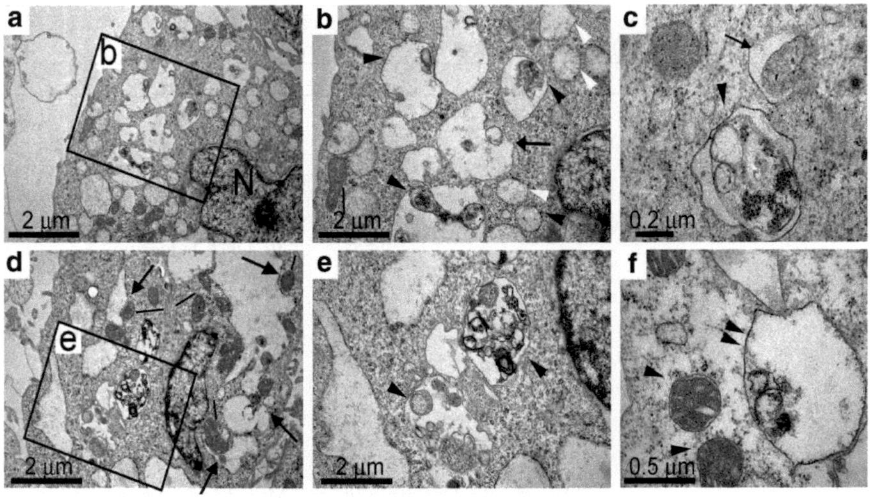

Fig. 4 Transmission electron microscopic images of rCHO cells (Hwang and Lee, 2008b). Scale bars, as indicated in the individual figures. Typical autophagic morphologies are shown within the cells after 7 days of batch culture. ***(a–e)***, *Arrowheads*, representative of autophagic vacuoles including residual digested material within the cytoplasm, *arrows*, sequestering cytoplasm by double membranous structure, *open arrowheads*, empty vacuoles, *lines*, intact mitochondria trapped in vacuoles; *double arrowheads*, a large late autophagic vacuole containing partially degraded material

analysis and multiple autophagic vacuoles within the cell by the transmission electron microscope (TEM) (Hwang and Lee, 2008b). The most convincing and standard method of detecting autophagy in the cell today is TEM (Gozuacik and Kimchi, 2004; Mizushima, 2004). Autophagic structures are characterized by multiple autophagic vacuoles, which are double membranous vacuoles engulfing cytoplasmic materials (Hwang and Lee, 2008b; Mizushima et al., 2001) as shown in Fig. 4.

LC3 is associated with the autophagosomes and is converted from the precursor form (LC3-I) to the processed form (LC3-II) upon autophagy induction (Levine and Klionsky, 2004). LC3 conversion is a commonly used marker for measuring autophagic activity, since the levels of LC3-II correlate with the number of autophagosomes (Kabeya et al., 2000; Mizushima et al., 2004).

In addition, green fluorescent protein (GFP)–LC3 is also a useful tool to quantify the level of autophagy by fluorescent microscopy (Mizushima et al., 2004). Two forms of LC3, LC3-I and LC3-II, show distinct localization as a cytosolic form and membrane bound form in cells, respectively (Kabeya et al., 2000). When autophagy is induced, GFP-LC3, seen as a diffused form in the cytoplasm, is converted to a punctate form associated with the autophagosomal membrane. The top panel in Fig. 5 shows a typical Western blot image

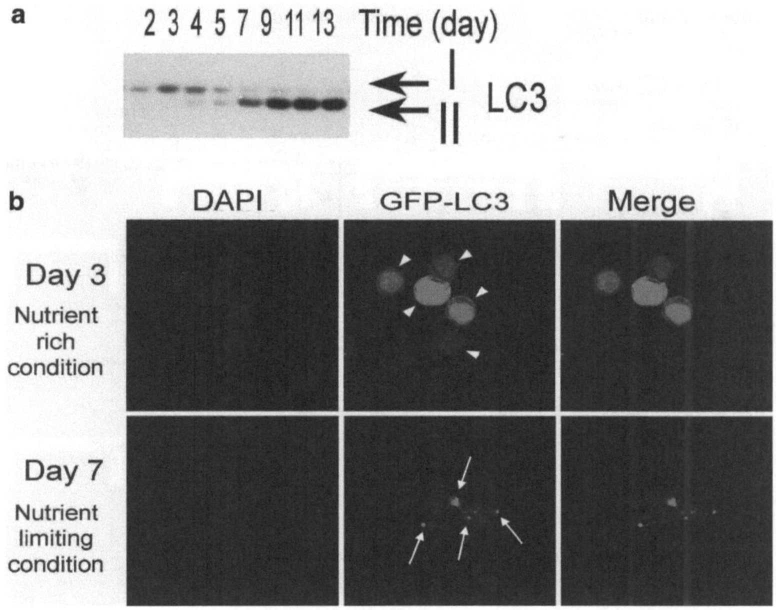

Fig. 5 Approaches to monitor autophagy in cells (Hwang and Lee, 2008b). (**a**) Western blot image of an rCHO cell line during batch culture. The LC3-I form is converted into the LC3-II form as the culture advances to nutrient limiting condition. (**b**) Confocal microscopic images of cells transfected with the GFP-LC3 vector. *Arrowheads*, diffused patterns of GFP-LC3-I; *arrows*, punctate structures of GFP-LC3-II. DAPI was used for nucleus staining

for LC3 expression in a CHO cell line producing an antibody over time in batch culture. The expression of LC3-I decreases due to its conversion into the membranous LC3-II form. The bottom panel of Fig. 5 shows the GFP-LC3 staining data. It can be seen that under nutrient limitation the cytoplasmic LC3-I is converted to the membranous LC3-II punctuate form (Hwang and Lee, 2008b).

3 Apoptosis Engineering

Apoptosis engineering is mainly aimed at preventing or alleviating the occurrence, or progression of the process. Strategies to prevent or down-regulate apoptosis can be broadly grouped under the following two categories: manipulation of the external cellular environment through media supplementation and employing genetic engineering strategies to manipulate the intracellular environment. Few of the commonly used strategies to tackle programmed cell death are illustrated in Fig. 6.

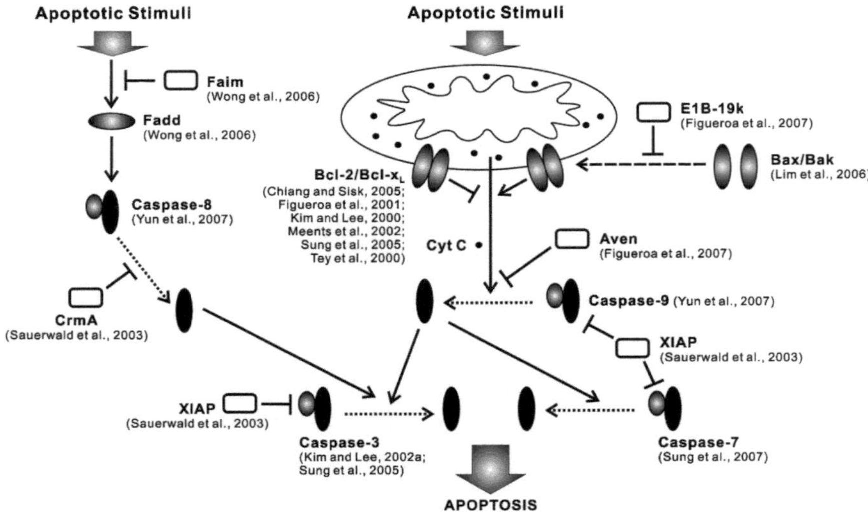

Fig. 6 Various strategies for anti-apoptosis cell engineering

3.1 Manipulation of External Cellular Environment

Apoptosis in CHO cells has been known to be induced by nutrient or serum deprivation during culture (Goswami et al., 1999; Hwang and Lee, 2008b; Singh et al., 1994). Periodic nutrient feeding after determining the time at which nutrients are depleted, can help in delaying the onset of apoptosis. Additionally, use of supplements like serum, suramin (Zanghi et al., 1999; Zanghi et al., 2000), N-acetylcysteine (NAC) (Chang et al., 1999; Oh et al., 2005), silkworm hemolymph (Choi et al., 2005), or insulin and IGF-1 (Adamson and Walum, 2007)is known to offer protection against apoptosis under serum-free conditions. Chemical caspase inhibitors offer protection against apoptosis in CHO-K1 and HEK-293 cells thereby enhancing the culture viabilities and protein titers (Arden et al., 2007a) but their cost might be formidable for use in large scale cell culture processes. Though the use of media supplements/additives are easily implemented in the cell culture set-up because they do not require the generation of new cell lines, which is usually laborious and time-consuming, their limitations and cost efficiency are factors to be considered before putting them into commercial usage.

3.2 Genetic Strategies to Manipulate the Intracellular Environment

Genetic manipulations to prevent/alleviate apoptosis mainly focus on the overexpression of anti-apoptotic genes, down regulation of pro-apoptotic genes and caspases – the actual executors of apoptosis.

3.2.1 Down Regulation of Caspases

Caspases belong to a family of highly conserved cysteine-dependent aspartate-specific acid proteases that use a cysteine residue as the catalytic nucleophile and share a stringent specificity for cleaving their substrates after aspartic acid residues in target proteins. The caspase gene family consists of 15 mammalian members that are grouped into two major sub-families, namely inflammatory caspases and apoptotic caspases. The apoptotic caspases are further subdivided into two sub-groups, initiator caspases and executioner caspases. The caspases form a caspase-cascade system that plays the central role in the induction, transduction and amplification of intracellular apoptotic signals for cell fate determination, regulation of immunity, and cellular proliferation and differentiation. The substrates of apoptotic caspases have been associated with cellular dismantling, while inflammatory caspases mediate the proteolytic activation of inflammatory cytokines (extensively reviewed in Chowdhury et al., 2008; Cohen, 1997; Earnshaw et al., 1999; Salvesen and Riedl, 2008; Siegel and Lenardo, 2002).

The three apoptotic pathways initiated from the death receptor, mitochondria and the ER converge on caspase-3 and caspase-7 (Arden and Betenbaugh, 2004). Hence numerous research articles have focused on the inhibition of these caspase activities by either blocking them or by down regulating them genetically. Since the cost of chemical caspase inhibitors make them unaffordable for use in large-scale cultures, the focus has been mainly towards genetically manipulating cells for silencing caspase gene expression.

XIAP, considered as the most potent caspase inhibitor was studied along with two of its deletion mutants in CHO and HEK-293 cell lines. On subjecting to apoptotic insults such as serum deprivation, Sindbis virus infection and etoposide treatment, it was observed that XIAP and the mutant containing three tandem baculovirus IAP repeat (BIR) domains provided protection against apoptosis whereas the other mutant containing only the C-terminal RING domain offered no protection and was pro-apoptotic in transient studies (Sauerwald et al., 2002).

Silencing of caspase-3 using the antisense technology, in an antibody-producing cell line, significantly suppressed sodium butyrate (NaBu)-induced apoptosis and extended the culture longevity. However, this technique did not translate into increased volumetric productivity because of the loss of metabolic capability due to depolarization of mitochondrial membrane (Kim and Lee, 2002a). Caspase-3 siRNA silencing on thrombopoietin (TPO)-producing cell line could not effectively inhibit NaBu-induced apoptosis because the cells seemed to compensate for the lack of caspase-3 by increasing the active caspase-7 levels (Sung et al., 2005). To circumvent this, Sung et al. also attempted to co-down-regulate caspase-3 and caspase-7 to inhibit apoptosis under NaBu treatment and serum-free conditions; however, the effect on TPO production was modest, about a 55% increase (Sung et al., 2007). Stable dominant negative mutants of caspase-2, caspase-8 or caspase-9 were overexpressed in CHO cells and it was observed that the inhibition of caspase-8 or caspase-9 enhanced the viability of CHO cell in batch and fed-batch culture whereas the inhibition of caspase-2 had minimal effects (Yun et al., 2007). Studies on genetic caspase inhibitors, XIAP and CrmA under apoptotic insults including, spent medium, Sindbis virus, and etoposide suggest that the apoptosis pathways induced and the level of protection afforded by a particular caspase inhibitor may vary with the insult considered (Sauerwald et al., 2003). The combined use of caspase (XIAP-BIRs) and mitochondrial dysfunction inhibitors (Bcl-x_L, Aven) in stable CHO cell lines illustrates that the utilization of a mitochondrial dysfunction inhibitor in combination with a caspase inhibitor is more effective in preventing the progression of apoptosis than either inhibitor expressed individually (Sauerwald et al., 2006).

Another strategy that was reported for anti-apoptosis engineering was the use of a ribozyme targeted at caspase-3 in CHO cells. The caspase-3 in cells was cleaved by this U6 Sn RNA promoter-driven ribozyme, resulting in enhanced cell viability, cell density and interferon-beta production in low serum cultures (Lai et al., 2004).

Hence, it has been observed that the blocking/down-regulation of caspases may be beneficial; however, it depends on the apoptotic insults, and the strategy of usage. It has also been seen that cells compensate for the lack of some effectors by up-regulating the others. Thus, when genetic engineering is considered for apoptosis inhibition employing caspase silencing, the interrelationship between molecules in a cell and the cell's competence to cope with changes needs to be considered.

3.2.2 Overexpression of Anti-apoptotic Genes

The B cell CLL/lymphoma 2 (Bcl-2) family consists of both pro- and anti-apoptotic members grouped into three families: (i) anti-apoptotic Bcl-2 homologs containing Bcl-2 homology (BH) domains 1, 2, 3 and 4; (ii) pro-apoptotic members containing BH1, 2, and 3 homology domains; and (iii) BH3 only pro-apoptotic members (Harris and Thompson, 2000). Upon exposure to intrinsic or extrinsic death signals, pro-apoptotic proteins such as Bax and Bak undergo structural modifications in which they alter the mitochondrial membrane integrity, causing the release of cytochrome c and other pro-apoptotic molecules (Cheng et al., 1996; Goswami et al., 1999; Scorrano and Korsmeyer, 2003; Subramanian and Chinnadurai, 2003). These molecules are released in response to apoptotic stimuli, such as toxin exposure or DNA damage to promote activation of caspases, responsible for initiating cellular apoptosis signaling. The aim of apoptosis engineering to deal with the Bcl-2 family proteins has been mainly to up-regulate the anti-apoptotic genes and/or silence the pro-apoptotic ones.

The establishment of apoptosis-resistant dhfr–CHO cell line by the incorporation of Bcl-2 gene (Lee and Lee, 2003) expedites the developmental process of establishing apoptosis-resistant CHO cell lines producing therapeutic proteins. Overexpression of Bcl-2 in a CHO cell line producing a chimeric antibody resulted in a tremendous enhancement in culture viability (over 80%) compared to the control cultures. Ammonium toxicity and growth arrest using thymidine also resulted in enhanced culture viabilities in the Bcl-2 overexpressing cells. However, there was no major difference in the antibody titer between the two cell lines (Tey et al., 2000b). Coexpression of c-Myc, the expression of which generally correlates with cell proliferation and Bcl-2, the anti-apoptotic protein, in a CHO cell line resulted in higher proliferation rates and maximum cell numbers, with a decrease in apoptosis (Ifandi and Al-Rubeai, 2005). CHO cells are extensively researched with respect to Bcl-2 overexpression. Bcl-2 protected cells against sodium butyrate (NaBu)-induced apoptosis (Kim and Lee, 2001), hyperosmotic pressure (Kim and Lee, 2002b), and in serum-free suspension culture (Sung and Lee, 2005). Under all these conditions, the culture viabilities were enhanced resulting in an increase in the culture longevity by over 2-days and the protein titer was found to be enhanced by over twofold. A comparative study between wild-type and a Bcl-2 mutant lacking the nonstructured loop domain in BHK and CHO cells revealed that the mutant provided increased protection as compared to wild-type against Sindbis virus infection and serum deprivation (Figueroa et al., 2001).

Another widely studied Bcl-2 family protein is Bcl-x_L. Coexpression and coamplification of a target protein (soluble intercellular adhesion molecule 1, sICAM) and Bcl-2 or Bcl-x_L resulted in higher sICAM yields in cells containing Bcl-x_L. However, the death-protective impact of Bcl-2 and Bcl-x_L in engineered CHO-DG44 was not significant under typical batch-mode operation. Bcl-2 and Bcl-x_L displayed their anti-apoptosis potential only following dhfr-based amplification in sICAM-producing CHO-DG44 cell lines. In all cases, Bcl-x_L outperformed Bcl-2 in its cell death-protective capacity (Meents et al., 2002). Bcl-2 and Bcl-x_L were also found to limit apoptosis upon alphavirus infections used in viral expression systems in BHK and CHO cells lines resulting in enhancement of protein titer by over twofold. However, this effect was found to be cell line specific (Mastrangelo et al., 2000a). Studies on Bcl-x_L-overexpressing antibody-producing CHO cell lines consistently displayed enhanced cell viabilities, specific productivities, and product titers in serum-free and chemically defined medium (Chiang and Sisk, 2005). Combinatorial anti-apoptosis cell engineering strategies employing Aven and Bcl-x_L (Figueroa et al., 2004) and E1B-19 K and Aven (Figueroa et al., 2007) resulted in an overall improvement in the culture viabilities, maximum cell densities, and product titers. The anti-apoptotic capabilities of these genes were found to be higher when co-overexpressed than when done singly. As with Bcl-2, wild type Bcl-x_L was compared to its mutant lacking the non-conserved unstructured loop domain and it was observed that the mutant displayed superior anti-apoptotic properties, resulting in enhanced viabilities and protein titers in CHO cells using viral expression systems (Figueroa et al., 2003). Recently, Bcl-x_L was also found to inhibit both forms of cell death, apoptosis and autophagy, in a CHO cell line producing erythropoietin by interacting with Bax and Bak and Beclin-1. This resulted in enhanced cell viabilities and culture longevity. However, the specific productivity was unaffected (Kim et al., 2009).

3.2.3 Down-Regulation of Pro-apoptotic Genes

The Bcl-2 family of proteins represents a critical checkpoint upstream of mitochondria where the life-or death decision is made. Bcl-2 family members possess up to 4 conserved BH domains. The anti-apoptotic members, such as Bcl-2, Bcl-x_L, Mcl-1 and A1, contain all 4 BH domains. Some of the pro-apoptotic members

(Bax, Bak and Bok) contain 3 BH domains (BH1, BH2, and BH3), whereas other pro-apoptotic members (Bid, Bad, Bim, Bik, Noxa, and Puma) contain only the BH3 domain and therefore are called "BH3-only" members. Recently, few studies have shifted focus on down regulating the pro-apoptotic genes because as such, over-expression of Bcl-2 or Bcl-x_L could not ensure a complete block of the mitochondrial death pathway.

Stable CHO cell line producing IFN-γ in adherent from on microcarrier beads and suspension culture were subjected to knockdown of two pro-apoptotic genes, namely Bax and Bak. On subsequent insult with cytotoxic lectins, UV irradiation, nutrient depletion and high osmolality, the knockouts showed a better viability profile with a 35% increase in product titer (Lim et al., 2006). Transcriptional profiling of CHO cell culture using microarray has generated immense knowledge and one such research which exploited these findings was reported by Wong et al. Four key early apoptosis signaling genes, Fadd, Faim (anti-apoptotic), Alg-2, Requiem (pro-apoptotic) were identified and overexpressed or knocked down, respectively in CHO cells producing batch and fed-batch cultures demonstrated an increase in the cell viability and a much higher maximum cell densities. The IFN-γ titer was found to be enhanced by over 2.5-fold with the molecule being more highly sialylated (Wong et al., 2006).

3.2.4 Others

Other strategies to tackle apoptosis include the use of the 30Kc6 gene of silkworm hemolymph resulting in a significant enhancement in the cell density (5-fold) and erythropoietin (EPO) titer (10-fold) (Choi et al., 2006).

Overexpression of MDM2, an E3 ubiquitin ligase for p53, was employed to inhibit upstream apoptotic pathways. The overexpressing cells not only showed greater cell viabilities but also when compared to Bcl-2 overexpressing cells, MDM2 expression showed better protection against apoptosis in various insults such as passaged culture, spent medium, and transient p53 overexpression (Arden et al., 2007b).

Human telomerase reverse transcriptase (hTERT) catalytic subunit was overexpressed in CHO-K1 cells and differences in the growth profile and apoptosis levels between the cells over-expressing hTERT (Telo) and the cells containing mock vector were observed with the Telo cells being superior. They also showed lower levels of apoptosis, greater attachment tendency and higher viable cell density under serum-deprived conditions compared to the control cell line (Crea et al., 2006). Table 2 summarizes the strategies commonly used in anti-apoptotic cell engineering.

4 Autophagy Engineering

Until now, only one type of programmed cell death, apoptosis, has drawn attention in CHO cell culture. Recently, our group suggested the importance of autophagic cell death in CHO cells (Hwang and Lee, 2008a,b). Since the two kinds of PCD act

Table 2 Strategies employed to tackle apoptosis in mammalian cell cultures

Approach	Example	Reference
Continued activity of survival signaling pathways	IGF-action on cultures	D'Ambrosio et al. (1997)
	Matrix attachment factors	Zhang et al. (1995)
Genetic modification of regulatory or execution pathways	Inhibition of proteases	Cotter and Al-Rubeai (1995), Kim and Lee (2002a), and Liston et al. (1996)
	Bcl-2	Farrow and Brown (1996), and Itoh et al. (1995), Kim and Lee (2001), Mastrangelo et al. (2000a,b), Reed (1994), Simpson et al. (1997), Terada et al. (1997), Tey et al. (2000b), Vaux et al. (1994), and Zanghi et al. (1999)
	Bcl-x(L)	Chiang and Sisk (2005), Figueroa et al. (2003), Figueroa et al. (2003), Mastrangelo et al. (2000a,b), Meents et al. (2002)
	Manipulation of signaling pathways	Dickens et al. (1997) and Kauffmann-Zeh et al. (1997)
Prevention of nutrient deprivation and waste accumulation	Growth factor addition	deZengotita et al. (2000) and Morris and Schmid (2000)
	Fed-batch culture	Zhou et al. (1997)

independently, inhibiting apoptotic cell death does not guarantee the inhibition of autophagic cell death, suggesting that it is not sufficient to inhibit apoptosis without considering autophagic cell death for blocking PCD. For example, autophagic cell death occurs in cells defecting apoptosis machinery or cultured in the presence of caspase inhibitors (Shimizu et al., 2004; Yu et al., 2004).

4.1 Manipulation of External Cellular Environment for Autophagy Cell Engineering

A non-toxic compound 3-methyladenine (3-MA), a chemical inhibitor of autophagy, blocked PI3-K activity, indicating that it might be helpful to inhibit autophagy occurring at the end of CHO culture, as described earlier (Blommaart et al., 1997; Fuertes et al., 2003; Petiot et al., 2000; Seglen and Gordon, 1982).

The second possible method is feeding essential amino acid mixtures to culture medium to suppress the induction of autophagy (Schworer and Mortimore, 1979). Similarly, there has been a report that amino acid re-addition inhibited autophagic vacuole formation (Kovacs et al., 1981). Also, adding low concentrations of amino acids inhibited autophagy via stimulating the phosphorylation of S6K (Blommaart et al., 1995). Altogether, it is expected that autophagic cell death in rCHO cells can be prevented by the addition of nutrients including amino acids during batch culture.

4.2 Genetic Strategies to Manipulate the Intracellular Environment for Autophagy Cell Engineering

Despite the increasing importance of autophagic cell death, there is, to date, no attempts to improve the cell growth and protein production by inhibiting the autophagic cell death in CHO cells (Hwang and Lee, 2008a). One of the possible methods to overcome autophagic cell death in CHO cells is to engineer Bcl-2 or Bcl-x_L genes that are known to negatively regulate both apoptosis and autophagy. Recently, Bcl-x_L has been considered as a crucial protein in autophagic cell death, proposed by studies regarding the pharmacological suppression of Bcl-2 function, antisense-mediated down-regulation of Bcl-2 expression, and siRNA-mediated down-regulation of Bcl-2 expression (Akar et al., 2008; Kessel and Reiners, 2007; Pattingre et al., 2005; Saeki et al., 2000). Furthermore, there have been many studies about the overexpression of Bcl-2 and Bcl-x_L inhibit autophagic cell death by binding with Beclin-1 (Ku et al., 2008; Liang et al., 2006; Maiuri et al., 2007b; Pattingre et al., 2005). Regulating Atg genes that are known to be essential proteins in autophagic process can be a good approach. Beclin-1, also called Atg6, has been isolated and identified as a binding protein of Bcl-2 (Liang et al., 1998) and is known as an essential protein in promoting autophagy (Liang et al., 1999). Interestingly, although Beclin-1 shows no overall sequence homology with the Bcl-2 family proteins, the recent crystallographic studies revealed that Beclin-1 possesses a novel BH3 domain that mediates interactions with Bcl-x_L (Feng et al., 2007; Oberstein et al., 2007). Also, it was confirmed that a BH3-mimetic compound and BH3-only protein disrupt the interaction between Beclin-1 and Bcl-x_L or Bcl-2 by competitive interaction (Maiuri et al., 2007a,b).

Another approach inhibiting autophagy in CHO cells is to engineer downstream molecules of the PI3-K pathway for regulating mTOR which is known to be an important protein in autophagic cell death (Lum et al., 2005; Miron and Sonenberg, 2001; Oldham and Hafen, 2003). A previous report was shown that activation of p53 inhibits mTOR activity, which resulted in the regulation of autophagic death (Feng et al., 2005). Similarly, activation of S6 by S6K, a downstream molecule of mTOR, could inhibit autophagy (Blommaart et al., 1995). Also, inhibition of PI3-K/Akt signaling pathway by specific inhibitors augmented rapamycin-induced autophagic cell death (Kuo et al., 2006; Takeuchi et al., 2005). However, overexpression of CA-Akt, the constitutively active form of Akt, resulted in reduced levels of autophagy (Arico et al., 2001).

5 Conclusions

The ability to regulate programmed cell death in mammalian cell cultures represents one approach to developing more economical and efficient processes. Enhanced product yields, reduction in throughput time, improved cost-effectiveness and product quality are a few examples of benefits achieved by alleviating or preventing programmed cell death in bioreactors.

References

Adamson L, Walum E (2007) Insulin and IGF-1 mediated inhibition of apoptosis in CHO cells grown in suspension in a protein-free medium. Altern Lab Anim 35:349–352

Akar U, Chaves-Reyez A, Barria M et al (2008) Silencing of Bcl-2 expression by small Interfering RNA induces autophagic cell death in MCF-7 breast cancer cells. Autophagy 4:669–679

Al-Rubeai M, Singh RP (1998) Apoptosis in cell culture. Curr Opin Biotechnol 9:152–156

Al-Rubeai M, Singh RP, Goldman MH et al (1995) Death mechanisms of animal cells in conditions of intensive agitation. Biotechnol Bioeng 45:463–472

Alvarez B, Garrido E, Garcia-Sanz JA et al (2003) Phosphoinositide 3-kinase activation regulates cell division time by coordinated control of cell mass and cell cycle progression rate. J Biol Chem 278:26466–26473

Arden N, Betenbaugh MJ (2004) Life and death in mammalian cell culture: strategies for apoptosis inhibition. Trends Biotechnol 22:174–180

Arden N, Ahn SH, Vaz W et al (2007a) Chemical caspase inhibitors enhance cell culture viabilities and protein titer. Biotechnol Prog 23:506–511

Arden N, Majors BS, Ahn SH et al (2007b) Inhibiting the apoptosis pathway using MDM2 in mammalian cell cultures. Biotechnol Bioeng 97:601–614

Arico S, Petiot A, Bauvy C et al (2001) The tumor suppressor PTEN positively regulates macroautophagy by inhibiting the phosphatidylinositol 3-kinase/protein kinase B pathway. J Biol Chem 276:35243–35246

Blommaart EF, Luiken JJ, Blommaart PJ et al (1995) Phosphorylation of ribosomal protein S6 is inhibitory for autophagy in isolated rat hepatocytes. J Biol Chem 270:2320–2326

Blommaart EF, Krause U, Schellens JP et al (1997) The phosphatidylinositol 3-kinase inhibitors wortmannin and LY294002 inhibit autophagy in isolated rat hepatocytes. Eur J Biochem 243:240–246

Bowen ID, Lockshin RA (1981) Cell death in biology and pathology. Chapman and Hall, London

Brown EJ, Schreiber SL (1996) A signaling pathway to translational control. Cell 86:517–520

Buja LM, Eigenbrodt ML, Eigenbrodt EH (1993) Apoptosis and necrosis. Basic types and mechanisms of cell death. Arch Pathol Lab Med 117:1208–1214

Bursch W (2001) The autophagosomal-lysosomal compartment in programmed cell death. Cell Death Differ 8:569–581

Chang KH, Kim KS, Kim JH (1999) N-acetylcysteine increases the biosynthesis of recombinant EPO in apoptotic Chinese hamster ovary cells. Free Radic Res 30:85–91

Cheng EH, Levine B, Boise LH et al (1996) Bax-independent inhibition of apoptosis by Bcl-x_L. Nature 379:554–556

Chiang GG, Sisk WP (2005) Bcl-x(L) mediates increased production of humanized monoclonal antibodies in Chinese hamster ovary cells. Biotechnol Bioeng 91:779–792

Choi SS, Rhee WJ, Park TH (2005) Beneficial effect of silkworm hemolymph on a CHO cell system: Inhibition of apoptosis and increase of EPO production. Biotechnol Bioeng 91:793–800

Choi SS, RheeWJ KEJ, Kim EJ et al (2006) Enhancement of recombinant protein production in Chinese hamster ovary cells through anti-apoptosis engineering using 30Kc6 gene. Biotechnol Bioeng 95:459–467

Chowdhury I, Tharakan B, Bhat GK (2008) Caspases–An update. Comp Biochem Physiol B. doi:10.1016/j.cbpb.2008.05.010.

Cohen GM (1997) Caspases: The executioners of apoptosis. Biochem J 326:1–16

Cotter TG, Al-Rubeai M (1995) Cell death (apoptosis) in cell culture systems. Trends Biotechnol 13:150–155

Coultas L, Strasser A (2003) The role of the Bcl-2 protein family in cancer. Semin Cancer Biol 13:115–123

Crea F, Sarti D, Falciani F et al (2006) Over-expression of hTERT in CHO K1 results in decreased apoptosis and reduced serum dependency. J Biotechnol 121:109–123

D'Ambrosio C, Valentinis B, Prisco M et al (1997) Protective effect of the insulin-like growth factor I receptor on apoptosis induced by okadaic acid. Cancer Res 57:3264–3271

deZengotita VM, Miller WM, Aunins JG et al (2000) Phosphate feeding improves high-cell-concentration NS0 myeloma culture performance for monoclonal antibody production. Biotechnol Bioeng 69:566–576

Dickens M, Rogers JS, Cavanagh J et al (1997) A cytoplasmic inhibitor of the JNK signal transduction pathway. Science 277:693–696

Earnshaw WC, Martinsm LM, Kaufmann SH (1999) Mammalian caspases: Structure, activation, substrates, and functions during apoptosis. Annu Rev Biochem 68:383–424

Farrow SN, Brown R (1996) New members of the Bcl-2 family and their protein partners. Curr Opin Genet Dev 6:45–49

Feng Z, Zhang H, Levine AJ et al (2005) The coordinate regulation of the p53 and mTOR pathways in cells. Proc Natl Acad Sci USA 102:8204–8209

Feng W, Huang S, Wu H et al (2007) Molecular basis of Bcl-xL's target recognition versatility revealed by the structure of Bcl-xL in complex with the BH3 domain of Beclin-1. J Mol Biol 372:223–235

Figueroa B Jr, Sauerwald TM, Mastrangelo AJ et al (2001) Comparison of Bcl-2 to a Bcl-2 deletion mutant for mammalian cells exposed to culture insults. Biotechnol Bioeng 73:211–222

Figueroa B Jr, Sauerwald TM, Oyler GA et al (2003) A comparison of the properties of a Bcl-xL variant to the wild-type anti-apoptosis inhibitor in mammalian cell cultures. Metab Eng 5:230–245

Figueroa B Jr, Chen S, Oyler GA et al (2004) Aven and Bcl-xL enhance protection against apoptosis for mammalian cells exposed to various culture conditions. Biotechnol Bioeng 85:589–600

Figueroa B Jr, Ailor E, Osborne D et al (2007) Enhanced cell culture performance using inducible anti-apoptotic genes E1B–19 K and Aven in the production of a monoclonal antibody with Chinese hamster ovary cells. Biotechnol Bioeng 97:877–892

Fuertes G, Martin De Llano JJ, Villarroya A et al (2003) Changes in the proteolytic activities of proteasomes and lysosomes in human fibroblasts produced by serum withdrawal, amino-acid deprivation and confluent conditions. Biochem J 375:75–86

Fulda S, Debatin KM (2006) Extrinsic versus intrinsic apoptosis pathways in anticancer chemotherapy. Oncogene 25:4798–4811

Goldman MH, James DC, Ison AP et al (1997) Monitoring proteolysis of recombinant human interferon-γ during batch culture of Chinese hamster ovary cells. Cytotechnology 23:103–111

Goswami J, Sinskey AJ, Steller H et al (1999) Apoptosis in batch cultures of Chinese hamster ovary cells. Biotechnol Bioeng 62:632–640

Gozuacik D, Kimchi A (2004) Autophagy as a cell death and tumor suppressor mechanism. Oncogene 23:2891–2906

Harris MH, Thompson CB (2000) The role of the bcl-2 family in the regulation of outer mitochondrial membrane permeability. Cell Death Differ 7:1182–1191

Hay N, Sonenberg N (2004) Upstream and downstream of mTOR. Genes Dev 18:1926–1945
Hwang SO, Lee GM (2008a) Autophagy and apoptosis in Chinese hamster ovary cell culture. Autophagy 4:70–72
Hwang SO, Lee GM (2008b) Nutrient deprivation induces autophagy as well as apoptosis in Chinese hamster ovary cell culture. Biotechnol Bioeng 99:678–685
Ifandi V, Al-Rubeai M (2005) Regulation of cell proliferation and apoptosis in CHO-K1 cells by the coexpression of c-Myc and Bcl-2. Biotechnol Prog 21:671–677
Ishaque A, Al-Rubeai M (1998) Use of intracellular pH and annexin-V flow cytometric assays to monitor apoptosis and its suppression by bcl-2 over-expression in hybridoma cell culture. J Immunol Methods 221:43–57
Itoh Y, Ueda H, Suzuki E (1995) Overexpression of bcl-2, apoptosis suppressing gene: Prolonged viable culture period of hybridoma and enhanced antibody production. Biotechnol Bioeng 48:118–122
Kabeya Y, Mizushima N, Ueno T et al (2000) LC3, a mammalian homologue of yeast Apg8p, is localized in autophagosome membranes after processing. EMBO J 19:5720–5728
Kauffmann-Zeh A, Rodriguez-Viciana P, Ulrich E et al (1997) Suppression of c-Myc-induced apoptosis by Ras signalling through PI(3) K and PKB. Nature 385:544–548
Kerr JFR, Wyllie AH, Currie AR (1972) Apoptosis: A basic biological phenomenon with wide ranging implications in tissue kinetics. Br J Cancer 26:239–257
Kessel D, Reiners JJ Jr (2007) Initiation of apoptosis and autophagy by the Bcl-2 antagonist HA14–1. Cancer Lett 249:294–299
Kim NS, Lee GM (2001) Overexpression of bcl-2 inhibits sodium butyrate-induced apoptosis in Chinese hamster ovary cells resulting in enhanced humanized antibody production. Biotechnol Bioeng 71:184–193
Kim NS, Lee GM (2002a) Inhibition of sodium butyrate-induced apoptosis in recombinant Chinese hamster ovary cells by constitutively expressing antisense RNA of caspase-3. Biotechnol Bioeng 78:217–228
Kim NS, Lee GM (2002b) Response of recombinant Chinese hamster ovary cells to hyperosmotic pressure: effect of Bcl-2 overexpression. J Biotechnol 95:237–248
Kim YG, Kim JY, Mohan C, et al (2009) Effect of Bcl-x_l overexpression on apoptosis and autophagy in Recombinant chinese Hamster overy cells under nutrient-deprived condition. Biotechnol Bioeng
Kovacs AL, Grinde B, Seglen PO (1981) Inhibition of autophagic vacuole formation and protein degradation by amino acids in isolated hepatocytes. Exp Cell Res 133:431–436
Ku B, Woo JS, Liang C et al (2008) Structural and biochemical bases for the inhibition of autophagy and apoptosis by viral BCL-2 of murine gamma-herpesvirus 68. PLoS Pathog 4:e25
Kuo PL, Hsu YL, Cho CY (2006) Plumbagin induces G2-M arrest and autophagy by inhibiting the AKT/mammalian target of rapamycin pathway in breast cancer cells. Mol Cancer Ther 5:3209–3221
Lai D, Fu L, Weng S et al (2004) Blocking caspase-3 activity with a U6 SnRNA promoter-driven ribozyme enhances survivability of CHO cells cultured in low serum medium and production of interferon-beta. Biotechnol Bioeng 85:20–28
Lee SK, Lee GM (2003) Development of apoptosis-resistant dihydrofolate reductase-deficient Chinese hamster ovary cell line. Biotechnol Bioeng 82:872–876
Levine B (2005) Eating oneself and uninvited guests: autophagy-related pathways in cellular defense. Cell 120:159–162
Levine B, Klionsky DJ (2004) Development by self-digestion: molecular mechanisms and biological functions of autophagy. Dev Cell 6:463–477
Levine B, Yuan J (2005) Autophagy in cell death: an innocent convict? J Clin Invest 115:2679–2688
Liang XH, Kleeman LK, Jiang HH et al (1998) Protection against fatal Sindbis virus encephalitis by beclin, a novel Bcl-2-interacting protein. J Virol 72:8586–8596
Liang XH, Jackson S, Seaman M et al (1999) Induction of autophagy and inhibition of tumorigenesis by beclin 1. Nature 402:672–676

Liang C, Feng P, Ku B et al (2006) Autophagic and tumour suppressor activity of a novel Beclin1-binding protein UVRAG. Nat Cell Biol 8:688–699
Lim SF, Chuan KH, Liu S et al (2006) RNAi suppression of Bax and Bak enhances viability in fed-batch cultures of CHO cells. Metab Eng 8:509–522
Liston P, Roy N, Tamai K et al (1996) Suppression of apoptosis in mammalian cells by NAIP and a related family of IAP genes. Nature 379:349–353
Lockshin RA, Williams CM (1964) Programmed cell death. II. Endocrine potentiation of the breakdown of the intersegmental muscles of silkmoths. J Insect Physiol 10:643–649
Lum JJ, DeBerardinis RJ, Thompson CB (2005) Autophagy in metazoans: cell survival in the land of plenty. Nat Rev Mol Cell Biol 6:439–448
Maiuri MC, Criollo A, Tasdemir E et al (2007a) BH3-only proteins and BH3 mimetics induce autophagy by competitively disrupting the interaction between beclin 1 and Bcl-2/Bcl-X(L). Autophagy 3:374–376
Maiuri MC, Le Toumelin G, Criollo A et al (2007b) Functional and physical interaction between Bcl-X(L) and a BH3-like domain in Beclin-1. EMBO J 26:2527–2539
Mastrangelo AJ, Hardwick JM, Bex F et al (2000a) Part I Bcl-2 and Bcl-x(L) limit apoptosis upon infection with alphavirus vectors. Biotechnol Bioeng 67:544–554
Mastrangelo AJ, Hardwick JM, Zou S et al (2000b) Part II overexpression of bcl-2 family members enhances survival of mammalian cells in response to various culture insults. Biotechnol Bioeng 67:555–564
Meents H, Enenkel B, Eppenberger HM et al (2002) Impact of coexpression and coamplification of sICAM and antiapoptosis determinants bcl-2/bcl-x(L) on productivity, cell survival, and mitochondria number in CHO-DG44 grown in suspension and serum-free media. Biotechnol Bioeng 80:706–716
Meijer AJ, Codogno P (2004) Regulation and role of autophagy in mammalian cells. Int J Biochem Cell Biol 36:2445–2462
Mercille S, Massie B (1999) Apoptosis-resistant E1B–19 K-expressing NS/0 myeloma cells exhibit increased viability and chimeric antibody productivity under perfusion culture conditions. Biotechnol Bioeng 63:529–543
Miron M, Sonenberg N (2001) Regulation of translation via TOR signaling: insights from *Drosophila melanogaster*. J Nutr 131:2988S–29893S
Mizushima N (2004) Methods for monitoring autophagy. Int J Biochem Cell Biol 36:2491–2502
Mizushima N, Yamamoto A, Hatano M et al (2001) Dissection of autophagosome formation using Apg5-deficient mouse embryonic stem cells. J Cell Biol 152:657–68
Mizushima N, Yamamoto A, Matsui M et al (2004) In vivo analysis of autophagy in response to nutrient starvation using transgenic mice expressing a fluorescent autophagosome marker. Mol Biol Cell 15:1101–1111
Moore A, Donahue CJ, Hooley J et al (1995) Apoptosis in CHO cell batch cultures: examination by flow cytometry. Cytotechnology 17:1–11
Morris AE, Schmid J (2000) Effects of insulin and LongR(3) on serum-free Chinese hamster ovary cell cultures expressing two recombinant proteins. Biotechnol Prog 16:693–697
Mortimore GE, Pösö AR (1988) Amino acid control of intracellular protein degradation. Methods Enzymol 166:461–476
Munzert E, Muthing J, Buntemeyer H et al (1996) Sialidase activity in culture fluid of Chinese hamster ovary cells during batch culture and its effect on recombinant human antithrombin III integrity. Biotechnol Prog 12:559–563
Nivitchanyong T, Martinez A, Ishaque A et al (2007) Anti-apoptotic genes Aven and E1B–19 K enhance performance of BHK cells engineered to express recombinant factor VIII in batch and low perfusion cell culture. Biotechnol Bioeng 98:825–841
Oberstein A, Jeffrey PD, Shi Y (2007) Crystal structure of the Bcl-XL-Beclin 1 peptide complex: Beclin 1 is a novel BH3-only protein. J Biol Chem 282:13123–13132
Oh HK, So MK, Yang J et al (2005) Effect of *N*-acetylcystein on butyrate-treated Chinese hamster ovary cells to improve the production of recombinant human interferon-beta-1a. Biotechnol Prog 21:1154–1164

Oldham S, Hafen E (2003) Insulin/IGF and target of rapamycin signaling: A TOR de force in growth control. Trends Cell Biol 13:79–85

Pattingre S, Tassa A, Qu X et al (2005) Bcl-2 antiapoptotic proteins inhibit Beclin 1-dependent autophagy. Cell 122:927–939

Petiot A, Ogier-Denis E, Blommaart EF et al (2000) Distinct classes of phosphatidylinositol 3′-kinases are involved in signaling pathways that control macroautophagy in HT-29 cells. J Biol Chem 275:992–998

Raught B, Gingras AC, Sonenberg N (2001) The target of rapamycin (TOR) proteins. Proc Natl Acad Sci USA 98:7037–7044

Reed JC (1994) Bcl-2 and the regulation of programmed cell death. J Cell Biol 124:1–6

Saeki K, Yuo A, Okuma E et al (2000) Bcl-2 down-regulation causes autophagy in a caspase-independent manner in human leukemic HL60 cells. Cell Death Differ 7:1263–1269

Salvesen GS, Riedl SJ (2008) Caspase mechanisms. Adv Exp Med Biol 615:13–23

Sauerwald TM, Betenbaugh MJ, Oyler GA (2002) Inhibiting apoptosis in mammalian cell culture using the caspase inhibitor XIAP and deletion mutants. Biotechnol Bioeng 77:704–716

Sauerwald TM, Oyler GA, Betenbaugh MJ (2003) Study of caspase inhibitors for limiting death in mammalian cell culture. Biotechnol Bioeng 81:329–340

Sauerwald TM, Figueroa B Jr, Hardwick JM et al (2006) Combining caspase and mitochondrial dysfunction inhibitors of apoptosis to limit cell death in mammalian cell cultures. Biotechnol Bioeng 94:362–372

Schmelzle T, Hall MN (2000) TOR, a central controller of cell growth. Cell 103:253–262

Schworer CM, Mortimore GE (1979) Glucagon-induced autophagy and proteolysis in rat liver: Mediation by selective deprivation of intracellular amino acids. Proc Natl Acad Sci USA 76:3169–3173

Scorrano L, Korsmeyer SJ (2003) Mechanisms of cytochrome c release by proapoptotic bcl-2 family members. Biochem Biophys Res Commun 304:437–444

Scott PH, Brunn GJ, Kohn AD et al (1998) Evidence of insulin-stimulated phosphorylation and activation of the mammalian target of rapamycin mediated by a protein kinase B signaling pathway. Proc Natl Acad Sci USA 95:7772–7777

Seglen PO, Gordon PB (1982) 3-Methyladenine: Specific inhibitor of autophagic/lysosomal protein degradation in isolated rat hepatocytes. Proc Natl Acad Sci USA 79:1889–1892

Shimizu S, Kanaseki T, Mizushima N et al (2004) Role of Bcl-2 family proteins in a non-apoptotic programmed cell death dependent on autophagy genes. Nat Cell Biol 6:1221–1228

Siegel RM, Lenardo MJ (2002) Apoptosis signaling pathways. Curr Protoc Immunol. Chapter 11: 44:Unit 11.9C

Simpson NH, Milner AE, Al-Rubeai M (1997) Prevention of hybridoma cell death by bcl-2 during suboptimal culture conditions. Biotechnol Bioeng 54:1–16

Singh R, Al-Rubeai M, Gregory C et al (1994) Cell death in bioreactors-a role for apoptosis. Biotechnol Bioeng 4:720–726

Singh RP, Finka G, Emery AN et al (1997) Apoptosis and its control in cell culture systems. Cytotechnology 3:87–93

Subramanian T, Chinnadurai G (2003) Pro-apoptotic activity of transiently expressed bcl-2 occurs independent of BAX and BAK. J Cell Biochem 89:1102–1114

Sung YH, Lee GM (2005) Enhanced human thrombopoietin production by sodium butyrate addition to serum-free suspension culture of bcl-2-overexpressing CHO cells. Biotechnol Prog 21:50–57

Sung YH, Hwang SJ, Lee GM (2005) Influence of down-regulation of caspase-3 by siRNAs on sodium-butyrate-induced apoptotic cell death of Chinese hamster ovary cells producing thrombopoietin. Metab Eng 7:457–466

Sung YH, Lee JS, Park SH et al (2007) Influence of co-down-regulation of caspase-3 and caspase-7 by siRNAs on sodium butyrate-induced apoptotic cell death of Chinese hamster ovary cells producing thrombopoietin. Metab Eng 5–6:452–464

Takeuchi H, Kondo Y, Fujiwara K et al (2005) Synergistic augmentation of rapamycin-induced autophagy in malignant glioma cells by phosphatidylinositol 3-kinase/protein kinase B inhibitors. Cancer Res 65:3336–3346

Teige M, Weidemann R, Kretzmer G (1994) Problems with serum-free production of antithrombin III regarding proteolytic activity and product quality. J Biotechnol 34:101–105
Terada S, Fukuoka K, Fujita T et al (1997) Anti-apoptotic genes, bag-1 and bcl-2, enabled hybridoma cells to survive under treatment for arresting cell cycle. Cytotechnology 25:17–23
Tey BT, Singh RP, Piredda L et al (2000a) Bcl-2 mediated suppression of apoptosis in myeloma NS0 cultures. J Biotechnol 79:147–159
Tey BT, Singh RP, Piredda L et al (2000b) Influence of bcl-2 on cell death during the cultivation of a Chinese hamster ovary cell line expressing a chimeric antibody. Biotechnol Bioeng 8:31–43
Thornberry NA, Lazebnik Y (1998) Caspases: enemies within. Science 281:1312–1316
Tokunaga C, Yoshino K, Yonezawa K (2004) mTOR integrates amino acid- and energy-sensing pathways. Biochem Biophys Res Commun 313:443–446
Trump BF, Berezesky IK, Osornio-Vargas AR (1981) Cell death and the disease process. The role of calcium. In: Bowen ID, Lockshin RA (eds) Cell death in biology and pathology. Chapman and Hall, New York
Tsukada M, Ohsumi Y (1993) Isolation and characterization of autophagy-defective mutants of *Saccharomyces cerevisiae*. FEBS Lett 333:169–174
Vaux DL, Haecker G, Strasser A (1994) An evolutionary perspective on apoptosis. Cell 76:777–779
Wong DC, Wong KT, Nissom PM et al (2006) Targeting early apoptotic genes in batch and fed-batch CHO cell cultures. Biotechnol Bioeng 95:350–361
Wyllie AH, Kerr JF, Currie AR (1980) Cell death: the significance of apoptosis. Int Rev Cytol 68:251–306
Yorimitsu T, Klionsky DJ (2005) Autophagy: Molecular machinery for self-eating. Cell Death Differ 12:1542–1552
Yu L, Alva A, Su H et al (2004) Regulation of an ATG7-beclin 1 program of autophagic cell death by caspase-8. Science 304:1500–1502
Yun CY, Liu S, Lim SF et al (2007) Specific inhibition of caspase-8 and -9 in CHO cells enhances cell viability in batch and fed-batch cultures. Metab Eng 9:406–418
Zanghi JA, Fussenegger M, Bailey JE (1999) Serum protects protein-free competent Chinese hamster ovary cells against apoptosis induced by nutrient deprivation in batch culture. Biotechnol Bioeng 64:108–119
Zanghi JA, Renner WA, Bailey JE et al (2000) The growth factor inhibitor suramin reduces apoptosis and cell aggregation in protein-free CHO cell batch cultures. Biotechnol Prog 16:319–325
Zhang Z, Vuori K, Reed JC et al (1995) The alpha 5 beta 1 integrin supports survival of cells on fibronectin and up-regulates Bcl-2 expression. Proc Natl Acad Sci USA 92:6161–6165
Zhou W, Chen CC, Buckland B et al (1997) Fed-batch culture of recombinant NS0 myeloma cells with high monoclonal antibody production. Biotechnol Bioeng 55:783–792

Glycoengineering and Modeling of Protein N-Glycosylation

Sandra V. Bennun, Frederick J. Krambeck, and Michael J. Betenbaugh

Abstract Glycoproteins for treating human diseases have revolutionized the health care industry. However, controlling glycosylation has been a challenge as small variations in glycan structure can be responsible for significant changes in key therapeutic properties. Manipulation of glycan biosynthesis can be particularly complex since the process is not directly encoded on the genome but depends on multiple variables such as enzymes' activity, selectivity, localization, expression host, and process parameters and conditions. Furthermore, a particular glycoprotein may include many different glycan structures due to differences in processing that occur for each individual molecule. The present chapter focuses on experimental and computational approaches to direct N-glycosylation in expression systems for generation of biotherapeutics of superior value. Glycoengineering-based manipulations of glycan structures using glycosyltransferases, modification of precursor biosynthetic pathways, and predictions of glycosylation patterns using mathematical models are described including examples from the literature as a means of optimizing glycoform distributions in cells.

1 Introduction

Glycans structures constitute a dominant feature of the cell; they bind proteins conferring on them important biological functions such as mediation of cellular processes associated with immune regulation, tissue development, and cell signaling (Kornfeld and Kornfeld, 1985; Kornfeld, 1998). Importantly, the majority of the proteins are glycosylated, they are found in the cell surface as well as in the intra-cellular and extracellular enviroments (Kornfeld and Kornfeld, 1985; Apweiler et al., 1999). Small changes in glycan structure are the cause of wide variability in key

S.V. Bennun, F.J. Krambeck, and M.J. Betenbaugh (✉)
Department of Chemical and Biomolecular Engineering, Johns Hopkins University, 3400 North Charles Street – Baltimore, Maryland, 21218, USA
e-mail: beten@jhu.edu

M. Al-Rubeai (ed.), *Cell Line Development*, Cell Engineering 6,
DOI 10.1007/978-90-481-2245-5_11, © Springer Science+Business Media B.V. 2009

therapeutic properties of glycoproteins such as folding, solubility, immunogenicity, biological activity, and clearance rate (Jenkins et al., 1996). These attributes make glycosylation processing for therapeutics an important aspect of the biotechnology industry. Indeed, mammalian hosts such as Chinese Hamster Ovary (CHO) cells are used for production of glycoproteins because the glycosylation patterns of these systems are more similar to the human versions. In this chapter we describe modifications of the N-glycosylation pathway to improve glycoproteins properties for superior therapeutic value. We also present computational approaches to optimize oligosaccharide structural patterns towards a desired outcome. Computer tools for understanding and predicting the glycosylation processing will become more and more important to control and manipulate glycosylation pathways as the sophistication of analytical tools increases and the implications of glycosylation become more evident.

While many challenges exist, technologies for manipulating protein glycosylation have increasingly improved and demonstrated that modifications in the oligosaccharide structures of expression systems to mimic human type N-glycans are possible. What is especially important to that end is the development of means to control and predict the effects of these manipulations on the final glycosylation pattern. Because of the complexity of the mechanism of action of post-translational N-glycosylation the integration of both experimental and computational approaches will play a fundamental role in devising means to control N-glycosylation in expression systems that generate valuable biotherapeutics.

2 Processing Pathway for N-Glycosylation in Mammalian Expression Systems

Two major types of glycosylation are predominant in glycoproteins: N-glycosylation and O-glycosylation. These glycosylation types are defined by the type of linkage between the protein backbone to the glycan structure. In N-Glycosylation the N-acetylglucosamine (GlcNac) residue of the glycan structure is β-linked to the amide (N) group of asparagine of a Asn-Xxx-(Ser, Thr) motif, where Xxx is any amino acid except proline. In O-glycosylation usually the N-acetylgalactosamine (GalNac) residue of the glycan structure is α-linked to the hydroxyl (O) group of serine or threonine (Kornfeld and Kornfeld, 1985; Brockhausen et al., 1998). While N-glycosylation takes place in the endoplasmic reticulum (ER) and Golgi apparatus, O-glycosylation occurs in the Golgi. Since N-glycans structures are particularly relevant for biotherapeutics proteins of commercial relevance such as monoclonal antibodies, this review will focus on N-glycosylation. The reader interested in O-glycosylation is referred to other reviews in the literature (Goochee et al., 1991; Brockhausen et al., 1998; Peter-Katalinic, 2005).

N-glycosylation biosynthesis is the result of a complex interaction of glycosyltransferases and glycosidases enzymes with cellular metabolites. The assembly of N-glycans starts with the generation of Man_5-P-P-dol in the cytosol. Next, the lipid linked oligosaccharide moiety Man_5-P-P-dol structure flips from the cytosol to the lumen of the endoplasmic reticulum (ER) where the synthesis of the lipid-glycan moiety ($Glc_3Man_9GlcNAc_2$-P-P-dol) is completed. This step is followed by the en bloc transfer of complex oligosacharide structure ($Glc_3Man_9GlcNAc_2$) precursor to the asparagine (Asn) residue of the Asn-Xxx-Ser/Thr sequence of the polypeptide chain of the nascent protein by the action of oligosaccharyltransferase (OST) (Kornfeld and Kornfeld, 1985). After trimming of terminal glucose structures, which depend on the proper folding of the polypeptide, the proteins are folded and moved to the Golgi apparatus, where additional glycosylation post translational modifications of the glycoproteins take place sequentially in the cis-Golgi network (CGN), the cis-, medial- and trans- Golgi cisternae (CGC, MGC and TGC), and the trans-Golgi network (TGN). A schematic representation for mammalian N-glycosylation in the ER and Golgi apparatus is depicted in (Fig. 1), where glycosidases facilitate removal of sugars, while glycotransferases facilitate transfer of the sugars from a sugar-nucleotide donor molecule. In mammalian expression systems elimination of the majority of mannoses occurs before the addition of N-acetylglucosamine, galactose and N-acetylneuraminic acid (sialic acid) takes place.

The simplified action mannosidases I (ER Man I) and (Golgi Man I) to remove the four α(1,2)-mannose residues from $Man_{8-9}GlcNAc$ to yield $Man_5GlcNAc_2$ (Tabas and Kornfeld, 1979) is more complex than what was depicted in (Fig. 1). For instance, Class I Golgi α(1,2)-mannosidases (Golgi Man I) belong to the glycosylhydrolase family 47, which includes three class I mammalian α(1,2)-mannosidases: Man IA, Man IB, and Man IC (Herscovics, 2001). Glycan processing of

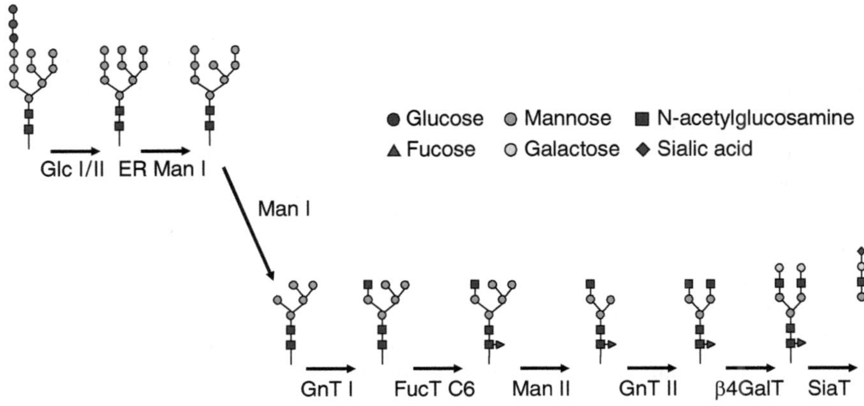

Fig. 1 Schematic pathway representation for N-glycosylation in the endoplasmic reticulum (ER) (*left*) and Golgi apparatus (*right*) for mammalian expression systems. The key enzymes' processings shown result in glycan structures with N-acetylglucosamine, fucose, galactose and sialic acid residues. The enzymes' processings include trimming of glucose residues in the ER by α-glucosidase I (Glc I) and α-glucosidase II (Glc II), elimination of a specific terminal mannose residue by ER α-mannosidase I (ER Man I), and subsequent action of Golgi α(1,2)-mannosidase I (Man I), β-N-acetylglucosaminyltrasferase I (GnTI), core α(1,6)fucosyltransferase (FucTC6), Golgi α(1,3) and α(1,6)-mannosidase II (Man II), β-N-acetylglucosaminyltrasferase II (GnTII), β(1,4)-galactosyltransferase (β4GalT), and sialyltransferase (SiaT).

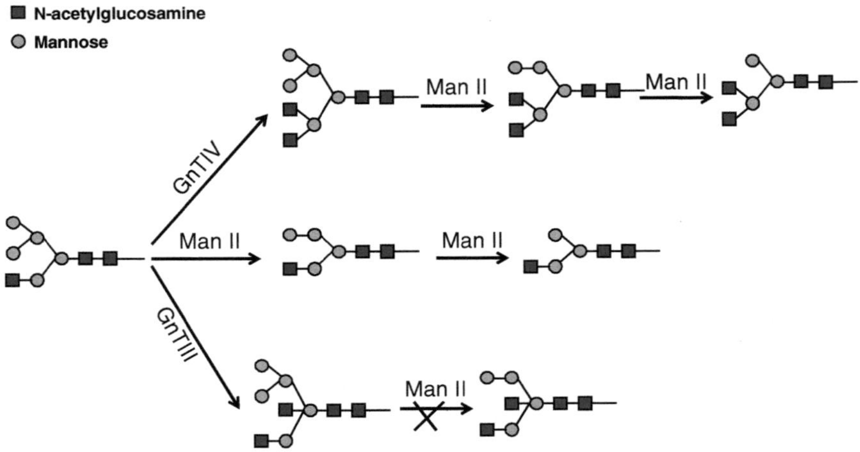

Fig. 2 Initial buildup of glycan structures by considering the action of β-N-acetylglucosaminyl-trasferase IV (GnTIV); Golgi α-mannosidase II (Man II); and β-N-acetylglucosaminyltrasferase III (GnTIII) enzymes.

class I Golgi α(1,2)-mannosidases IA ,IB, and IC results in various high-mannose isomers (Herscovics, 2001). After the high-mannose glycan structure $Man_5GlcNAc_2$ is formed (Fig. 1) GnTI adds a GlcNAc residue through an α(1,3)- linkage to $Man_5GlcNAc_2$, which results in the formation of $GlcNAcMan_5GlcNAc_2$, a hybrid structure in which one antenna is processed. This structure can follow the different pathways (Fig. 2 – first step), where GnTIV can add a GlcNAc β(1,4)-linked to the lower core mannose group or Mannosidase II can eliminate the α(1,3)-mannose residue from the $GnMan_5GlcNAc_2$ glycan structure. If GnTIII is considered in the initial buildup of the N-glycosylation pathway, it can add the bisecting GlcNAc group. The resulting structures are of the hybrid type glycans. Subsequent processing steps of these glycans in these three pathways result in the elimination of terminal mannose α(1,3)-residue, followed by elimination of the α(1,6)-mannose residue by the action of Mannosidase II to generate the $Man_3GlcNAc_2$ type oligosaccharide structure (Fig. 2 – second and third steps).

After the initial buildup on (Fig. 2), additional processing by GnTII and GnTV add GlcNAc groups to the upper core mannose groups and GnTIV to add GlcNac to the lower core mannosegroup, generating bi-antennary, tri-antennary and tetra-antennary complex glycan structures as shown in (Fig. 3 – top row). In following steps, GalT may add galactose to all the terminal GlcNAc groups except the bisecting GlcNAc group. Then two directions are possible: one is the action of SiaT to add sialic acid residues to the galactose groups or alternatively the action of iGnT to add additional GlcNAc groups, which can be further galactosylated and then sialylated. Representative glycan structures of these types are shown in (Fig. 3 – bottom row).

Once these posttranslational modifications take place secreted and membrane glycoproteins will leave the Golgi apparatus to carry out their biological functions.

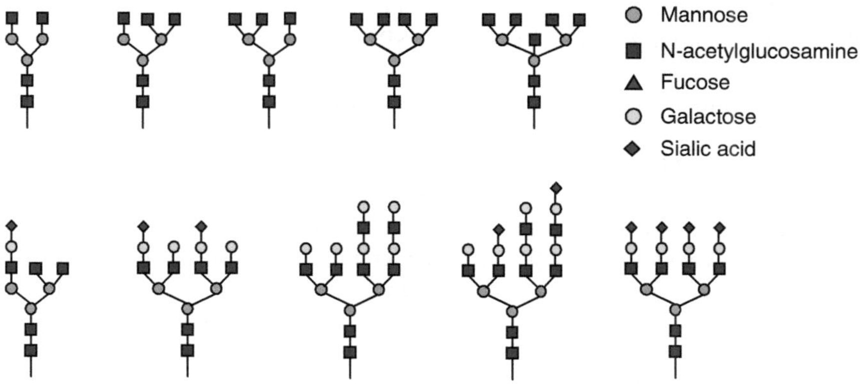

Fig. 3 Following the initial buildup of glycans (Fig. 2). *Top row* shows: characteristic glycan configurations after addition of GlcNac groups to the upper core mannose residues by β-N-acetylglucosaminyltrasferase II (GnTII) and β-N-acetylglucosaminyltrasferase V (GnTV) and by addition of the GlNac group to the lower core mannose by the action of GnTIV. *Bottom row* shows: bi, tri and tetra branched oligosaccharides structures resulting from additional processing of β(1,4)-galactosyltransferase (GalT), sialyltransferase (SiaT) and β-1,3-N-acetylglucosaminyltransferase (iGnT).

Additional enzyme activities can be found in mammalian expression systems (Stanley et al., 1996; Brockhausen et al., 1998), including fucosyltransferases (α3-Fuc-transferase, α2-Fuc-transferase, α3/4-Fuc-transferaseIII); galactosyltransferases (β3-Gal-transferase, Gal 3-sulfotransferase), sialyltransferases (α6-Sialyltransferase) and N-acetylglucosaminyltransferases (GlcNAc-transferase VI, Blood group i β3-GlcNAc-transferase, Blood group I β6-GlcNAc-transferase). These and other types of genetic modifications can expand the number of human like type N-glycan structures obtained from mammalian sources. Importantly, only a subset of N-glycosylation enzymes are present in mammalian cell lines such as Chinese Hamster Ovary cells (CHO), which are used for the production of recombinant therapeutic proteins. For example CHO include α6-Fuc-transferase but lack α6-Sialyltransferase.

3 Glycoengineering of Mammalian Expression Systems

Production of recombinant proteins represented a breakthrough in medicine, as proteins previously were obtained from animals such as bovine/porcine or isolated from human sources. A large number of recombinant proteins have been developed since 1977 when for the first time a synthetic gene for the human hormone somatostatin was cloned and expressed in *Escherichia coli* (Itakura et al., 1977). In 2003, somatostatin got profits accounting for approximately $1.4 billion in global sales. Current manufacturing methods prefer mammalian cell cultures, including Chinese

Hamster Ovary (CHO), NS0 Mouse Myeloma, Human Embryonic Kidney (HEK 293), Baby Hamster Kidney (BHK), and SP2/0 Mouse Myeloma, because they express similar glycan structures to the ones in native human glycoproteins. For instance recombinant human erythropoietin (rhEPO) from mammalian cell lines represent the largest market of biopharmaceuticals with more than $10 billion in the global market as 2007 (Jelkmann, 2007). However, in mammalian expression systems it is difficult to achieve complete control over glycosylated structures. For that reason several approaches to optimize mammalian cell glycosylation are of significant interest. A widely used approach is alteration of the media to favor certain structures. In addition the emergence of genetic engineering tools has allowed researches to make significant progress towards manipulating and controlling N-glycosylation in mammalian cell expression systems in order to obtain more uniform and humanized distributions of glycan structures that preserve key properties for valuable biopharmaceuticals.

Glycan charge is one of the variables that affect the clearance rates of proteins in vivo, for example sialic acid with is negative charge favors slower in vivo clearance kinetics due to avoidance of the asialoglycoprotein receptor (Weiss et al., 1989). The presence of endogenous CHO cell sialidase activity, which cleaves the CHO naturally occurring α(2,3)-sialyl residues from the glycoproteins giving a mixture of desialylated glycans structures in the extracellular medium have been demostrated (Gramer and Goochee, 1993). Changes in process parameters were attempted to decrease sialidase action (Munzert et al., 1996) as well as suppressing sialidase activity by inhibitors (Gramer et al., 1995). Furthermore, partial reduction in the gene expression for sialidase was achieved by developing a CHO cell line that expresses sialidase antisense RNA (Ferrari et al., 1998). To improve the quality of glycans in CHO cell cultures, α(2,6)-sialyltransferase gene has been expressed (Minch et al., 1995) in order to confer NeuAc α-2,6 linkage that is naturally present in human glycans but absent in native CHO cells, on recombinant tissue plasminogen activator (tPA). Alternatively, the overexpression of sialuria mutated UDP-GlcNAc 2-epimerase/ManAc kinase (GNE), which generates more intracellular sialic acid, resulted in increased content of sialic acid on EPO (Bork et al., 2007).

Whereas glycoproteins from the most common mammalian cell lines often present N-acetylneuraminic acid (Neu5Ac) and N-glycolylneuraminic acid (Neu5Gc) humans lack the enzyme to produce Neu5Gc (Bardor et al., 2005). The fact that humans have antibodies against Neu5Gc (Zhu and Hurst, 2002) is of major concern when glycoproteins therapeutics with high Neu5Gc content are produced. Inactivation of the mouse CMAH gene, which is responsible for biosynthesis of the sialic acid N-glycolylneuraminic acid (Neu5Gc), was done to study the in vivo consequences in mice with deficiency of Neu5Gc (Hedlund et al., 2007). This type of approach could be of importance to alleviate the immune responses that Neu5Gc may cause in humans and the elimination of glycans containing Neu5Gc, which reduces drug effectiveness.

Other manipulations to change the number of N-glycans attached, called site occupancy, have been done to alter protein production rates and bioactivity. For example it was show that removal of glycosylation sites in baby hamster kidney

(BHK) cells expressing tranferrin decreased protein secretion five times compared to wild type transferrin (Mason et al., 1993). Importantly, improper site occupancy in human transferrin is one of the glycosylation defects of clinical relevance to Congenital Disorders of Glycosylation (CDG) (Wada et al., 1992; Wada, 2006). Another example concerns recombinant human erythropoietin (EPO), a highly glycosylated complex protein with four carbohydrate side chains and three N-linked glycosylation sites (Lai et al., 1986; Cindric et al., 2006). Elimination of the three native N-glycosylation sites of EPO in baby hamster kidney (BHK) cells confirmed a 90% reduction in production (Yamaguchi et al., 1991). Other studies further support that N-linked oligosaccharides are essential for in vivo activity (Yamaguchi et al., 1991; Higuchi et al., 1992; Elliott et al., 2003). Yamaguchi et al. (1991) found that while in vitro activity in unglycosylated EPO derivatives is maintained, the elimination of the glycosylation sites reduced in vivo activity. More recently, the EPO gene was modified with the addition of two additional N-glycosylation sites to the protein, resulting in improvement of in vivo activity and longevity of action (Elliott et al., 2003; Elliott et al., 2004). These studies show that significant changes in the protein properties can be achieved by altering the number of N-glycans (Elliott et al., 2003; Elliott et al., 2004; Borman, 2006) through genetically reengineering.

As the therapeutic antibodies market is expected to growth to $30.3 billion by 2010, which will account for approximately 60% of the recombinant protein market revenues, monoclonal antibodies are one of the most important products of this industry (Pavlou and Reichert, 2004; Datamonitor Report, 2005). Currently pharmaceutical antibodies are expressed in mammalian cell lines (CHO, mouse myeloma cell line NS0 and SP2/0). Genetic engineering has allowed modifying the antibody carbohydrate moiety located in the Fragment crystallizable (Fc) region of the antibody; resulting in bioactivity increases and improvement of the strength of the FcγRIIIa-receptor specific binding relative to the immune system cells (Rothman et al., 1989; Umana et al., 1999; Ferrara et al., 2006). Antibody Dependent Cellular Cytotoxicity (ADCC) was improved by manipulation of the glycosylation pathway through elimination of fucose (Yamane-Ohnuki et al., 2004) or by expressing β(1,4)-N-acetylglucosaminyltransferase III (GnTIII), enzyme that catalyzes addition of bisecting GlcNAc, a non-existant glycan structure in wild type CHO cells (Umana et al., 1999). Importantly, in vivo cytotoxicity of antibodies against tumors, an Fc-receptor-dependent mechanism, was improved in commercial antibodies such as Rituximab (Cartron et al., 2002) by expressing GnTIII. It was found that the increase in activity correlates with level of GnTIII over expression and that GnTIII outcompetes endogenous FucT to inhibit the addition of fucose (Ferrara et al., 2006). Strong binding to the human receptor FcγRIIIa and a 100-fold improvement of ADCC was obtained in defucosylated antibodies with respect to Rituxan. Knock out of FUT8 that encodes an α-1,6-fucosyltransferase that catalyze the transfer of fucose from GDP-fucose to N-acetylglucosamine (GlcNAc) in Chinese hamster ovary cells resulted in complete defucosylated antibodies (Yamane-Ohnuki et al., 2004). To achieve further improvements both expression of GnTIII and fucose elimination have been explored. This manipulation of the glycosylation pathway provided a 50-fold increase in binding affinities of Non-fucosylated bisected IgGs to FcγRIIIa receptor (Ferrara et al., 2006). These genetic glycoengineering modifications when used alone or in combination with changes in process parameters can significantly improve efficiency and activity of protein therapeutics.

4 Computer Aided Glycoengineering

In order to obtain glycan structures of desired therapeutic efficiency various methods have been employed from genetic manipulations (Lee et al., 1989; Stanley et al., 1996; Umana et al., 1999; Lee et al., 2001) to variations in the cell culture processing parameters (Goochee et al., 1991; Baker et al., 2001). The fact that glycan structures are not directly encoded by the genome as in the case of protein structures makes it difficult to predict a priori the effects of genetic and cell culture manipulations on glycan structure profiles of therapeutic proteins. Indeed, glycosylation is the product of a complex sequence of enzyme catalyzed reactions, where glycosydases and glycosyltransferases interact with other metabolites to add or remove sugars to the glycan chains and generate a wide diversity of glycan structures. Predicting the result of the action of those glycosylation enzymes working in a cooperative mode is not always intuitive and straight forward as exemplified by a number of contributing factors such as the localization of the enzymes to specific Golgi compartments, multiple enzyme action on common substrates, the non-linear nature of enzyme kinetics and the restrictions on substrate accessibility for these enzymes. Computers programs able to predict glycosylation have emerged as promising tools to help understanding the complexity of the glycosylation processing we outlined. Here on, we will briefly address some developments in computational modeling, with emphasis on the work from our group.

A mathematical model for the N-glycosylation process in mammalian cells was developed by Umana and Bailey, 1997. This model predicts 33 N-glycan structures and is limited to the first galactosylation steps of the N-glycosylation pathway. Glycosylation of proteins is assumed to take place in four compartments, the cis-, medial- and trans-Golgi cisternae (CGC, MGC and TGC), and the trans-Golgi network (TGN). The compartments are considered 'well mixed reactors' with reactions following Michaelis–Menten kinetics. Values for the model parameters were obtained or estimated from literature information. The solution for the set of 33 reactions gave galactosylated glycan structures that followed the experimental glycan distributions for recombinant CHO cells (tPA, EPO, β-interferon). Importantly, simulations of β(1,4)-N acetylglucosaminyltransferase III (GnTIII) predicted maximum levels of complex bisected oligosaccharides at intermediate values of GnTIII activity. This prediction was later tested experimentally by the same group (Umana et al., 1999) for production of monoclonal antibodies of higher in vivo efficacy resulting from overexpression of GnTIII.

In order to predict the degree of sialylation Monica et al. (1997) developed a mathematical model that assumes like the Umana and Bailey model Michaelis–Menten reaction with the N-linked sialyltransferases localized in the trans Golgi network. The compartment is treated as a homogeneous reaction environment with bulk transport of metabolites moving in and out. The predicted values for the extent of sialylation were consistent with experimental data obtained for CD4 and t-PA glycoproteins (Spellman et al., 1989; Spellman et al., 1991).

An extension of the Umana and Bailey model was developed by Krambeck and Betenbaugh (2005). The model was designed to be easily and iteratively refined to include additional enzymes for galactosylation, fucosylation, extension of N-acetyllactosamine repeat and sialylation. As with the Umana and Bailey system, the model assumes that the cis, medial and trans-Golgi cisternae (CGC, MGC and TGC), and the trans-Golgi network (TGN) compartments are 'well-mixed reactors' with enzyme catalyzed glycosylation reactions following Michaelis–Menten kinetics. The substrate specificity of the glycosylation enzymes in the Golgi apparatus is defined by restriction rules on the enzymes that follow the biosynthetic N-glycosylation pathway. Once this framework is set, the reaction network is derived and the equations are solved for the concentration of each glycan in each Golgi compartment. The 33 glycan structures and 33 reactions generated by the Umana and Bailey model were extended to 7,565 glycan structures and a network of 22,871 reactions. Solution of the model equations requires parameters including reaction kinetic parameters, compartment residence times, enzyme distributions between compartments, compartment volumes, total glycan concentration, and donor cosubstrate concentrations, which were obtained from literature sources.

Different from the Umana and Bailey model the Krambeck and Betenbaugh model incorporates methods for not only solving the complete glycan distribution profile but also methods for adjusting the model parameters to match experimental glycan distribution data. In addition, the Krambeck and Betenbaugh model considers the donor concentration effects by including the dissociation constants for donor cosubstrates (K_{md}) while the Umana and Bailey model assumed that enzymes were saturated with cosubstrates. Similar to the Umana and Bailey model a mixture of high mannose glycans of nine and eight mannose residues (Man9 and Man8) are the input to the first compartment of the model (cis-Golgi cisternae). The reactions that remove some mannoses in the endoplasmic reticulum (ER) and in the cis-Golgi network (CGN) to form the high mannose glycan mixture (Man8 and Man9) are neglected. Importantly, the model predictions of glycosylation profile for recombinant human thrombopoietin (rhTPO) in CHO cells were validated with experimental evidence (Inoue et al., 1999). Almost all of the structures obtained by the model were fucosylated with mostly one to four sialic acid residues per glycan structure as show by the sialic acid distribution of Table 1.

Table 1 Distribution of sialic acid groups on N-glycans of recombinant human TPO (adapted from Krambeck and Betenbaugh, 2005)

Number of sialic acid groups	0	1	2	3	4
Percent of N-glycans (measured) (Inoue et al., 1999)	1	12	47	27	12
Calculated (unadjusted)	3	15	68	13	1
Calculated (adjusted)	3	10	48	26	13

In Table 1, the calculated 'adjusted sialylated glycan distributions', were obtained by making adjustments to the overall enzyme concentrations in the model from their assumed values using an optimization algorithm to match the experimental data. For the 'unadjusted sialylated glycan distributions' in Table 1, the distribution of sialic acid groups was obtained without adjusting the model parameters to match the experimental data. In all cases no adjustment was made to the distribution of enzymes among the Golgi compartments. The distribution of glycan structures following in vitro elimination of the sialic acid residues by sialidase (Inoue et al., 1999) are included in Table 2, the model predicted oligosaccharide distributions were in good agreement with the measured distributions for each glycan structure.

After adjustment of the overall enzyme concentrations, the model was used to analyze how glycan structures change as protein productivity is increased (i.e. increase of total glycan concentration, which is proportional to protein productivity). For instance, as shown in Fig. 4, a general decrease in the degree of sialylation is observed as the total glycan concentration increases. This decline in sialic acid content with higher yield implies a bottleneck in the sialylation processing. Similar trends are observed for N-acetyllactosamine repeat groups, which decrease as total glycan concentration increases. This trend is related to the effect of the denominator term in the Michaelis–Menten type equation, where the summation term over all

Table 2 Distribution of desialylated structures for recombinant human TPO. Measured data was obtained from Inoue et al. (1999) (adapted from Krambeck and Betenbaugh, 2005)

Structure	Measured Amount	Calculated (unadj.)	Calculated (adjusted)
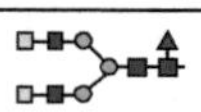	27.9	77.8	27.4
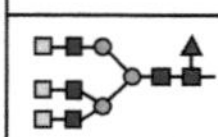	5.9	1.6	4.2
	16.2	11.8	17.6
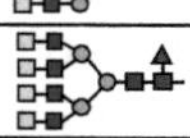	22.6	1.6	22.5
	3.4	0.8	3.4
	4.0	0.8	3.5
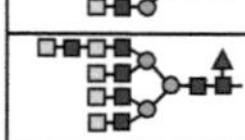	5.2	0.1	6.7
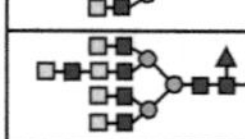	5.7	0.1	6.8
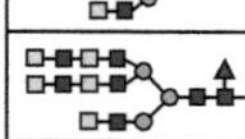	2.7	0.1	1.0
	5.2	0.0	2.9

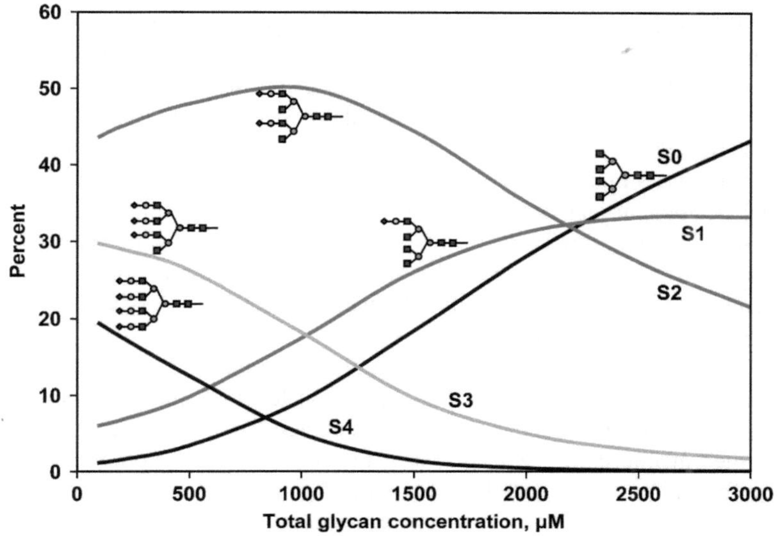

Fig. 4 Effect of N-glycan concentration on sialylation level. S0 to S4 correspond to glycan structures with 0 to 4 sialic acid residues (adapted from Krambeck and Betenbaugh, 2005).

the substrates-K_m enzyme combinations, may result in substrate competition for the same enzyme. This substrate competition prevents yield increases of branched sialylated glycan structures with glycan concentration. In effect enzymes with lower K_m values such as SiaT will saturate quickly as concentration increases.

The authors also show how one could manipulate model parameters to increase protein productivity but maintain glycan distribution within the original carbohydrate structural distribution. The model identified intuitive and non intuitive changes in enzyme concentration and donor cosubstrate that could be implemented in order to decrease the shift in glycan distribution as the protein productivity is increased.

Another mathematical model (Lau et al., 2007) was developed to identify the effects of N-glycan branching ultrasensitivity to UDP-GlcNAc. Conditions for branching ultrasensitivity were identified, showing agreement with experiments. The model assumes parameters from Umana and Bailey (1997) and Krambeck and Betenbaugh (2005). Different from the Krambeck and Betenbaugh model an additional reversible reaction represents the dissociation of the enzyme-substrate-cosubstrate complex into product, UDP and free enzyme, implying that UDP competes for the donor site. In the Krambeck and Betenbaugh model the dissociation of the enzyme-substrate-cosubstrate complex into enzyme-product and UDP was neglected by assuming an irreversible reaction. This may be justified by considering low concentration of UDP in vivo, as UDP is rapidly hydrolyzed to UMP. Nevertheless, the term of reversibility can be included in the Krambeck and Betenbaugh model for use of the model for in vitro applications. Also, the Lau et al. (2007) model assumes that the reverse rate constant is four times the forward rate constant for unsaturated systems. This model includes several branching enzymes except sialyltransferases and fucosyltransferases.

Lately, a software package named: In silico Biochemical Reaction Network Analysis (IBRENA) was developed (Liu and Neelamegham, 2008). This package for analysis and simulation of reaction networks facilitates sensitivity analysis, principal component analysis as well as singular value decomposition and model reduction. Computer modeling results for O-glycosylation were validated with experiments. The model predicted the variation in expression of sialyl Lewis-X (sLe^x) epitope by perturbation of glycotransferases α2,3ST3Gal-IV and α2,3ST3 Gal-I/II expression (Liu et al., 2008) .

In an effort to link expression data with glycan structure Kawano et al. (2005) developed a method that assumes that expression levels of the genes are proportional to glycan linkage abundances. The second version of this model (Suga et al., 2007) aimed at improving two aspects of the scheme: gene expression based on binary information was replaced by real-valued gene expression intensity and the number of predictable glycan structure candidates was increased by estimating missing glycans from a global glycan structure map. Interestingly, increases of Lewis-a, Lewis-x, or sialyl-Lewis-x structures, which are known to be related to cancer, were consistent with changes in expression data from human cancer cells with respect to normal cells. However, several aspects that restrict enzyme action were not considered using this bioinformatics approach.

Future developments to link expression data with structural data will require more systematic efforts to define how changes in glycosylation enzyme concentrations can be realized by expressing the genes that encode those enzymes. Importantly, increasing amounts of information is available including detailed Glycomics web resources (von der Lieth et al., 2006; Lutteke, 2008). Other resources include the BRaunschweig ENzyme Database (BRENDA) which collects glycosylation enzyme data in terms of reaction rate parameters and specificity (Schomburg et al., 2004; Barthelmes et al., 2007).

5 Conclusions

With the improvement of analytical and computational methods, manipulation and control of desired structural glycan alterations will continue to improve in cell culture systems. As glycan structures are not directly encoded in the genome, a thorough understanding of the enzymology and cellular localization of glycosyltransferases and glycosidases involved in the biosynthesis of N-glycosylation and their relation with gene expression will be fundamental to manipulate and control N-glycosylation. Although a large amount of resources is available, data interpretation and integration of this knowledge using complex computation methods will be essential in order to optimize the profiles of N-glycans using cell and process engineering strategies.

Acknowledgements This work was supported by Award Number 5R41CA127885-02 from the National Cancer Institute.

References

Apweiler R, Hermjakob H et al (1999) On the frequency of protein glycosylation, as deduced from analysis of the SWISS-PROT database. Biochim Biophys Acta – General Subjects 1473(1):4–8

Baker KN, Rendall MH et al (2001) Metabolic control of recombinant protein N-glycan processing in NS0 and CHO cells. Biotechnol Bioeng 73(3):188–202

Bardor M, Nguyen DH et al (2005) Mechanism of uptake and incorporation of the non-human sialic acid N-glycolylneuraminic acid into human cells. J Biol Chem 280(6):4228–4237

Barthelmes J, Ebeling C et al (2007) BRENDA, AMENDA and FRENDA: the enzyme information system in 2007. Nucleic Acids Res 35:D511–D514

Bork K, Reutter W et al (2007) Enhanced sialylation of EPO by overexpression of UDP-GlcNAc 2-epimerase/ManAc kinase containing a sialuria mutation in CHO cells. FEBS Lett 581(22):4195–4198

Borman S (2006) Glycosylation engineering: controlling personalities tame wild sugars on proteins and natural products. Chem Eng News 84(36):13–22

Brockhausen I, Schutzbach J et al (1998) Glycoproteins and their relationship to human disease. Acta Anat 161(1–4):36–78

Cartron G, Dacheux L et al (2002) Therapeutic activity of humanized anti-CD20 monoclonal antibody and polymorphism in IgG Fc receptor Fc gamma RIIIa gene. Blood 99(3):754–758

Cindric M, Bindila L et al (2006) Mass spectrometry-based glycoproteomic approach involving lysine derivatization for structural characterization of recombinant human erythropoietin. J Proteome Res 5(11):3066–3076

Datamonitor Report (2005) Monoclonal Antibody Therapies: Evolving into a $30 billion Market.

Elliott S, Lorenzini T et al (2003) Enhancement of therapeutic protein in vivo activities through glycoengineering. Nat Biotechnol 21(4):414–421

Elliott S, Egrie J et al (2004) Control of rHuEPO biological activity: the role of carbohydrate. Exp Hematol 32(12):1146–1155

Ferrara C, Brunker P et al (2006a) Modulation of therapeutic antibody effector functions by glycosylation engineering: influence of Golgi enzyme localization domain and co-expression of heterologous beta1, 4-N-acetylglucosaminyltransferase III and Golgi alpha-mannosidase II. Biotechnol Bioeng 93(5):851–861

Ferrara C, Stuart F et al (2006b) The carbohydrate at Fc gamma RIIIa Asn-162 – an element required for high affinity binding to non-fucosylated IgG glycoforms. J Biol Chem 281(8):5032–5036

Ferrari J, Gunson J et al (1998) Chinese hamster ovary cells with constitutively expressed sialidase antisense RNA produce recombinant DNase in batch culture with increased sialic acid. Biotechnol Bioeng 60(5):589–595

Goochee CF, Gramer MJ et al (1991) The oligosaccharides of glycoproteins – bioprocess factors affecting oligosaccharide structure and their effect on glycoprotein properties. Bio/Technology 9(12):1347–1355

Gramer MJ, Goochee CF (1993) Glycosidase activities in Chinese-hamster ovary cell lysate and cell-culture supernatant. Biotechnol Prog 9(4):366–373

Gramer MJ, Goochee CF et al (1995) Removal of sialic-acid from a glycoprotein in Cho cell-culture supernatant by action of an extracellular Cho cell sialidase. Bio/Technology 13(7):692–698

Hedlund M, Tangvoranuntakul P et al (2007) N-glycolylneuraminic acid deficiency in mice: implications for human biology and evolution. Mol Cell Biol 27:4340–4346
Herscovics A (2001) Structure and function of Class I alpha 1, 2-mannosidases involved in glycoprotein synthesis and endoplasmic reticulum quality control. Biochimie 83(8):757–762
Higuchi M, Oh-eda M et al (1992) Role of sugar chains in the expression of the biological activity of human erythropoietin. J Biol Chem 267(11):7703–7709
Inoue N, Watanabe T et al (1999) Asn-linked sugar chain structures of recombinant human thrombopoietin produced in Chinese hamster ovary cells. Glycoconj J 16(11):707–718
Itakura K, Hirose T et al (1977) Expression in *Escherichia coli* of a chemically synthesized gene for hormone somatostatin. Science 198(4321):1056–1063
Jelkmann W (2007) Recombinant EPO production – points the nephrologist should know. Nephrol Dial Transplant 22(10):2749–2753
Jenkins N, Parekh RB et al (1996) Getting the glycosylation right: implications for the biotechnology industry. Nat Biotechnol 14(8):975–981
Kawano S, Hashimoto K et al (2005) Prediction of glycan structures from gene expression data based on glycosyltransferase reactions. Bioinformatics 21(21):3976–3982
Kornfeld S (1998) Diseases of abnormal protein glycosylation: an emerging area. J Clin Invest 101(7):1293–1295
Kornfeld R, Kornfeld S (1985) Assembly of asparagine-linked oligosaccharides. Annu Rev Biochem 54:631–664
Krambeck FJ, Betenbaugh MJ (2005) A mathematical model of N-linked glycosylation. Biotechnol Bioeng 92(6):711–728
Lai PH, Everett R et al (1986) Structural characterization of human erythropoietin. J Biol Chem 261(7):3116–3121
Lau KS, Partridge EA et al (2007) Complex N-glycan number and degree of branching cooperate to regulate cell proliferation and differentiation. Cell 129(1):123–134
Lee EU, Roth J et al (1989) Alteration of terminal glycosylation sequences on N-linked oligosaccharides of Chinese-hamster ovary cells by expression of beta-galactoside alpha-2, 6-sialyltransferase. J Biol Chem 264(23):13848–13855
Lee JH, Sundaram S et al (2001) Chinese hamster ovary (CHO) cells may express six beta 4-galactosyltransferases (beta 4GalTs) – consequences of the loss of functional beta 4GalT-1, beta 4GalT-6, or both in CHO glycosylation mutants. J Biol Chem 276(17):13924–13934
Liu G, Neelamegham S (2008) In silico biochemical reaction network analysis (IBRENA): a package for simulation and analysis of reaction networks. Bioinformatics 24:1109–1111
Liu G, Neelamegham S., et al. (2008) Systems level modeling of cellular glycosylation reaction networks: O-linked glycan formation on natural selectin ligands. Bioinformatics (submitted).
Lutteke T (2008) Web resources for the glycoscientist. Chembiochem 9(13):2155–2160
Mason AB, Miller MK et al (1993) Expression of glycosylated and nonglycosylated human transferrin in mammalian-cells – characterization of the recombinant proteins with comparison to 3 commercially available transferrins. Biochemistry 32(20):5472–5479
Minch SL, Kallio PT et al (1995) Tissue-plasminogen activator coexpressed in Chinese-hamster ovary cells with alpha(2, 6)-sialyltransferase contains neuac-alpha(2, 6) Gal-beta(1, 4) Glc-N-Acr linkages. Biotechnol Prog 11(3):348–351
Monica TJ, Andersen DC et al (1997) A mathematical model of sialylation of N-linked oligosaccharides in the trans-Golgi network. Glycobiology 7(4):515–521
Munzert E, Muthing J et al (1996) Sialidase activity in culture fluid of Chinese hamster ovary cells during batch culture and its effect on recombinant human antithrombin III integrity. Biotechnol Prog 12(4):559–563
Pavlou AK, Reichert JM (2004) Recombinant protein therapeutics – success rates, market trends and values to 2010. Nat Biotechnol 22:1513–1519
Peter-Katalinic J (2005) Methods in enzymology: O-glycosylation of proteins. Mass Spectrom: Modified Proteins and Glycoconjugates 405:139–171

Rothman RJ, Perussia B et al (1989) Antibody-dependent cytotoxicity mediated by natural-killer-cells is enhanced by castanospermine-induced alterations of Igg glycosylation. Mol Immunol 26(12):1113–1123

Schomburg I, Chang A et al (2004) BRENDA, the enzyme database: updates and major new developments. Nucleic Acids Res 32:D431–D433

Spellman MW, Basa LJ et al (1989) Carbohydrate structures of human-tissue plasminogen-activator expressed in Chinese-hamster ovary cells. J Biol Chem 264(24):14100–14111

Spellman MW, Leonard CK et al (1991) Carbohydrate structures of recombinant soluble human Cd4 expressed in Chinese-hamster ovary cells. Biochemistry 30(9):2395–2406

Stanley P, Raju TS et al (1996) CHO cells provide access to novel N-glycans and developmentally regulated glycosyltransferases. Glycobiology 6(7):695–699

Suga A, Yamanishi Y, Hashimoto K, Goto S, Kanehisa M (2007) An improved scoring scheme for predicting glycan structures from gene expressison data. Genome Inform 18:237–246

Tabas I, Kornfeld S (1979) Purification and characterization of a rat-liver golgi alpha-mannosidase capable of processing asparagine-linked oligosaccharides. J Biol Chem 254(22):1655–1663

Umana P, Bailey JE (1997) A mathematical model of N-linked glycoform biosynthesis. Biotechnol Bioeng 55(6):890–908

Umana P, Jean-Mairet J et al (1999) Engineered glycoforms of an antineuroblastoma IgG1 with optimized antibody-dependent cellular cytotoxic activity. Nat Biotechnol 17(2):176–180

von der Lieth CW, Lutteke T et al (2006) The role of informatics in glycobiology research with special emphasis on automatic interpretation of MS spectra. Biochim Biophys Acta – General Subjects 1760(4):568–577

Wada Y (2006) Mass spectrometry for congenital disorders of glycosylation, CDG. J Chromatogr B Analyt Technol Biomed Life Sci 838(1):3–8

Wada Y, Nishikawa A et al (1992) Structure of serum transferrin in carbohydrate-deficient glycoprotein syndrome. Biochem Biophys Res Commun 189(2):832–836

Weiss P, Tietze F et al (1989) Identification of the metabolic defect in sialuria. J Biol Chem 264(30):17635–17636

Yamaguchi K, Akai K et al (1991) Effects of site-directed removal of N-glycosylation sites in human erythropoietin on its production and biological properties. J Biol Chem 266(30):20434–20439

Yamane-Ohnuki N, Kinoshita S et al (2004) Establishment of FUT8 knockout Chinese hamster ovary cells: an ideal host cell line for producing completely defucosylated antibodies with enhanced antibody-dependent cellular cytotoxicity. Biotechnol Bioeng 87(5):614–622

Zhu A, Hurst R (2002) Anti-N-glycolylneuraminic acid antibodies identified in healthy human serum. Xenotransplantation 9(6):376–381

Engineering the Secretory Pathway in Mammalian Cells

Ren-Wang Peng and Martin Fussenegger

Abstract There has been rapid growth in the commercial market for biopharmaceuticals, most of which are produced in mammalian cells as secreted recombinant products. Thus, cellular engineering strategies, involving modulation of the secretory capacity of the producer cells, are highly desirable. Genetic and biochemical analyses of intracellular trafficking of proteins have generated a detailed portfolio of molecular mechanisms involved in cargo transport between organelles. Together with a large array of technical advances, this progress has enabled the rational identification of core machinery, key factors, complex controls and various regulatory mechanisms that define the directionality, specificity and efficiency of vesicular transport along the secretory pathway. The unprecedented availability of information has led to new goals for the development of novel strategies to improve the productivity of the secretory pathway in mammalian cells.

1 Introduction

Mammalian cells have dominated the production of biopharmaceuticals due to their superiority in protein folding, assembly and complex post-translational modifications (Wurm, 2004; Fussenegger and Hauser, 2007). Among the commercial biopharmaceutical products (e.g. hormones, monoclonal antibodies, interferons, etc.), most are secreted proteins. A variety of successful transcription and translation engineering strategies have been implemented over the past decade, which have driven the specific productivity of mammalian cells to an apparent limit. The secretion capacity of a producer cell or organism has now been considered as the major bottleneck preventing mammalian cells from fully exploiting their physiological production capacity in a biopharmaceutical manufacturing scenario. As part of the complex vesicle trafficking system that manages the precise and regulated distribution of

R.-W. Peng and M. Fussenegger (✉)
ETH Zurich, Department of Biosystems Science and Engineering (D-BSSE),
Mattenstrasse 26, CH-4058, Basel, Switzerland
e-mail: fussenegger@bsse.ethz.ch

M. Al-Rubeai (ed.), *Cell Line Development*, Cell Engineering 6,
DOI 10.1007/978-90-481-2245-5_12, © Springer Science+Business Media B.V. 2009

proteins, membranes and other molecular cargo between cellular compartments, the secretory pathway, which is used by proteins destined for secretion and for residence in the plasma membrane, consists of a series of consecutive and interconnected steps. After synthesis on the ribosome-bound endoplasmic reticulum (rough ER), secreted proteins are first translocated into the lumen of the ER before being transported to the cell exterior through the Golgi apparatus. This long march distance covered by secretory proteins occurs by means of vesicle intermediates, which bud from a donor compartment and fuse with an acceptor/target compartment. Vesicle budding and cargo selection in the donor compartment are mediated by coat protein complexes (COPI, COPII and clathrin), while vesicle fusion depends on a mechanism involving SNARE and Sec1/Munc18 proteins. The precise regulation of these two aspects of vesicular transport ensures efficient cargo transfer during which integrity of the organelles is maintained. This contribution focuses on advances that have been made in our understanding of the molecular mechanisms involved in vesicle trafficking and in the development of engineering strategies for enhanced secretion.

2 Overview of the Secretory Pathway

2.1 Protein Secretion: Concepts and Mechanisms

Newly synthesized secretory proteins pass through a series of membrane-enclosed compartments, including the endoplasmic reticulum (ER) and the Golgi apparatus on their way to the extracellular space (Palade, 1975; Fig. 1). In principle, there are two modes of secretion: constitutive and regulated. Constitutive secretion is the most common and occurs in a variety of cell types and organisms with proteins being secreted as fast as they are synthesized. Experiments with protein labeling have shown that it takes only about 10 min for secreted proteins to traverse the cellular compartments along the secretory pathway to reach their final destination (Kelly, 1985). Alteration of the rate of protein secretion by constitutive secretory cells is achieved by altering the rate of protein synthesis. In regulated secretion, on the other hand, secretory proteins are first stored in transport vesicles after being synthesized. Regulated secretory cells have evolved to release large amounts of cargo molecules in a few milliseconds, which far exceeds the rate of protein synthesis. The time for the storage of cargo in secretory vesicles varies with the type of cell and type of cargo molecules, and usually exceeds 10 h (Kelly, 1985). As a consequence, one of the morphological hallmarks of secretory cells is the accumulation of large numbers of secretory vesicles in the cytoplasm. These vesicles do not undergo exocytosis until the cellular level of a cytoplasmic messenger (e.g. Ca^{2+}) is modulated.

Since most FDA-licensed biopharmaceuticals are secreted proteins produced by mammalian cells, secretion-competent producer cell lines such as Chinese hamster ovary cells (CHO-K1) have become the preferred production host of the biopharmaceutical manufacturing industry. The overall production capacity and product quality

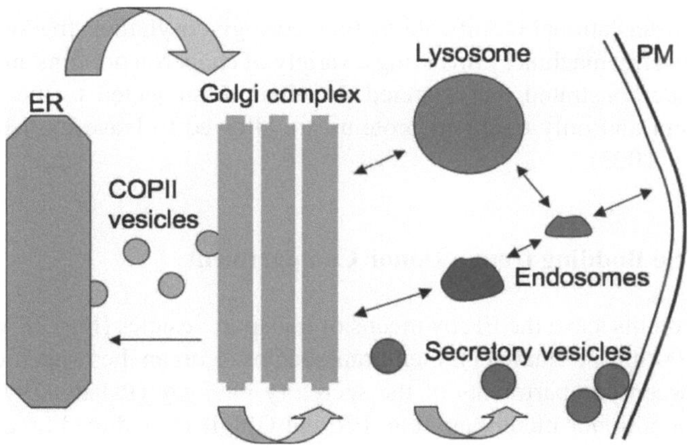

Fig. 1 Protein trafficking network and the secretory pathway in mammalian cells showing the compartment/organelle arrangement involved in the secretory pathway (*yellow, curved arrows*) and intracellular protein transport. Different classes of vesicles mediating distinct trafficking steps of the secretory pathway are indicated by different colors. PM, plasma membrane

of a biopharmaceutical manufacturing process is a well-balanced function of production gene transcription, translation and secretion of the protein pharmaceutical. Since tremendous progress has been achieved in recent decades to increase transcription as well as translation of product genes (Fussenegger et al., 1998a; Fussenegger et al., 2000; Meents et al., 2002a; Schlatter and Fussenegger, 2003; Underhill et al., 2003; Weber and Fussenegger, 2004), the secretion capacity of a mammalian production cell line has become the major bottleneck limiting protein production of mammalian cells (Dorai et al., 2006; Barnes et al., 2004; Barnes and Dickson, 2006). A potential solution of this challenge is secretion engineering, which aims at modulating the secretion capacity of secretory cells by genetic engineering.

2.2 *Key Steps Along the Secretory Pathway*

2.2.1 Translocation into the Endoplasmic Reticulum

A common feature of secretory proteins is that they contain an N-terminal signal sequence (SS), which guides the protein-making ribosomes to the ER surface. This process is mediated by a protein-protein interaction that involves the signal sequence recognition particle (SRP) which binds the SS and SRP receptors on the ER membrane. Secretory proteins, synthesized on the ribosome-bound ER (rough ER), are then translocated into the lumen of the ER by a cotranslational mechanism.

The endoplasmic reticulum is the first station of the secretory pathway. There, secretory proteins fold, are assembled into competent conformations and undergo various post-translational modifications (e.g. core-glycosylation) that are mediated by the ER-specific machinery involving a variety of chaperon proteins and enzymes. It has been demonstrated that secreted proteins are subjected to the ER quality control system and only qualified proteins are allowed to leave the ER (Ellgaard and Helenius, 2003).

2.2.2 Vesicle Budding from a Donor Compartment

Secretory proteins leave the ER by means of transport vesicles (Fig. 2). Vesicles are small (60–100 nm in diameter), membrane-enclosed organelles specific for cargo transfer between compartments of the secretory pathway (Palade, 1975). Vesicle budding from a donor membrane (e.g. ER and Golgi) is mediated by coat protein complexes (Kirchhausen, 2000, Antonny and Schekman, 2001; Bonifacino and Lippincott-Schwartz, 2003), which are soluble proteins and recruited from the cytoplasm onto donor membranes. Binding of coat proteins is believed to deform the membrane and trigger the genesis of the curvature to form a coated bud (Bi et al., 2002). There are different classes of coat proteins that bud vesicles from distinct compartments (Waters et al., 1991; Barlowe et al., 1994). Of these coat proteins, the coat protein complex II (COP II), generating vesicles from the ER, has been studied extensively. COPII is a supramolecular assembly consisting of the GTPase Sar1 and two protein subcomplexes, Sec23/Sec24 and Sec13/Sec31. Assembly of the functional COPII complex from individual components occurs in a stepwise manner. First, cytosolic, GDP bound Sar1 is converted to membrane and GTP bound Sar1, which then recruits the Sec23/Sec24 subcomplex. The Sec13/Sec31 subcomplex reaches the ER membrane last (Antonny and Schekman, 2001; Bi et al., 2002).

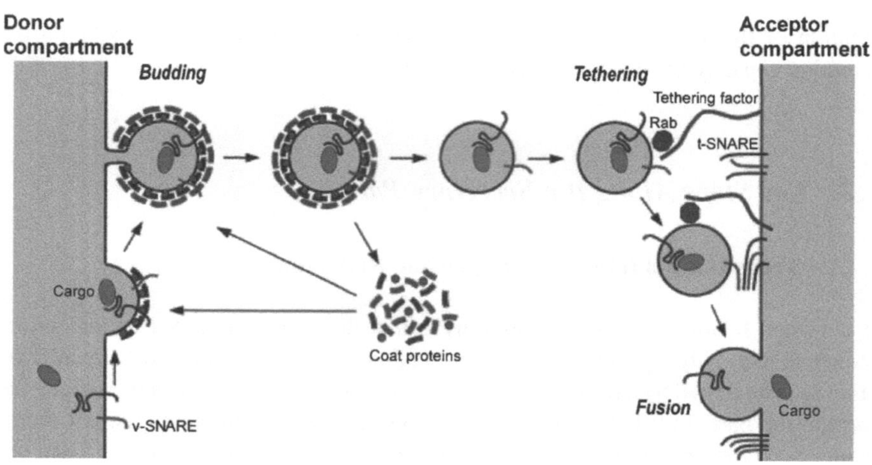

Fig. 2 The molecular machinery involved in vesicle budding and fusion. Note that Sec1/Munc18 proteins, the essential components of the vesicle fusion machinery, are not shown (adapted from article of Bonifacino and Glick, 2004)

In vitro experiments with purified proteins and artificial membranes (liposomes) have demonstrated that COPII is required to deform membranes and generate vesicles (Matsuoka et al., 1998).

The assembly of coat complex is precisely regulated by two protein families with contrasting functions: the guanine nucleoside exchange factors (GEF), which convert the GDP bound GTPase (e.g. Sar1) to the GTP bound one, thus activating the GTPase and triggering vesicle budding, and GTPase-activating proteins (GAP), which accelerate the intrinsic GTPase activity of GTPases, thus inactivating the GTPase by converting the GTP bound GTPase to the GDP bound again. The GTPase cycle (in either GTP-binding or GDP-binding) determines the activity of GTPase and, therefore, controls vesicle formation from a donor membrane.

2.2.3 Tethering and Fusion of Vesicles with Acceptor Membranes

Vesicle Tethering

Once detached from a donor compartment, vesicles must identify their cognate acceptor compartments for the unloading of cargo molecules. This is achieved by tethering and fusion of vesicles (Fig. 2; Rothman, 1994). As many other cellular processes, docking and fusion of transport vesicles also operate in a sequential manner to guarantee the high fidelity of transport. Prior to membrane fusion, tethering is the initial contact between a transport vesicle and the target membrane. This process is mediated by multi-component heteromeric protein complexes (called tethering factors), and by evolutionarily conserved Rab GTPases (Whyte and Munro, 2002). Different tethering factors and Rab GTPases are involved in tethering of distinct classes of vesicles. For example, the multimeric TRAPP complex has been suggested to tether ER-derived vesicles to the Golgi (Whyte and Munro, 2002), while "Exocyst", a hexametric protein complex, links the Golgi-derived secretory carriers to the plasma membrane (Guo et al., 1999; Whyte and Munro, 2002). By interaction with vesicles and targeting membranes, tethering factors and Rab GTPases ensure that tethering and the ensuing membrane fusion occur at the correct time and place.

Mechanisms of Membrane Fusion

Membrane Fusion and SNARE Proteins

A key step in vesicle trafficking is membrane fusion, which defines the merging of two opposing membranes and results in the unloading of cargo molecules in acceptor or target compartments. In the regulated secretory pathway, fusion is a major target for temporal regulation because it occurs only when a cellular signal is transmitted. Central to the mechanisms of membrane fusion are the two most important protein families, namely SNARE (soluble N-ethylmaleimide-sensitive-factor attachment receptor) and Sec1/Munc18 proteins. SNARE proteins are a super family of small molecules, with more than 35 found in mammals (Bock et al., 2001). Most SNAREs are type II membrane proteins bearing a single transmembrane domain (TMD) in their C termini. Although the size and primary structure of SNARE proteins vary widely, they all share a homologous domain, the SNARE motif, adjacent to the TMD. SNARE motifs, consisting of 60 to 70 amino acids, and prone to form coiled coils, are the defining feature of all SNARE proteins (Jahn et al., 2003; Jahn and Scheller, 2006).

One of the striking properties of SNARE proteins is that they bind to each other and assemble into complexes (Fig. 3; Sutton et al., 1998). Found on both vesicles (v-SNARE) and target compartments (t-SNAREs), SNARE proteins engage each other when two membranes are in close proximity, which is achieved by the tethering process. Biochemical and structural studies have indicated that the SNARE complex, between the cognate v- and t-SNAREs, is a very stable four-helix bundle, with one α-helix contributed by the monomeric v-SNARE and the other three by the oligomeric t-SNAREs (Sutton et al., 1998). Of the SNARE proteins, syntaxin1, SNAP-25 and VAMP2, mediating the fusion of synaptic vesicles with the plasma membrane, are the best characterized. In the assembled synaptic SNARE complex, syntaxin 1 and VAMP2 each provide one SNARE motif, while the other two SNARE motifs are donated by SNAP-25. All the SNARE complexes in the cell follow this rule, although in most cases the two SNARE motifs donated by SNAP-25 are from two separate proteins (Jahn et al., 2003; Bonifacino and Glick, 2004).

Reconstituted experiments have demonstrated that purified SNARE proteins promote the fusion of liposomes, provided that v- and t-SNAREs are present on different membranes (Weber et al., 1998; Jahn et al, 2003). Furthermore, when engineered into mammalian cells and expressed with the topology that the SNARE

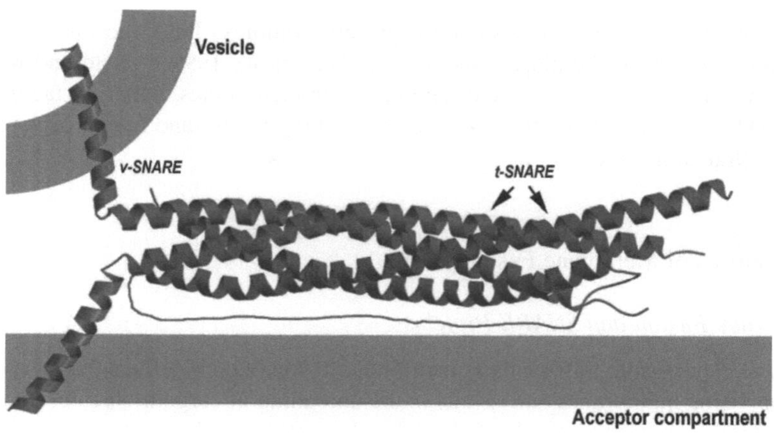

Fig. 3 Structural properties of the functional SNARE complex. One SNARE from a transport vesicle (v-SNARE) is engaged with two or three other SNAREs from the target membrane (t-SNARE) and a 4-α-helix bundle forms between the SNARE motifs (adapted from Sutton et al., 1998)

motif faces the outside rather than the cytoplasm of the cell, these flipped SNARE proteins specifically drive the fusion of cells expressing v- and t-SNAREs (Hu et al., 2003). These experiments provide compelling evidence that by forming SNARE complexes, SNAREs are the nano-machinery that drives the fusion reaction.

Membrane Fusion and SM Proteins

Fusion of liposomes with purified SNARE proteins is much slower than physiological fusion, indicating that other factors are required for in vivo reactions. Among others, the Sec1/Munc18 (SM) proteins are the most suitable for modulating SNARE activity (Fig. 4). First identified in yeast (Sec1) and in mammals (Munc18), this evolutionarily conserved protein family is especially known for its capacity to control membrane fusion and secretion. By interacting with SNAREs, SM proteins are involved in every fusion step of protein trafficking pathway. There are seven Sec1/Munc18 proteins encoded by mammalian genomes, four of which (Sly1, Munc18a-c) are directly involved in the secretory pathway. Sly1 is required for the fusion of ER-derived vesicles with the Golgi apparatus and Munc18 proteins with the plasma membrane. The knockout or mutation of SM proteins inevitably leads to the blocking of membrane fusion or secretion. For example, neither stimulated nor non-stimulated exocytosis occurred in Munc18a knockout mice (Verhage et al., 2000). This phenotype is even stronger than that of the VAMP2 knockout, indicating the importance of SM proteins in membrane fusion (Jahn et al., 2003).

In contrast to the widely accepted function of SNAREs in membrane fusion, the mechanism by which SM proteins regulate this process is still unclear. This is due in part to the diverse modes involved in the binding of SM proteins to SNAREs (Rizo and Sudhof, 2002; Gallwitz and Jahn, 2003; Toonen and Verhage, 2003; Peng, 2005). Recent findings, however, suggest that SM proteins might play a

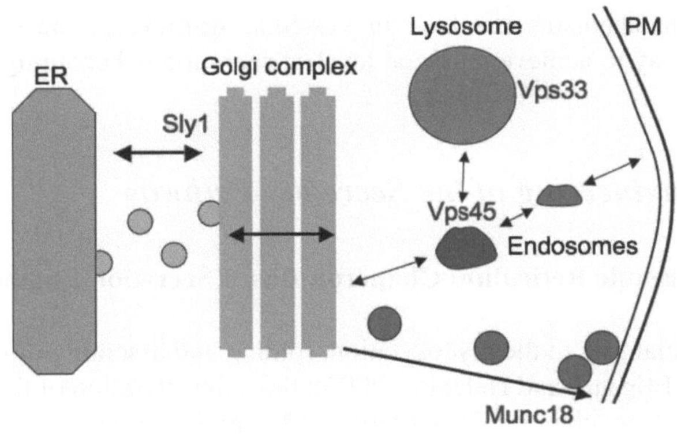

Fig. 4 Geographic scheme of Sec1/Munc18 proteins in the intracellular protein trafficking network of mammalian cells. Sly1 and Munc18 are the SM proteins directly involved in the secretory pathway. Note that there are three isoforms of Munc18 (a, b and c) and two of Vps33 (a, b) in vertebrates

general role in stimulating the assembly of cognate SNARE complexes (Jahn et al., 2003; Peng 2005), as demonstrated for Sly1 (Peng and Gallwitz, 2002; Peng and Gallwitz, 2004), Munc18a (Shen et al., 2007) and Munc18c (Latham et al., 2006).

3 Secretion Engineering

3.1 Secretory Pathway and Biopharmaceutical Production

The first biopharmaceutical product of mammalian cells was put on the market in the mid-1980s (Wurm, 2004). This ushered in a prosperous period of the production of biopharmaceuticals by mammalian systems. Today hundreds of natural and manipulated gene products are being manufactured and marketed (Wurm, 2004).

Significant effort has been directed at the high-level production of recombinant proteins in transgenic mammalian cells. These attempts have primarily focused on vector design, DNA delivery and integration, promoter engineering, mRNA processing, generic host cell engineering and optimization of cell culture media (Wurm, 2004; Barnes and Dickson, 2006; Fussenegger and Hauser, 2007; Weber and Fussenegger, 2007).

The production of secreted recombinant proteins depends not only on the transcription and translation of transgenes, but also on the secretory capacity of host cells. For cells producing recombinant proteins, saturated secretion is the major factor that hinders mammalian cells from fully exploiting their physiologic capacity in the production of biopharmaceuticals (Barnes and Dickson, 2006; Tigges and Fussenegger, 2006; Ku et al., 2008). Although a variety of transcription and translation engineering strategies have been implemented over the past decade that have driven the specific productivity of mammalian cells to an apparent limit, restricted post-translation competence has been poorly addressed. With the rapid elucidation of molecular mechanisms involved in vesicular trafficking, the engineering of secretory pathway to achieve maximal level of secretion is becoming feasible.

3.2 The Engineering of the Secretory Pathway

3.2.1 Endoplasmic Reticulum Chaperon-Based Secretion Engineering

ER plays a crucial role in the glycosylation, folding and assembly of newly synthesized proteins (Ellgaard and Helenius, 2003); thus, the alteration of the ER environment might induce changes in secretion. The most common approach in the engineering of secretion is to modify the expression level of ER chaperons and enzymes. It has been shown that overexpression of the immunoglobulin-binding protein (BiP) leads to more efficient secretion of heterologous proteins in yeast (Shusta et al., 1998; Smith et al., 2004) and in insect cells (Hsu and Betenbaugh,

1997; Ailor and Betenbaugh, 1998). In mammalian cells, proteomic approaches have been employed to identify proteins responsible for the secretion capacity. By applying cell extracts of an antibody-producing cell line (NSO) on two-dimensional polyacrylamide gel electrophoresis, Smales et al. (2004) found that expression levels of the chaperon protein BiP and the disulfide isomerase (PDI) are relevant to the productivity of antibodies. This was confirmed by Borth et al. (2005), who examined the effect of BiP and protein disulfide isomerase (PDI) on the secretion of recombinant antibodies by Chinese hamster ovary (CHO) cells. Although the elevated expression of PDI significantly increased the secretion of antibodies, this was observed only when the expression of BiP was downregulated, indicating that BiP mechanism differed in mammalian, in yeast and in insect cells (Shusta et al., 1998; Smith et al., 2004; Hsu and Betenbaugh, 1997; Ailor and Betenbaugh, 1998). In line with this, the negative role played by BiP in the secretion of mammalian cells was also reported by Kitchin and Flickinger (1995) and Davis et al. (2000).

3.2.2 Xbp-1-Based Organelle Engineering

The X-box binding protein 1 (XBP-1), a transcription activator, is the first such master regulator affecting multiple key genes and organelles in the secretory pathway. Originally isolated as a protein that binds to a cyclic AMP-response element (CRE) of the gene encoding the major histocompatibility complex (MHC) class II molecule DRα (Liou et al., 1990), Xbp-1 was later shown to be essential for the unfolded protein response (UPR) (Yoshida et al., 2001; Dinnis and James, 2005) and a key regulator of the differentiation of B cells into immunoglobulin-secreting plasma cells (Reimold et al., 2001).

There are two isoforms of Xbp-1, Xbp-1 (u) and Xbp-1 (s), encoded by the unspliced and spliced Xbp-1 mRNA, respectively (Yoshida et al., 2001; Mori, 2003). Shaffer et al. have demonstrated that the spliced form of Xbp-1 upregulated the expression of an array of ER chaperon proteins and resulted in increased contents of a subset of secretion-related organelles such as ER, Golgi (Shaffer et al., 2004). Xbp-1(s) upregulated the secretion capacity of plasma cells (Iwakoshi et al., 2003a); overexpression of Xbp-1(s) led to increased secretion of immunoglobulin M (IgM) in primary B cells, while knockout of the gene resulted in a significant decrease in IgM production (Iwakoshi et al., 2003b), clearly demonstrating the essential role of Xbp-1 in the secretory pathway of B cells.

Capitalizing on the impact of Xbp-1 on the global secretion machinery in mammalian cells, we have designed an Xbp-1-based secretion engineering strategy to increase the secretion and overall production capacity of CHO-K1-derived production cell lines. Through the ectopic expression of human-derived Xbp-1, we specifically expanded the endoplasmic reticulum and the Golgi of transgenic Chinese hamster ovary (CHO-K1)-derived cell lines with a resulting increase in overall production capacity of the mammalian cells (Figs. 5 and 6; Tigges and Fussenegger, 2006). Xbp-1-based engineering of secretion was compatible with a variety of different promoter-product gene configurations, suggesting that Xbp-1 induces

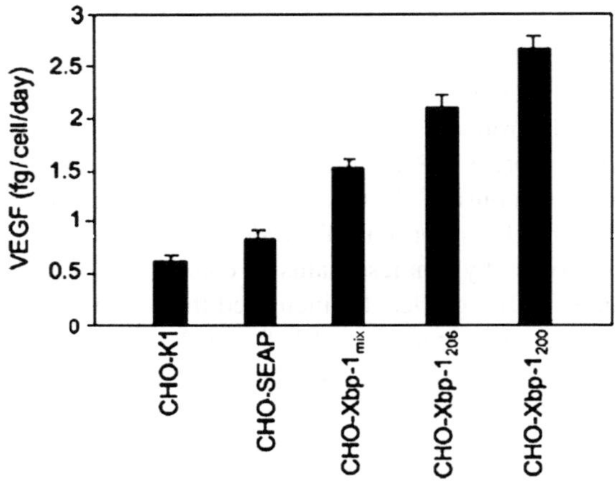

Fig. 5 Engineering Xbp-1-based secretion enhances the production of human vascular endothelial growth factor 121 ($VGEF_{121}$). CHO-K1, CHO-SEAP, which stably expresses the human placental secreted alkaline phosphatase (SEAP), and its derivatives, stably expressing the human Xbp-1(s), were transiently transfected by a VEGF-expressing construct. Secreted VEGF was profiled 72 h after transfection(adapted from Tigges and Fussenegger, 2006)

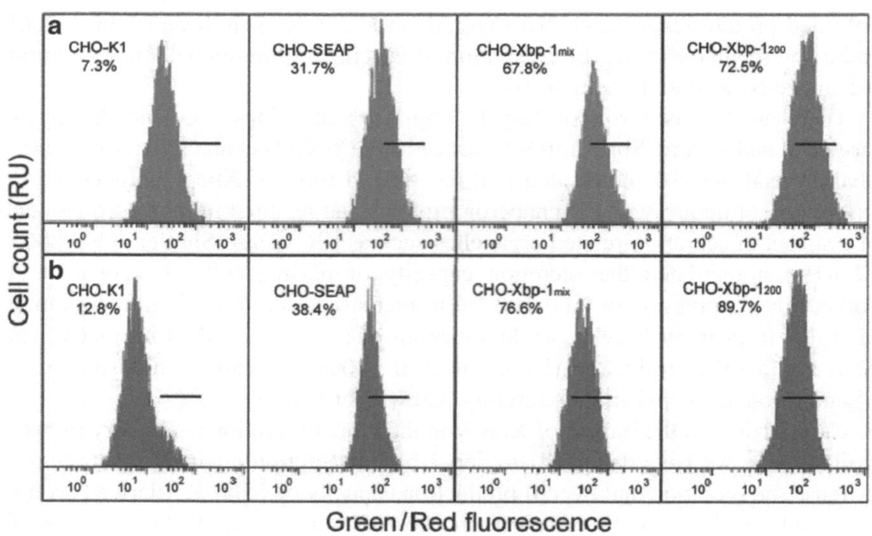

Fig. 6 Engineering Xbp-1 overexpression expands the size of the ER and the Golgi. CHO-derived stable cell lines constitutively expressing Xbp-1 were stained by the ER- and Golgi-specific fluorescent dye BODIPY FL Cy5-ceramide and analyzed by FACS. (**a**), (**b**) Quantification of ER/Golgi-specific structures (515 nm; *green*) and Golgi apparatus (620 nm; *red*). RU, random units (adapted from Tigges and Fussenegger, 2006)

generic production increases in CHO-K1 cell derivatives (Tigges and Fussenegger, 2006). This latest secretion engineering, as illustrated by Xbp1-based reprogramming of the post-translational processing machinery, provides new insights into the major bottlenecks, which impact the manufacture of secreted biopharmaceutical proteins.

3.2.3 Vesicle Trafficking-Based Engineering

A complex vesicle trafficking system manages the precise and regulated distribution of proteins, membranes and other molecular cargo between cellular compartments as well as the secretion of (heterologous) proteins in mammalian cells. Effective reprogramming of this complex trafficking machinery to achieve bioprocess goals, such as increased secretion of high-quality protein therapeutics, requires detailed information on vesicle-target membrane fusion to define production-boosting interventions along the exocytic pathway. Compared to other metabolic engineering approaches that aim at increased productivity of proteins, molecular intervention of vesicle trafficking machinery has long been elusive. Recent progress in our understanding of the molecular mechanisms involved in vesicle trafficking have revealed key factors in the control of the flow of client proteins in the secretory pathway.

Sec1/Munc18 (SM) proteins are key to controlling both the specificity and the rate of membrane fusion (Peng and Gallwitz, 2002; Dulubova et al., 2007; Shen et al., 2007). Based on a detailed physiological analysis of the function of two Sec1/Munc18 proteins, Sly1 and Munc18c, in Chinese hamster ovary (CHO-K1) and human embryonic kidney (HEK-293) cells, we have shown that both SM proteins stimulate specific vesicle traffic by interaction with particular SNAREs and have designed a straightforward multigene-based secretion engineering strategy which boosts the secretion significantly (Fig. 7; Peng and Fussenegger, 2008).

Furthermore, the engineering of Sly1- and Munc18c-based vesicle trafficking complements with Xbp-1-based organelle engineering in accelerating the secretory efficiency, highlighting that the multifaceted engineering provides a powerful tool to obtain optimal production by mammalian cells (Peng and Fussenegger, 2008). This first vesicle trafficking-based secretion engineering has set the standard for targeted interventions in the trafficking of exocytic vesicles and may foster advances in biopharmaceutical manufacturing of difficult-to-produce protein therapeutics, or to improve manufacturing capacities and process economics (Fig. 7).

4 Conclusions and Perspectives

Efficient conversion of transgenic information into desired protein therapeutics requires a well orchestrated sequence of events, including transcription, translation and secretion. Although mammalian cell culture has become the dominant system for biopharmaceutical manufacturing of protein therapeutics due to superior protein

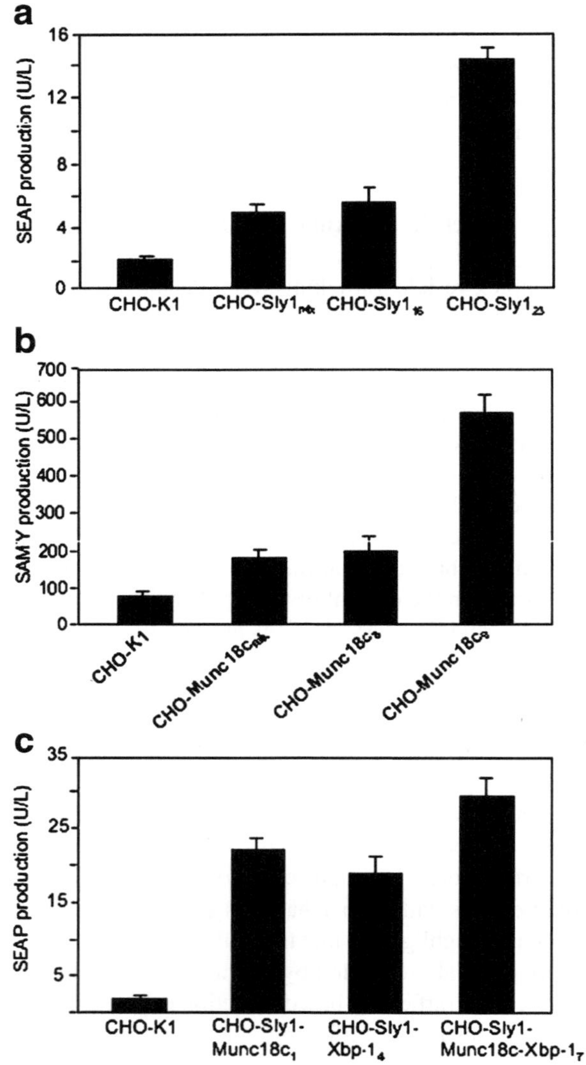

Fig. 7 Engineering of Sly1- and Munc18-based vesicle trafficking significantly increase the secretory capacity of CHO cells. (**a**) CHO-K1 and its derivatives constitutively expressing Sly1 were transiently transfected with a SEAP-encoding plasmid and the secretion profile of SEAP was monitored after 48 h. (**b**) SAMY secretion was measured in CHO-derived cell lines stably expressing Munc18c. (**c**) SEAP profiling was monitored in stable cell lines constitutively expressing Sly1, Munc18c and Xbp-1 (adapted from Peng and Fussenegger, 2008)

folding, assembly and post-translational modification capacities and despite an over 100-fold increase in volumetric productivity achieved since the first commercial production process was licensed in the mid-1980s (Wurm, 2004), the difference in specific productivity between native professional secretory cells and mammalian production cell lines used for biopharmaceutical manufacturing remains significant (Fussenegger and Hauser, 2007). To bridge this gap, different metabolic engineering strategies have been developed to reduce production bottlenecks and improve the specific productivity of mammalian producer cells or the quality of the product. Most successful examples include (i) transcription engineering using trigger-inducible expression systems (Fussenegger et al., 1998a; Meents et al., 2002a; Weber and Fussenegger, 2004), (ii) translation engineering consisting of improving ribosomal entry and translation-initiation (Schlatter and Fussenegger, 2003; Underhill et al., 2003), (iii) glycoengineering improving the ADCC activity of therapeutic antibodies (Umaña et al., 1999; Prati et al., 2002; Yamane-Ohnuki et al., 2004), (iv) controlled proliferation technology to redirect metabolic energy from cell growth to product formation (Fussenegger et al., 1998b; Meents et al., 2002b) and (v) anti-apoptosis engineering, which is based on the ectopic expression of survival genes (Ishaque and Al-Rubeai, 2002; Meents et al., 2002a; Arden and Betenbaugh, 2004).

It is now widely recognized that, when mRNA levels are high, the specific productivity of host cells or organisms is bottlenecked by the secretory capacity. The success of orchestrated reprogramming of exocytic control circuits in expanding secretory organelles (ER, Golgi) by Xbp-1 and optimizing the trafficking of secretory vesicles between the ER and the Golgi by Sly1 and between the Golgi and the plasma membrane by Munc18c showed the power of a multifaceted strategy in secretion engineering (Tigges and Fussenegger, 2006; Peng and Fussenegger, 2008). Future goals in this field include the thorough understanding of the molecular mechanisms involved in the production of heterologous proteins and the overcoming of the restricted post-translation events. Comprehensive techniques such as proteomics and state-of-the-art RNAi technology will surely give momentum to the identification of key components that control high-level protein production. Hence, in order to fully define the regulatory profile(s) of the secretory machinery, techniques that exploit the power of systems biology and data-rich analyses are becoming increasingly important.

References

Arden N, Betenbaugh MJ (2004) Life and death in mammalian cell culture: strategies for apoptosis inhibition. Trends Biotechnol 22:174–180

Ailor E, Betenbaugh MJ (1998) Overexpression of a cytosolic chaperone to improve solubility and secretion of a recombinant IgG protein in insect cells. Biotechnol Bioeng 58:196–203

Antonny B, Schekman R (2001) ER export: public transportation by the COPII coach. Curr Opin Cell Biol 13:438–443

Barnes LM, Bentley CM, Dickson AJ (2004) Molecular definition of predictive indicators of stable protein expression in recombinant NS0 myeloma cells. Biotechnol Bioeng 85:115–121

Barnes LM, Dickson AJ (2006) Mammalian cell factories for efficient and stable protein expression. Curr Opin Biotechnol 17:381–386

Barlowe C, Orci L, Yeung T, Hosobuchi M, Hamamoto S, Salama N, Rexach MF, Ravazzola M, Amherdt M, Schekman R (1994) COPII: a membrane coat formed by sec proteins that drive vesicle budding from the endoplasmic reticulum. Cell 77:895–907

Bi X, Corpina RA, Goldberg J (2002) Structure of the Sec23/24-Sar1 pre-budding complex of the COPII vesicle coat. Nature 419:271–277

Bock JB, Matern HT, Peden AA, Scheller RH (2001) A genomic perspective on membrane compartment organization. Nature 409:839–841

Bonifacino JS, Glick BS (2004) The mechanisms of vesicle budding and fusion. Cell 116:153–166

Bonifacino JS, Lippincott-Schwartz J (2003) Coat proteins: shaping membrane transport. Nat Rev Mol Cell Biol 4:409–414

Borth N, Mattanovich D, Kunert R, Katinger H (2005) Effect of increased expression of protein disulfide isomerase and heavy chain binding protein on antibody secretion in a recombinant CHO cell line. Biotechnol Prog 21:106–111

Davis R, Schooley K, Rasmussen B, Thomas J, Reddy P (2000) Effect of PDI overexpression on recombinant protein secretion in CHO cells. Biotechnol Prog 16:736–743

Dinnis DM, James DC (2005) Engineering mammalian cell factories for improved recombinant monoclonal antibody production: lessons from nature? Biotechnol Bioeng 91:180–189

Dorai H, Csirke B, Scallon B, Ganguly S (2006) Correlation of heavy and light chain mRNA copy numbers to antibody productivity in mouse myeloma production cell lines. Hybridoma 25:1–9

Dulubova I, Khvotchev M, Liu S, Huryeva I, Südhof TC, Rizo J (2007) Munc18-1 binds directly to the neuronal SNARE complex. Proc Natl Acad Sci USA 104:2697–2702

Ellgaard L, Helenius A (2003) Quality control in the endoplasmic reticulum. Nat Rev Mol Cell Biol 4:181–191

Fussenegger M, Hauser H (2007) Protein expression by engineering of yeast, plant and animal cells. Curr Opin Biotechnol 18:385–386

Fussenegger M, Mazur X, Bailey JE (1998a) pTRIDENT, a novel vector family for tricistronic gene expression in mammalian cells. Biotechnol Bioeng 57:1–10

Fussenegger M, Schlatter S, Datwyler D, Mazur X, Bailey JE (1998b) Controlled proliferation by multigene metabolic engineering enhances the productivity of Chinese hamster ovary cells. Nat Biotechnol 16:468–472

Fussenegger M, Morris RP, Fux C, Rimann M, von Stockar B, Thompson CJ, Bailey JE (2000) Streptogramin-based gene regulation systems for mammalian cells. Nat Biotechnol 18:1203–1208

Gallwitz D, Jahn R (2003) The riddle of the Sec1/Munc-18 proteins – new twists added to their interactions with SNAREs. Trends Biochem Sci 28:113–116

Guo W, Roth D, Walch-Solimena C, Novick P (1999) The exocyst is an effector for Sec4p, targeting secretory vesicles to sites of exocytosis. EMBO J 18:1071–1080

Hsu TA, Betenbaugh MJ (1997) Coexpression of molecular chaperone BiP improves immunoglobulin solubility and IgG secretion from Trichoplusia ni insect cells. Biotechnol Prog 13:96–104

Hu C, Ahmed M, Melia TJ, Söllner TH, Mayer T, Rothman JE (2003) Fusion of cells by flipped SNAREs. Science 300:1745–1749

Ishaque A, Al-Rubeai M (2002) Role of vitamins in determining apoptosis and extent of suppression by bcl-2 during hybridoma cell culture. Apoptosis 7:231–239

Iwakoshi NN, Lee AH, Glimcher LH (2003a) The X-box binding protein-1 transcription factor is required for plasma cell differentiation and the unfolded protein response. Immunol Rev 194:29–38

Iwakoshi NN, Lee AH, Vallabhajosyula P, Otipoby KL, Rajewsky K, Glimcher LH (2003b) Plasma cell differentiation and the unfolded protein response intersect at the transcription factor XBP-1. Nat Immunol 4:321–329

Jahn R, Scheller RH (2006) SNAREs – engines for membrane fusion. Nat Rev Mol Cell Biol 7:631–643

Jahn R, Lang T, Sudhof TC (2003) Membrane fusion. Cell 112:519–533

Kelly RB (1985) Pathways of protein secretion in Eukaryotes. Science 230:25–32

Kirchhausen T (2000) Three ways to make a vesicle. Nat Rev Mol Cell Biol 1:187–198

Kitchin K, Flickinger MC (1995) Alteration of hybridoma viability and antibody secretion in transfectomas with inducible overexpression of protein disulfide isomerase. Biotechnol Prog 11:565–574

Ku SC, Ng DT, Yap MG, Chao SH (2008) Effects of overexpression of X-box binding protein 1 on recombinant protein production in Chinese hamster ovary and NS0 myeloma cells. Biotechnol Bioeng 99:155–164

Latham CF, Lopez JA, Hu SH, Gee CL, Westbury E, Blair DH, Armishaw CJ, Alewood PF, Bryant NJ, James DE, Martin JL (2006) Molecular dissection of the Munc18c/syntaxin4 interaction: implications for regulation of membrane trafficking. Traffic 7:1408–1419

Liou HC, Boothby MR, Finn PW, Davidon R, Nabavi N, Zeleznik-Le NJ, Ting JP, Glimcher LH (1990) A new member of the leucine zipper class of proteins that binds to the HLA DR alpha promoter. Science 247:1581–1584

Matsuoka K, Orci L, Amherdt M, Bednarek SY, Hamamoto S, Schekman R, Yeung T (1998) COPII-coated vesicle formation reconstituted with purified coat proteins and chemically defined liposomes. Cell 93:263–275

Meents H, Enenkel B, Eppenberger HM, Werner RG, Fussenegger M (2002a) Impact of coexpression and coamplification of sICAM and antiapoptosis determinants bcl-2/bcl-x(L) on productivity, cell survival, and mitochondria number in CHO-DG44 grown in suspension and serum-free media. Biotechnol Bioeng 80:706–716

Meents H, Enenkel B, Werner RG, Fussenegger M (2002b) p27Kip1-mediated controlled proliferation technology increases constitutive sICAM production in CHO-DUKX adapted for growth in suspension and serum-free media. Biotechnol Bioeng 79:619–627

Mori K (2003) Frame switch splicing and regulated intramembrane proteolysis: key words to understand the unfolded protein response. Traffic 4:519–528

Palade G (1975) Intracellular aspects of the process of protein synthesis. Science 189:347–358

Peng R (2005) Decoding the interactions of SM proteins with SNAREs. ScientificWorldJournal 5:471–477

Peng R, Fussenegger M (2009) Molecular engineering of exocytic vesicle traffic enhances the productivity of Chinese hamster ovary cells. Biotechnol Bioeng 102:1170–1181

Peng R, Gallwitz D (2002) Sly1 protein bound to Golgi syntaxin Sed5p allows assembly and contributes to specificity of SNARE fusion complexes. J Cell Biol 157:645–655

Peng R, Gallwitz D (2004) Multiple SNARE interactions of an SM protein: Sed5p/Sly1p binding is dispensable for transport. EMBO J 23:3939–3949

Prati EG, Matasci M, Suter TB, Dinter A, Sburlati AR, Bailey JE (2002) Engineering of coordinated up- and down-regulation of two glycosyl-transferases of the O-glycosylation pathway in Chinese hamster ovary (CHO) cells. Biotechnol Bioeng 79:580–585

Reimold AM, Iwakoshi NN, Manis J, Vallabhajosyula P, Szomolanyi-Tsuda E, Gravallese EM, Friend D, Grusby MJ, Alt F, Glimcher LH (2001) Plasma cell differentiation requires the transcription factor XBP-1. Nature 412:300–307

Rizo J, Sudhof TC (2002) SNAREs and Munc18 in synaptic vesicle fusion. Nat Rev Neurosci 3:641–653

Rothman JE (1994) Mechanisms of intracellular protein transport. Nature 372:55–63

Schlatter S, Fussenegger M (2003) Novel CNBP- and La-based translation control systems for mammalian cells. Biotechnol Bioeng 81:1–12

Shaffer AL, Shapiro-Shelef M, Iwakoshi NN, Lee AH, Qian SB, Zhao H, Yu X, Yang L, Tan BK, Rosenwald A, Hurt EM, Petroulakis E, Sonenberg N, Yewdell JW, Calame K, Glimcher LH, Staudt LM (2004) XBP1, downstream of Blimp-1, expands the secretory apparatus and other organelles, and increases protein synthesis in plasma cell differentiation. Immunity 21:81–93

Shen J, Tareste DC, Paumet F, Rothman JE, Melia TJ (2007) Selective activation of cognate SNAREpins by Sec1/Munc18 proteins. Cell 128:183–195

Shusta EV, Raines RT, Pluckthun A, Wittrup KD (1998) Increasing the secretory capacity of *Saccharomyces cerevisiae* for production of single-chain antibody fragments. Nat Biotechnol 16:773–777

Smales CM, Dinnis DM, Stansfield SH, Alete D, Sage EA, Birch JR, Racher AJ, Marshall CT, James DC (2004) Comparative proteomic analysis of GS-NS0 murine myeloma cell lines with varying recombinant monoclonal antibody production rate. Biotechnol Bioeng 88:474–488

Smith JD, Tang BC, Robinson AS (2004) Protein disulfide isomerase, but not binding protein, overexpression enhances secretion of a non-disulfide-bonded protein in yeast. Biotechnol Bioeng 85:340–350

Sutton RB, Fasshauer D, Jahn R, Brunger AT (1998) Crystal structure of a SNARE complex involved in synaptic exocytosis at 2.4 Å resolution. Nature 395:347–353

Tigges M, Fussenegger M (2006) Xbp1-based engineering of secretory capacity enhances the productivity of Chinese hamster ovary cells. Metab Eng 8:264–272

Toonen RF, Verhage M (2003) Vesicle trafficking: pleasure and pain from SM genes. Trends Cell Biol 13:177–186

Umaña P, Jean-Mairet J, Moudry R, Amstutz H, Bailey JE (1999) Engineered glyco-forms of an antineuroblastoma IgG1 with optimized antibody-dependent cellular cytotoxic activity. Nat Biotechnol 17:176–180

Underhill MF, Coley C, Birch JR, Findlay A, Kallmeier R, Proud CG, James DC (2003) Engineering mRNA translation initiation to enhance transient gene expression in Chinese hamster ovary cells. Biotechnol Prog 19:121–129

Verhage M, Maia AS, Plomp JJ, Brussaard AB, Heeroma JH, Vermeer H, Toonen RF, Hammer RE, van den Berg TK, Missler M, Geuze HJ, Südhof TC (2000) Synaptic assembly of the brain in the absence of neurotransmitter secretion. Science 287:864–869

Waters MG, Serafini T, Rothman JE (1991) Coatomer: a cytosolic protein complex containing subunits of non-clathrin-coated Golgi transport vesicles. Nature 349:248–251

Weber T, Zemelman BV, McNew JA, Westermann B, Gmachl M, Parlati F, Söllner TH, Rothman JE (1998) SNAREpins: minimal machinery for membrane fusion. Cell 92:759–772

Weber W, Fussenegger M (2004) Approaches for trigger-inducible viral transgene regulation in gene-based tissue engineering. Curr Opin Biotechnol 15:383–391

Weber W, Fussenegger M (2007) Inducible product gene expression technology tailored to bioprocess engineering. Curr Opin Biotechnol 18:399–410

Whyte JR, Munro S (2002) Vesicle tethering complexes in membrane traffic. J Cell Sci 115:2627–2637

Wurm FM (2004) Production of recombinant protein therapeutics in cultivated mammalian cells. Nat Biotechnol 22:1393–1398

Yamane-Ohnuki N, Kinoshita S, Inoue-Urakubo M, Kusunoki M, Iida S, Nakano R, Wakitani M, Niwa R, Sakurada M, Uchida K, Shitara K, Satoh M (2004) Establishment of FUT8 knockout Chinese hamster ovary cells: an ideal host cell line for producing completely defucosylated antibodies with enhanced antibody-dependent cellular cytotoxicity. Biotechnol Bioeng 87:614–622

Yoshida H, Matsui T, Yamamoto A, Okada K, Mori K (2001) XBP1 mRNA is induced by ATF6 and spliced by IRE1 in response to ER stress to produce a highly active transcription factor. Cell 107:881–891

Index

MIX
Papier aus verantwortungsvollen Quellen
Paper from responsible sources
FSC® C105338

If you have any concerns about our products,
you can contact us on
ProductSafety@springernature.com

In case Publisher is established outside the EU,
the EU authorized representative is:
Springer Nature Customer Service Center GmbH
Europaplatz 3, 69115 Heidelberg, Germany

Printed by Libri Plureos GmbH
in Hamburg, Germany